Handbook of Industrial Diamonds

Vol. 2

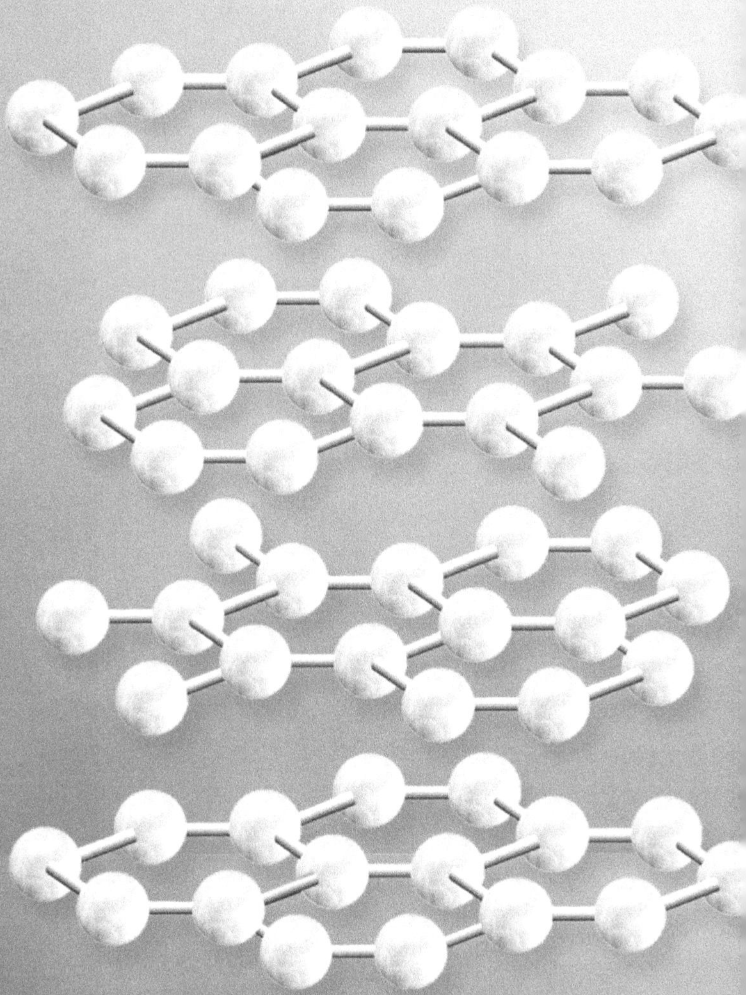

Handbook of Industrial Diamonds

Diamond Films and Carbon Coatings

Vol. 2

James Sung

JENNY STANFORD
PUBLISHING

Published by

Jenny Stanford Publishing Pte. Ltd.
Level 34, Centennial Tower
3 Temasek Avenue
Singapore 039190

Email: editorial@jennystanford.com
Web: www.jennystanford.com

British Library Cataloguing-in-Publication Data
A catalogue record for this book is available from the British Library.

Handbook of Industrial Diamonds: Diamond Films and Carbon Coatings

ISBN 978-981-4310-00-0 (Hardcover)
ISBN 978-1-003-18045-6 (eBook)

Contents

Preface

This volume covers diamond films and carbon coating, including principles and applications. Many industrial applications of diamond are described, such as cutting tools, heat spreaders, radiation windows, sound transmitters, diamond semiconductors, diamond electrodes, and various types of diamond-like carbon (DLC) coatings. The initial chapters connect diamond to the Universe at large and the origin of life in detail, as diamond is the most abundant solid in the Universe.

The chapter on ultrahard materials predicts new materials that may be harder than diamond. In 1984, the author originated the research of a hypothetical compound C_3N_4 and predicted that it could be harder than diamond. This prediction led to the publication of about 1000 research papers related to this subject in the 1990s. The CN coating on computer hard drives was derived from this flurry of research works. However, as unveiled in the chapter related to exotic ultrahard materials, there are many possible new lattice structures waiting to be discovered, such as a simple cubic oxygen lattice with six covalent bonds for every atom compared to tetrahedral coordinated carbon atoms in diamond.

The author pioneered one of the earliest diamond film companies in the world, Norton Diamond Film, which was the largest diamond film company around year 2000. The technology was based on Technion's arc jet space engine that was converted to make diamond films as one of the best cutting tool materials. Subsequently, the author established the largest DLC coating company at Kinik with a dozen of DC arc coating machines. The DLC coating was used to shield diamond pad conditioners from acid corrosion for CMP of semiconductor wafers, the first of such DLC applications worldwide starting in 1997. In 2013, the author collaborated with Ritek and offered LED chips cooled by DLC coatings with higher efficiency and longer life. In 2014, the author helped Hounghe Whirlwind establish Henan Graphene Company (Synthetic Element Six) based on physical exfoliation of graphite particles suspended in colliding supersonic water jets generated by diamond dies with micrometer holes. Beginning 2021, the author became the chief scientist of Shenzhen Graphene Institute, working on single crystals of graphite of wafer size for numerous applications (e.g., heat spreaders, noiseless antennae, interconnected circuits) for future.

Graphene is the stretched diamond's (111) face, and diamond is the puckered graphene stacked together. Hence, graphene is 2D diamond and diamond is 3D graphene. This century will enter the diamond age of human civilization with many new applications to make a debut. We are just about to enter the wonderland of diamond gadgets that will drive devices with ever-increasing degree of complexity. For example, smart diamond chips will dominate not only AI but also the interface with the human brain. As a keynote speaker at the International Conference on Graphene and Semiconductor held in Las Vegas in 2018, the author proposed graphene on diamond (GOD) for future quantum computers. The spin of extra neutrons in diamond's ^{13}C isotope atoms were proposed as qubits and diamond surface's seamlessly slanted graphene may be used as logic circuits. The rigid diamond lattice of ^{13}C atoms can shield the entanglement of neutron spins from quantum decoherence with environmental disturbance, so cryogenic cooling is not needed. Moreover, as diamond has the highest Debye temperature by far for all possible qubit atoms, the noise level is minimal, so complex corrections due to unavoided interference from outside may be mostly eliminated. Furthermore, the homoepitaxial interface of GOD can allow ultrasonic speed (Mach 100) of graphene's acoustic waves triggered by infrared light as input/output logic gates, so it is possible to replace massive sized quantum computing system based on light transmissions or metal circuits. Hence, palm-sized quantum cell phones may be realized in future. Let us wait to experience the dawn of the diamond age with awe.

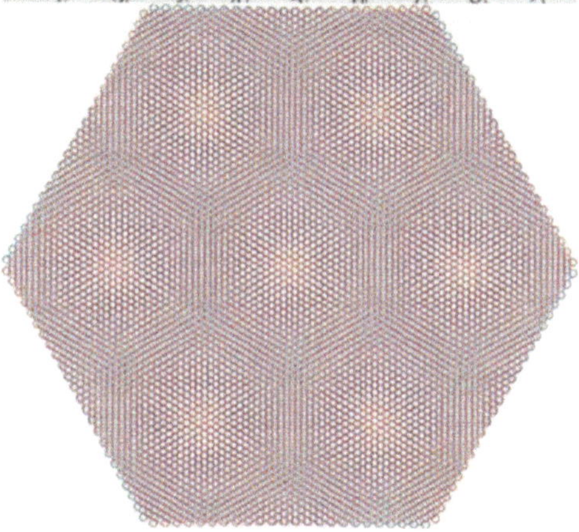

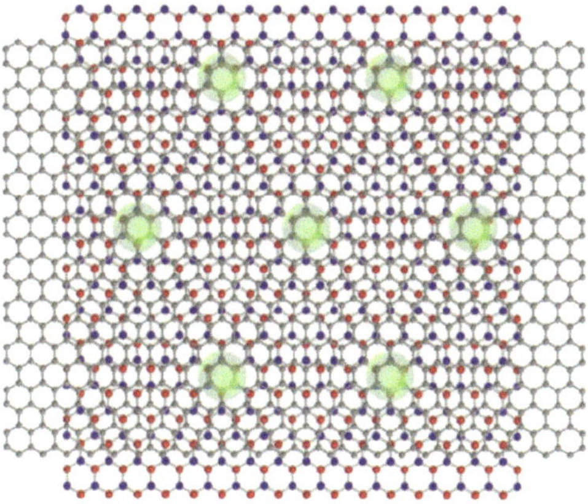

In 2008, the author invented single graphite films grown by recrystallization from molten alloy of carbon solvent metals (e.g., Invar Fe_2Ni), as shown in the diagram on the previous page (top) (refer to Kuen-Liang Chang et al, 2010, China Semiconductor Technology International Conference, *ECS Transactions*, 27(1), pp. 945–950). Such graphite films contain thousands of layers of single-crystal graphene with atoms in equilibrium sites of minimal lattice energy. Moreover, these single crystals of graphene layers are wafer sized suitable for making advanced gadgets of optic electronics. These large single crystals of graphene in place can be twisted to a magic angle of 1.1° so as to make twistronic devices (adjoining diagram, middle) with metal–insulator–superconductor transitions by inputting an electric current, as conceived by an MIT team in 2018. Moreover, the author also invented (e.g., US2011/0163298A1) epitaxial devices of graphene-hexagonal boron nitride, the only pair of 2D atomic devices with the conductor epitaxially coupled with the insulator. The bottom diagram illustrates the overlapping lattices of 2D graphene conductor sandwiched between 2D hexagonal boron nitride insulator with Moiré pattern of superlattice, as revealed in Berkeley Lab's news release in 2020. Here, the gray honeycomb depicts graphene lattice. Boron and nitrogen atoms are shown in red and blue, respectively and vice versa. The light green hexagons of stacking lattices may exhibit superconductivity. Just as a large single crystal of graphite can be grown by dissolving graphite powder in molten Ni and exsolved slowly, a large single crystal of hexagonal boron nitride can be grown by dissolving hexagonal boron nitride powder in Li_3N and exsolved slowly.

Chapter 1

Diamond in the Sky

Twinkle, twinkle, little star; how I wonder what you are?
Up above the world so high, like a diamond in the sky.

—Jane Taylor 1806

1.1 Diamond Stars

When Ms. Jane Taylor composed the famous lyric of "diamond in the sky" about 200 years ago, she could not have imagined that her figurative poetry would foretell the scientific discovery made in the 20th century. As it turns out that the real diamond is not only in the sky, but it may even be the most common solid in the universe. In addition, as will be introduced in the next chapter, diamond is also the most useful material. Thus, the most abundant material is also the most versatile. Of course, diamond has long been the king of all jewels; hence, diamond has all the three eternal virtues: true, good, and beautiful.

Carbon ranks number four in the universal abundances of elements, after hydrogen, helium, and oxygen. However, all the more abundant elements are gases, so carbon becomes the most abundant solid in the universe. The majority of carbon is locked up deep in stars where pressure is exceedingly high. As diamond is the stable form of carbon under pressure, naturally, diamond should be commonplace inside a large star.

Handbook of Industrial Diamonds: Diamond Films and Carbon Coatings
James Sung

ISBN 978-981-4310-00-0 (Hardcover), 978-1-003-18045-6 (eBook)
www.jennystanford.com

The stars are large nuclear reactors that generate heat by combining nuclei together. The first step of nuclear fusion is to merge hydrogen atoms to form helium atoms. For an ordinary star, such a process may take a long time to complete. For example, the Sun has been shining for 5 billion years and it will continue the process for another 5 billion years. During this tenure of hydrogen fusion, the star maintains its stability by balancing the two opposing forces; inward pulling gravity against outward rushing light. However, when a star of our Sun's size exhausts its nuclear fuel, it would shrink and become a white dwarf. Inside such dwarfs lie colossal mines of diamond. A typical white dwarf is about a million times more massive than the Earth; hence, chances are most white dwarfs contain diamonds more massive than the entire Earth.

There are only a handful of stars that lie within 20 light years from the Earth. In 1998, astronomer Steve Kawaler discovered that one of them was a diamond star (BPM37093). This star is located 17 light years from the Earth. It belongs to Centaurs Constellation, which is visible only in the southern hemisphere. This diamond star has a diameter of 12,800 km, which is almost the same size as the Earth. However, it is a white dwarf that is the ruminant of a sun-like star after it exhausted the fuels for nuclear fusion. Such a cinder of dead star is very dense; so a teaspoonful material may weigh over 1 ton.

The surface of this diamond star reaches 12,000°C, twice that of our Sun. It emits a greenish blue light that revealed characteristic absorption spectrum of diamond. It would appear that the diamond it contains might also be blue in color. As most diamonds are either colorless or yellowish on Earth, blue diamond is extremely scarce. The blue color is due to the absorption of yellow color by the incorporation of minute amount of boron atoms. The boron-doped diamond is also a semiconductor that may outperform the fastest silicon or other types of semiconductors.

The famous Hope diamond is the largest blue diamond known. It was originally embedded in the forehead of Indian goddess Sita in an Indian temple. The French adventurer J. J. Tavernier adventured to India many times, and at the age of 80, he made the last trip to India when he stole the blue diamond from Rama-Sita. Upon his return to France, he sold this 112 carats diamond to King Louis X IV. This blue diamond was eventually passed on to Louis X VI, who was beheaded during the French Revolution. In 1792, the blue diamond was robbed away from the palace during a mob. When it reappeared in 1839, it was re-cut into a much smaller (44.5-carat) gem. This blue diamond was said to carry a curse of Rama–Sita that induced tragedies to many of its owners, including the death of Louis X VI.

Eventually, the blue diamond fell into the hands of an English banker Henry Hope, and hence it bears his name since then. In 1958, the American gem collector Harry Winston presented it to the Smithsonian Museum in Washington, D.C. It has become the most sought-after attraction for tourists who visit the museum (Fig. 1.1). Although the Hope diamond may be the most glamorous gem that has ever existed, it could easily be dwarfed by a chip brought back from the blue diamond star of BMP37093.

Figure 1.1 The Hope diamond is the largest blue diamond known. It contains boron atoms that make it a positive semiconductor.

1.2 Diamond Planets

In addition to stars that may harbor diamonds of colossal size deep inside, many planets may also see diamonds raining down from the sky. This is because both carbon-containing gases methane (CH_4) and carbon dioxide (CO_2) are common constituents in the planetary atmosphere. Methane and carbon dioxide can be decomposed at high temperatures to form diamond. This is routinely practiced nowadays by a technique called chemical vapor deposition CVD) (Fig. 1.2). In addition to the high-temperature decomposition of methane, high pressure may also assist the diamond formation. Hence, diamond raining on planets' atmosphere may be as common as rains on the Earth.

Figure 1.2 Diamond grown from thermally decomposed methane (left) heated by tungsten filaments (Diamonex's sample); and from pyrolyzed carbon dioxide (right) heated by laser beams (QQC's sample).

In addition to the diamonds that are buried deep in stars, and that may rain down on planets, diamonds are ubiquitous in other places, too. Since carbon atoms may exist everywhere in the universe, the frequent impacts of these atoms may form diamonds inevitably.

There are many ways that carbon atoms may be squeezed together to become diamond. First, all stars and planets are moving relatively to one another, so their collisions are not uncommon. Each of such collisions will certainly convert much of carbon atoms into diamond. Second, each planet or satellite is constantly bombarded by asteroids or comets. This is why many satellites are covered with impact craters (e.g., Moon's surface). Again, each meteorite impact can produce diamonds as dust. Indeed, nano-sized diamond particles are present in meteorites as well as impact craters. Third, billions of asteroids and hundreds of billions of comets are constantly colliding. Hence, microscopic diamond may be formed in the solar system all the time. Fourth, many giant stars will explode after running out of nuclear fuel (e.g., supernovae) and the explosion can spew diamond debris in space. Fifth, the universe is soaked with cosmic rays, star winds, gamma ray bursts, and X-ray beams. These energetic particles and electromagnetic radiations can convert carbon atoms into diamond. Thus, atomic clusters of diamond have been found in stellar dusts as identified from their characteristic absorption spectrum.

Some coal mines could also be impregnated with diamonds. For example, an old meteorite hit a coal mine in Brazil a hundred million years ago. Some of the coal instantly melted and then quenched to form radiant structure of amorphous diamond known as ballas. The remaining less hot coal was converted into black aggregates of diamond known as carbonados. Certain uranium-containing coals also contain sub-nano-sized diamond. These clusters of hundreds of carbon atoms were located along the fission tracks of disintegrated uranium nuclei. Apparently, the impacts of this nuclear debris against carbon atoms in coal turned them into diamond.

In addition to various sources of diamond mentioned above, of course, we know diamond exists deep in the Earth as well. This is why volcanic mantle plumes such as kimberlite brought natural diamond up to the surface.

1.3 Nuclear Synthesis and Carbon Formation

According to the Big Bang theory, the orthodox teaching of the origin of the universe, we owe our existence to a proto universe of about 10 cm across that was formed when the universe was about 10^{-32} sec in existence. This proto universe has expanded since then to form today's space about 15 billion years later. The proto universe was extremely hot (10^{27} K), but the expansion of space has decreased its temperature to only 2.7 K now. During the early stage of quenching, the energy in the proto universe condensed into elementary particles that eventually coalesced to form protons and electrons. After expansion for about 300,000 years, the temperature was low enough (1000 K) to allow the formation of atoms. However, the cooling of the universe was so fast that only hydrogen atoms (93 atom% or 79 wt%) and helium atoms (6 atom% or 20 wt%) could form in time.

Hydrogen and helium atoms are less stable than heavier atoms, so they have a tendency to fuse together. But, this fusion may only take place at a high temperature that existed when the universe was still in infancy. However, although the ambient temperature has been too cold, the center of a star is hot enough to continue the fusion of light atoms. The stars are formed by the accumulation of sufficient amount of light atoms that attract one another by gravity. As soon as pressure and temperature reach a threshold level, nuclear synthesis by fusing light atoms into heavier ones will begin. Initially, like our Sun, hydrogen atoms will fuse into helium atoms. If the star is sufficiently massive (e.g., 3 times more massive than our Sun) helium atoms will also combine to form heavier atoms, and the trend will continue.

The binding energy of nuclei is the highest for iron atoms (Fig. 1.3). Accordingly, the nuclear synthesis may continue until iron atoms are formed. After then, any enlargement of nuclei will not release energy to sustain the reaction, but it will absorb energy so the nuclear synthesis will halt.

At high temperature, nucleons can move around so they tend to form the most stable nuclei possible; hence, four hydrogen atoms will fuse into one helium atom (two hydrogen atoms become two neutrons that join the nucleus, at the same time, two protons and two electrons from the other two atoms will combine in the same atom). Subsequently, helium atoms may also fuse into beryllium atoms. Beryllium atoms may combine with helium atoms to form carbon atoms. Carbon atoms may pack up helium atoms to form oxygen atoms, etc.

If the binding energy of nucleons are monotonously increasing with the addition of nucleons, then we would expect that the cosmic abundance of elements to decrease with the atomic size (atomic number N, proton number, or electron number). But the stabilities of nucleus size differ greatly with the number of protons present. Hence, the amounts of atoms formed are also highly variable. Although the general trend is to decrease the abundance of elements with the increasing size of atoms, but there are many exceptions. For example, although iron atoms are more than 10-fold heavier than beryllium atoms, but the former is millions times more abundant than the latter.

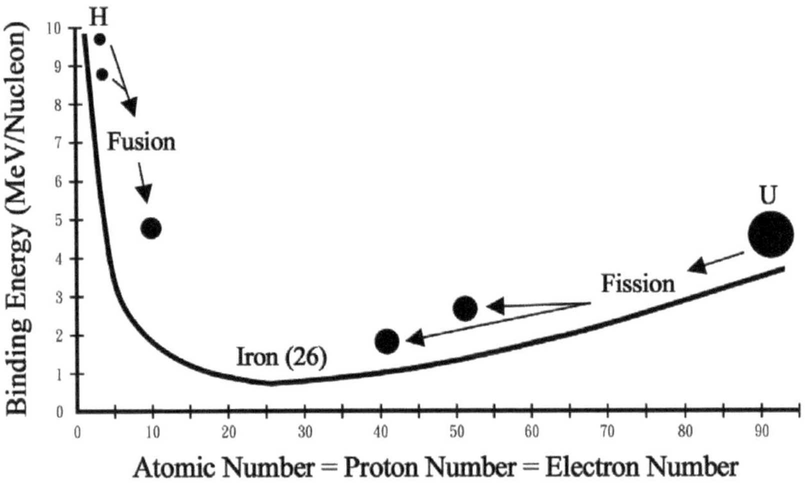

Figure 1.3 Iron (N = 26) has the lowest nuclear binding energy, so atoms lighter than iron atoms may combine (fusion) into larger atoms; but atoms heavier than iron may disintegrate (fission) to form smaller atoms.

The beryllium nuclei formed by the impact of two helium nuclei is unusually unstable with a half-life of only 10^{-17} sec. Hence, beryllium atoms are extremely scarce inside a star. However, beryllium atoms are the source for making heavier atoms; among them are carbon and oxygen, the two vital elements for us (e.g., our body may contain 12 atom% or 23 wt% of carbon atoms, and we need to breath the air that contains about 20 wt% oxygen). If there is no mechanism to boost the production of these essential elements, not only there would be no diamond stars, but also there would not be life in the universe.

In order to explain the ubiquitous presence of carbon and oxygen atoms in the universe, in 1953, the English astrophysicist Fred Hoyle proposed that carbon's nuclei could resonate at a frequency slightly (4%) higher than beryllium nuclei, so when the latter is hit by helium nuclei, the combined energy will make carbon nuclei tick. According to this hypothesis, although beryllium nuclei are extremely unstable, their chances to transmute into carbon nuclei are exceedingly high. As a result, carbon abundance can increase at the expense of beryllium.

However, in order to prevent the hardly formed carbon nuclei from being consumed too fast by fusing with helium nuclei,

Holye also proposed that the resonance frequency of oxygen nuclei be slightly (1%) less than carbon's. In this way, the collision of carbon and helium nuclei would not form oxygen nuclei readily. Hoyle's two predictions were confirmed latter by experimentation, a triumph to reason the cause based on effect. Thus, it is the subtle balance of the vibration frequencies between carbon and oxygen that balance the abundances of both critical elements. This is a miracle that eventually allowed life to prosper.

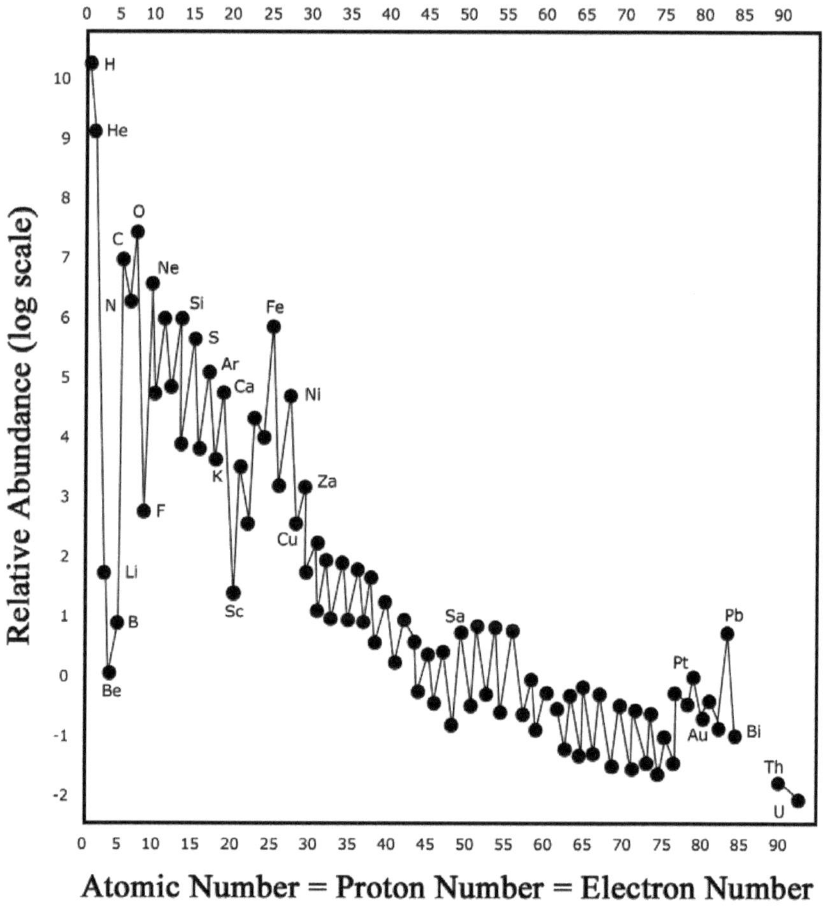

Figure 1.4 Cosmic abundance of elements. Note that carbon is the fourth abundant element and the most abundant solid element.

The abundance boosts of carbon and oxygen atoms have made them more than any other heavy element. As a consequence, the four most abundant elements in the universe are H (93 atom%), He (6 atom%), O (0.06 atom%), and C (0.03 atom%) (Fig. 1.4). Among them, C is the only solid element. All other heavy elements combined would account for only 0.1 atom% of the cosmic abundance; hence, carbon may account for 30% of all solid material in the universe. For example, the cosmic abundance of carbon is more than 10 times of silicon (the major semiconductor material and the essential constituent of rocks) or iron (the major industrial material and the core element of Earth). No wonder carbon is ubiquitous in space and diamond is everywhere in the cosmos.

1.4 Supernova Explosion

The Sun is now fusing hydrogen into helium. However, the Sun's mass is not high enough to cause further synthesis of heavier elements, so when its nuclear fuel is used up in about 5 billion years, it will die out as a white dwarf. If a star's mass is larger than 1.4 times of the Sun, its nuclear synthesis will proceed until just beginning to form iron. In this case, the dying white dwarf will contain concentric zones of increasingly heavy elements (Fig. 1.5).

However, if the star's mass is more than 1.4 times the solar mass, it will form a lot of iron during the nuclear synthesis. Upon the exhaustion of the nuclear fuel, the star cannot produce enough energy to sustain its size so the outer layer will collapse instantly. The implosion will smash against the iron core with the release of tremendous amount of energy. The sudden increase of pressure and temperature will break most iron atoms. As a consequence, outer electrons will be squeezed into the nucleus when they combine with protons to form neutrons. The result will be the formation of a gigantic soup of neutrons known as neutron star. The neutron star has a density as high as a nucleus (10^{14} g/cc), so a grape size of it will weigh over 100 million tons!

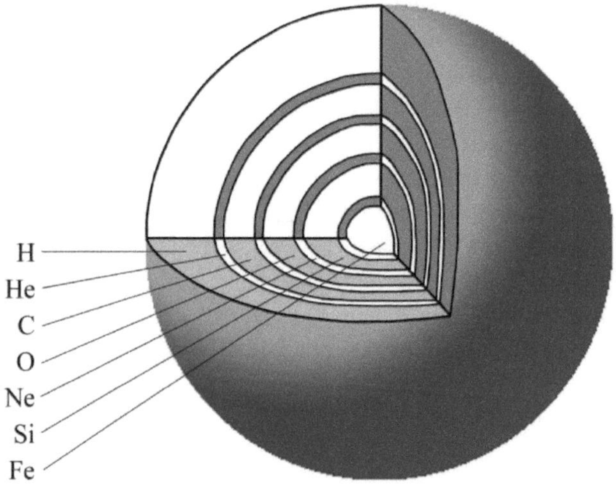

Figure 1.5 The onion structure of a large white dwarf. Note that carbon buried deep may form diamond. A typical white dwarf's mass is about one million times of that of the Earth, so the amount of diamond in such a star may easily be heavier than the entire Earth.

Hydrogen fusion in a star derives its energy by combining an electron with a proton to form a neutron one by one. The formation of a neutron star is like achieving nuclear fusion about 100 billion times faster than an ordinary star. Hence, once a neutron star is formed, the amount of energy must be released immediately. The result is a gigantic explosion that expels more than 90% of the outer materials into space to become stellar dusts. This spectacular astronomical phenomenon is known as "supernova" explosion. After the explosion, the supernova may outshine for weeks the entire galaxy that contains about 100 billion stars.

Supernova explosion occurs periodically in each galaxy (the human history has recorded hundreds of such cosmic glares). The explosion will release supernova's interior materials, such as carbon and oxygen that are otherwise locked up. The condensation of this expelled debris will make latter-generation stars richer in heavy elements. Our sun is one of these stars formed on the ashes of a previous supernova. The entire Earth is also built on such debris. Thus, we owe our existence to a parental supernova, so is diamond on the Earth.

1.5 Diamond Snow Flakes

If we exclude the noble gas He, which does not react with any other element, the three most abundant elements H, C, and O will form most common molecules of water (H_2O), methane (CH_4), and carbon dioxide (CO_2) (Fig. 1.6). Thus, these three compounds are the common gases in planets that harbor an atmosphere. Except for water, which does not contain carbon, both methane and carbon dioxide may decompose under high pressure, and the segregated carbon will form diamond.

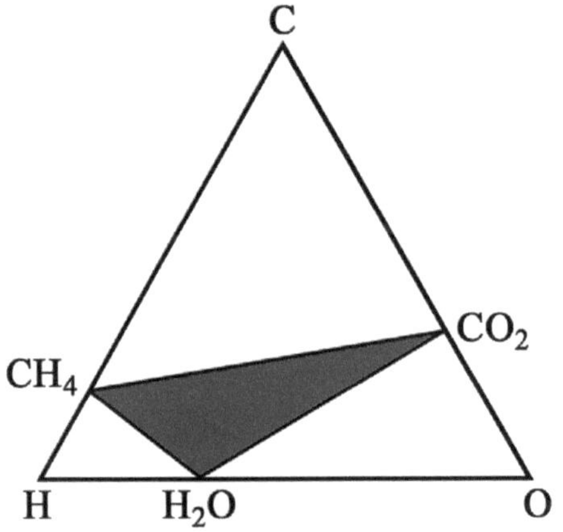

Figure 1.6 The common atmospheric gases are derived from the three most abundant cosmic elements.

Methane's carbon atom is surrounded by four hydrogen atoms that form a tetrahedron. The carbon's atom forms four bonds that extend to each of the four hydrogen atoms. This bonding structure (sp^3) is exactly that of diamond except that each of the bonds is slightly ionic due to the different bonding strength of electrons between carbon and hydrogen atoms. Hence, we may view methane as caged single-atom diamond (Fig. 1.7). Methane is the common ingredient in natural gas. It may be of interest to note that cooking is often done by burning floating diamond.

When methane is heated in the presence of hydrogen atoms, the single-atom diamonds can be coerced into linking together to form multiple atom diamonds. With time, micron diamonds may grow to microscopic size (Fig. 1.8), and eventually, they may link to form a continuous layer. This is a common practice of CVD growth of diamond film.

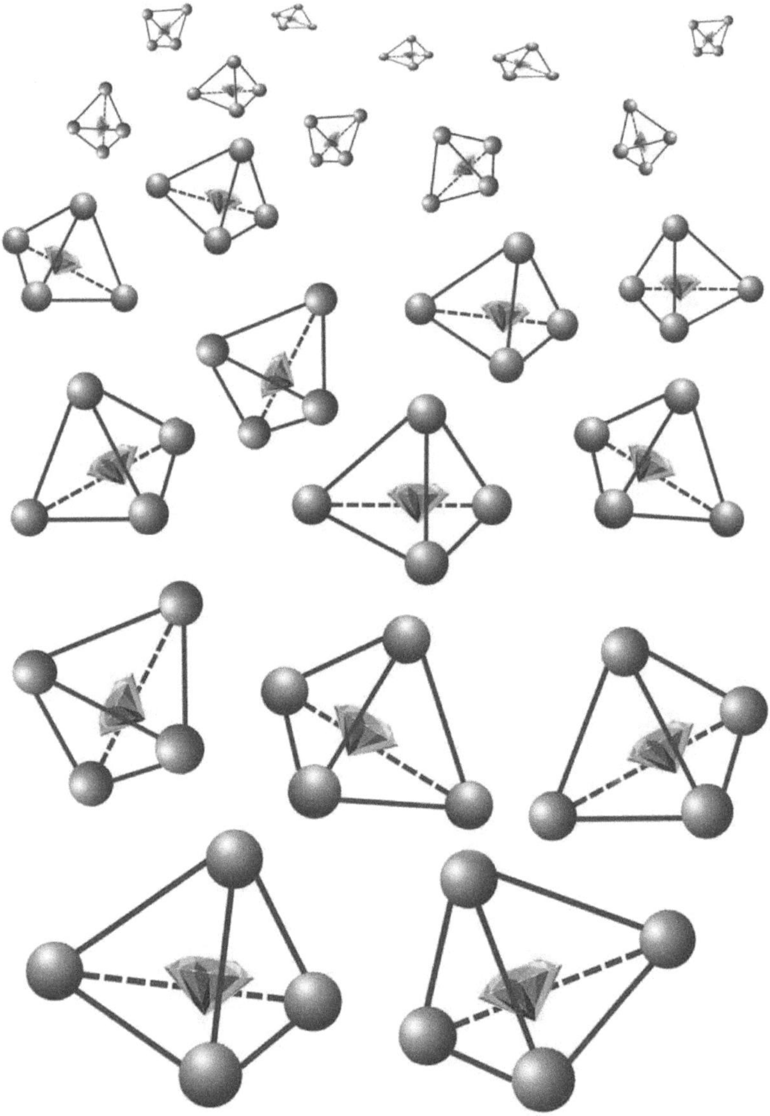

Figure 1.7 Methane is a gas of floating single-atom diamonds. Each of them is confined in a tetrahedron cage of hydrogen atoms.

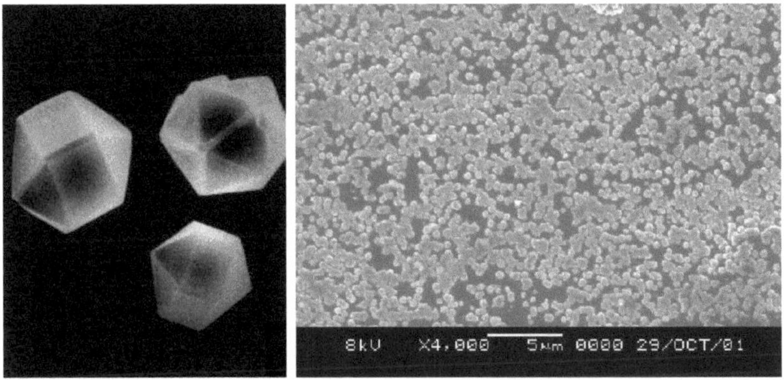

Figure 1.8 CVD growth of diamond micron powder (left) and the link of micron diamond to form a diamond film (right).

Methane may also be decomposed under high pressure to form diamond. Physicists have recovered such diamond by pressing methane solid to 50 GPa, or shot it by a bullet traveling at a speed of 30 mach (10 km/sec). So conceivably, diamond snowflakes may shower from sky in these giant planets. If the temperature is high enough (e.g., 4000°C), these snowflakes may melt to form diamond rain. At a much higher pressure, diamond rain may condense to form diamond ice or hail stones. At a great depth, diamond rain or diamond ice may accumulate to become diamond ocean or diamond continent, depending on the temperature there.

Of all gas giants in the solar system, Uranus and Neptune contain much methane in the atmosphere. Moreover, methane is under a high pressure without overheating, as in the case of Saturn and Jupiter (Fig. 1.9), such conditions are conducive to forming diamond. In 1981, the physicist Marvin Ross pointed out Uranus and Neptune may have accumulated a thick layer of diamond ice around the rocky core where the pressure may be as high as 60 GPa, and temperature 7000°C.

One place to look for spectacular diamond storm is deep in Neptune. Neptune has a blue appearance due to the presence of methane in the atmosphere that absorbs yellow light. Astronomers have found that this blue gaseous planet could form its own energy as it radiates 2.6 times of heat than it absorbed from the Sun. However, the source of such energy was unknown.

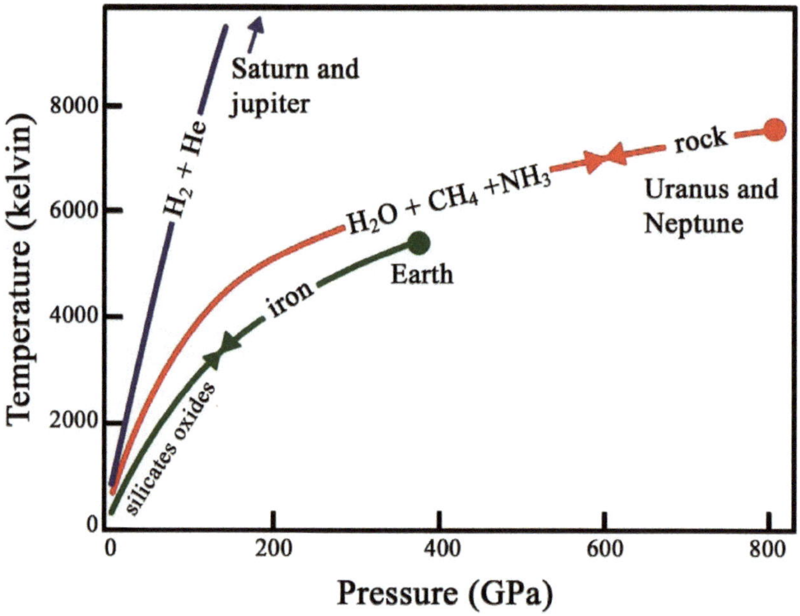

Figure 1.9 The pressure–temperature profiles of the four gas giants and the Earth in the solar system. Note that methane in Uranus and Neptune may decompose to form diamond.

In 1999, Robin Benedetti of the University of California at Berkeley squeezed methane in a pair of diamond anvils and heated it up with a laser to simulate the pressure and temperature conditions of about a third way toward the planet's center. He found that methane decomposed to form carbon atoms and these atoms bonded to one another to become diamond flakes. Hence, diamond snow shower from the sky may be common in Neptune (Fig. 1.10). The condensation of carbon to form diamond in Neptune would account for the mysterious heat source that baffled astronomers.

Not only methane could form diamond under high pressure, so is carbon dioxide. In 1995, metallurgist Pravin Mistry accidentally discovered that diamond could be formed by bombarding carbon dioxide with laser beams. Apparently, the decomposed carbon atoms banged against one another to link into a diamond lattice. Carbon dioxide is also common in gaseous giant planets. It is also the major constituent of Venus' atmosphere, which is 100 times thicker than the Earth's. Although the pressure (20 Mpa) and temperature (400°C) there may not high enough to form diamond rapidly, given the long period of geological time, a thin layer of diamond may have veneered Venus' surface.

Figure 1.10 The blue color of Neptune reflects the presence of abundant methane in its atmosphere. The interior of this methane atmosphere may shower diamond flakes down to its rocky core.

Chapter 2

The Dawn of Diamond Age

"Nothing can be too good to be true."

—James Clerk Maxwell (1831–1879)

2.1 The Miracle of Diamond

In the animal kingdom, humans (*Homo sapiens sapiens*) are the majority among all species, and they also possess the highest level of intelligence quality-wise. Hence, humans are the undisputable king of all animals. In the material world, diamond also enjoys the same royal status, as it is most abundant in the universe and also possesses the most useful properties.

Carbon ranks number 4 in the universal abundances of elements, after the three gas elements, hydrogen, helium, and oxygen. Hence, it is the most populous solid in the universe. Carbon forms the backbone of organic compounds that are 10 times more diversified than all other materials combined. Moreover, carbon is the essence of the biological cells (e.g., DNA and protein) that have sparked the life in the otherwise unanimated world. In the carbon family, diamond stands out as the supreme ruler; it has the most extreme properties of all materials.

Handbook of Industrial Diamonds: Diamond Films and Carbon Coatings
James Sung

ISBN 978-981-4310-00-0 (Hardcover), 978-1-003-18045-6 (eBook)
www.jennystanford.com

Although diamond may be rare on the Earth, it may be everywhere in the universe. It has been detected from the spectra of interstellar dusts. Nano-sized diamond particles were also recovered from plunging meteorites. Moreover, scientists have speculated that the interior of giant planets (e.g., Uranus or Neptune) are composed primarily of diamond formed by the decomposition of ubiquitous methane under tremendous pressure. Diamond may also form in many colossal stars as the high-pressure phase of carbon that is one of the terminal elements of nuclear synthesis. The deep buried diamond in these gregarious stars may have a mass millions times that of the Earth.

It is a miracle that the universe's omnipresent solid is also omnipotent. This impossible combination is a grace of God just like His human creation. The omnipotent diamond exhibits many extreme properties, which include aesthetical brilliance, mechanical strength, abrasion resistance, edge sharpness, thermal conductivity, thermal expansion, heat capacity, wave propagation, electrical insulation, hole mobility, chemical inertness, optical transparency, radiation hardness, electron emission, surface smoothness, friction coefficient, etc. These supreme properties, coupled with other unique attributes, such as extremely low thermal expansion coefficient, distinguish diamond from all other materials as a totally alien substance (Fig. 2.1). Thus, the queen of gems known in the past will become the king of materials in the future.

In the past, industrial diamonds were limited by size and geometry, so they served primarily in the mechanical industry as ultrahigh pressure anvils, surgical knives, and most commonly, as superabrasives (e.g., used in grinding wheels). However, with the availability of large-sized diamond films and complicated DLC coatings, many functional applications of diamond and its related materials are now either commercially available or under development. Thus, diamonds have become the fastest heat spreader available (e.g., for packaging 200 watt computer CPU or for bonding millions of semiconductor chips), the highest frequency resonator attainable (as 100 kHz tweeter diaphragm or 10 GHz surface acoustic wave filter), the most powerful semiconductor possible (e.g., recent Japanese initiative for developing ultracomputer with 20 times the speed of silicon chips), the most transparent radiation window achievable (e.g., for megawatt microwave transmission, or as six-mach missile radome), the most inert chemical barrier invented (e.g., for probing the corrosive atmosphere at high temperature), the most energetic particle detector possible (e.g., for monitoring cosmic rays), and the most efficient cold cathode known (e.g., for field emission display), to name just a few examples.

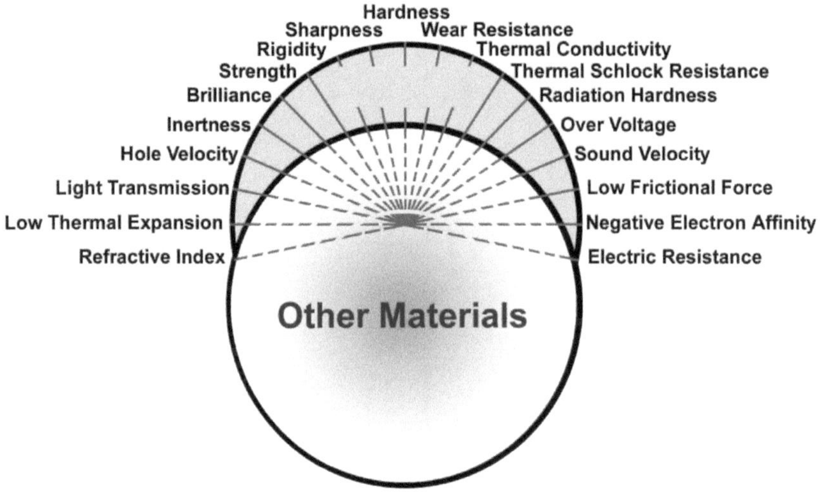

Figure 2.1 Diamond as an "alien substance" that possesses a cut above properties of all other materials.

In addition, diamond-like carbon (DLC) coatings have become a must for protecting computer hard drives and various chemical-mechanical parts (e.g., for thin film deposition on silicon wafers or acid slurry polishing of integrated circuitry). DLC-coated cassette disks and razor blades are also commonplace consumer products nowadays. DLC coating is also critical for sustained reliability of medical components (e.g., for heart valves, artificial joints, expandable stents, etc.).

With the availability of more shapes and sizes at affordable prices, diamond applications have increased steadily (Fig. 2.2). Diamond has also been tested constantly for new frontier applications; most of them assume the enabling function with no substitute possible. With the increasing commercial development underway, the arrival of the ultimate diamond age in the future with the climax of human civilization seems inevitable.

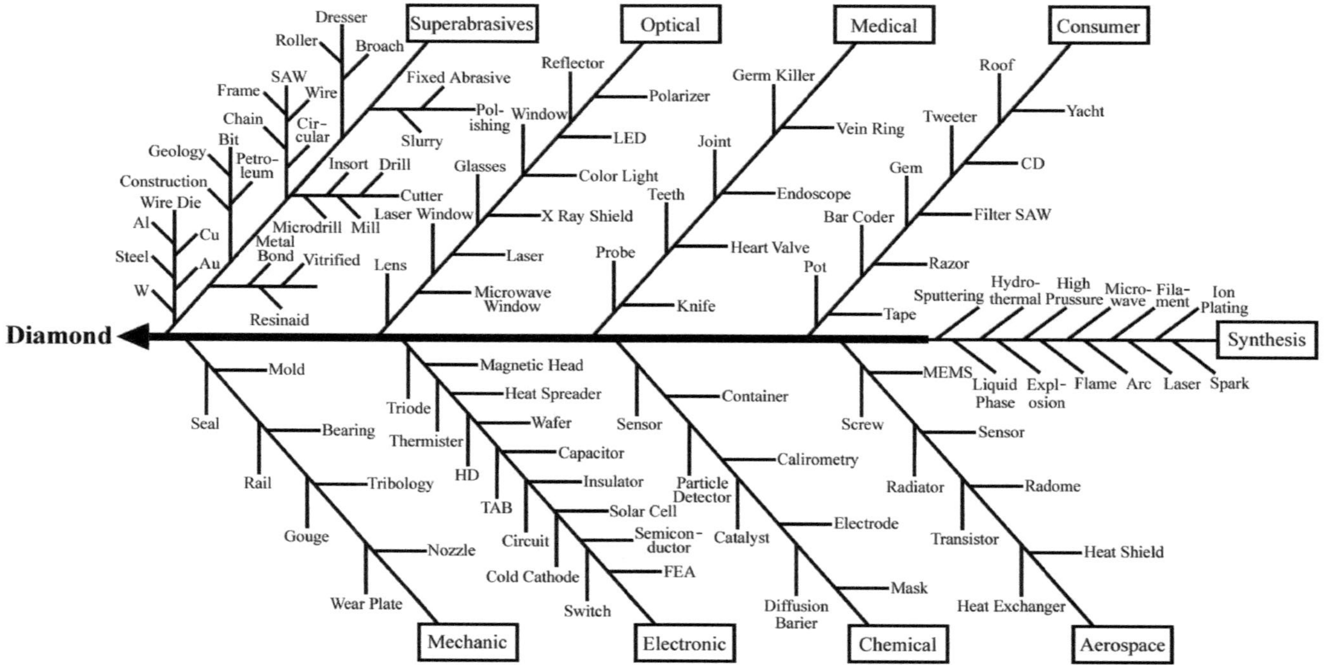

Figurer 2.2 Diamond has been used in almost every sector of industrial products. In many such applications, diamond is the enabler; its function cannot be substituted by other materials.

2.2 Diamond as the King Material

How did diamond acquire its royal quality above all materials? The secret lies in its ability to form four covalent bonds that define the simplest three-dimensional volume-tetrahedron. Each atom in a crystal lattice can be surrounded by 0, 1, 2, 3, 4, 6, 8, and 12 neighbors, the increasing trend of this "kissing number" (coordination number) marks the gain of the order (enthalpy) of a pile of atoms at the expense of the chaos (entropy). There is only one type of extreme order and one type of extreme chaos, so the compromise between them would allow diversification and complexity. The diversification of diamond's tetrahedral bonding has enriched organic chemistry, whereas its complexity has prompted the birth of biology.

The kissing number less than 4 is for point (e.g., gas molecules), line (e.g., polymer molecules), or plane (e.g., foliated molecules) relationship. This relationship either lacks a structure or defines a loosely attached structure. The low kissing number will highlight the local relationship and hence its exhibits insulating properties. On the other hand, with the increasing kissing number to go beyond 4, the solid structure will become tighter so the global property of metal becomes dominating. The compromise of loose and tight structures lies exactly at the kissing number of 4, and hence, it manifests well in the flexibility of tetrahedral bonding of diamond. This intermediate structure makes diamond's property falling between insulator and metal, and hence, it becomes a semiconductor.

Moreover, the packing between loose and tight can allow flexibility of atomic positions, which is not possible in insulators or metals. Thus, carbon atoms with the capability to form tetrahedral diamond bonding (e.g., methane and its polymers) and planar graphitic bonding (e.g., benzene and its derivatives) have become the cornerstone of organic chemistry. The open tetrahedral bonding also serves as the building block of many versatile semiconductors that power now indispensable computers and other electronic devices (e.g., lasers). Alternatively, the versatile tetrahedral bonding can be coupled with closer packed metal atoms to form most advanced composite materials for micro electro-mechanical devices (Fig. 2.3).

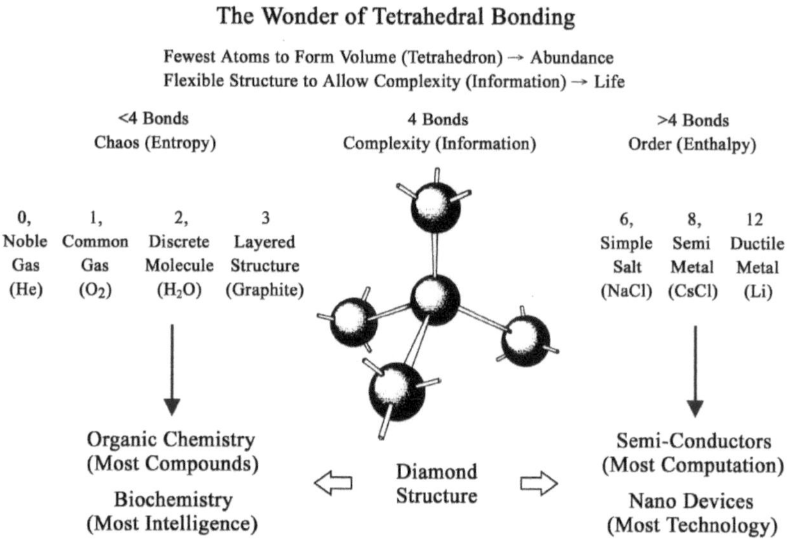

Figure 2.3 The locally connected atoms with kissing number less than 4 cannot form integral solid. On the other hand, the globally dominated structures with kissing number more than 4 do not allow many variations. The compromise between local and global allows versatile tetrahedral bonding that makes complex structures possible. This complexity has prompted the formation of organic compounds and the evolution of living organisms. The tetrahedral bonding has also enabled semiconductors for making integrated circuitry.

Although tetrahedral bonding is the common feature of all semiconductors, in comparison, diamond has the smallest atoms and hence, it possesses the highest atomic density (176 atoms/nm^3) of all semiconductors, and in fact, all materials. The combination of the highest atomic density with the largest number (4) of covalent bonds has greatly concentrated diamond's bond energy (7.4 eV). This strong bonding of atoms has zimparted diamond with the cut above properties that single diamond out as the king of materials.

Although diamond's "royal status" stems from its ability to form tetrahedral bonds, this ability is actually the result of carbon's "crown position" in the periodic table of elements (Fig. 2.4). Carbon is located at the center column (Group IV) and top period of all elements capable of forming chemically bonded solids. At the center column, carbon is impartial between valence electron deficient metals (kissing number >4) on the left side of the periodic table, and valence electron surplus insulators (kissing number < 4) on the right side. The atoms of left side elements (Group I, II, III) have many electron vacancies so they can attract more neighbors (kissing number = 6, 8, 12) to share their few valence electrons. The consequence of sharing few by many is forming isotropic metallic bonds. On the other hand, the atoms of the right side elements (Group V-VII) possess many valence electrons that repel one another. As a result, they can allow only few neighbors (kissing number = 1, 2, and 3) to share their valence electrons. The consequence of sharing many by few is forming directional covalent bonds. The compounds formed between indiscriminate metallic elements and selective covalent elements possess the intermediate ionic bonds.

Li	Be	B	C	N	O	F
Na	MG	Al	Si	P	S	Cl
K	Ca	Ga	Ge	As	Se	Br
Rb	Sr	In	Sn	Sb	Te	I
Cs	Ba	Tl	Pb	Bi	Po	At

Figure 2.4 Periodic table of solid materials with atomic bonding. This table stripped hydrogen and noble gases that may form solids only with weak metallic bonds (e.g., metallic H) or with polarized (van der Waals) electrostatic force (e.g., cryogenic He). The transition metals are inner filling elements of d and f orbitals so they are hidden in the parental Group II. This periodic table highlights carbon's crown position at the center column and top row. The center column gives carbon the highest (4) number of covalent bonds per atom. The top position makes its atoms the smallest among all center column elements. The combination of these two attributes imparts diamond with the highest concentration of bond energy, and thus the unsurpassed hardness and other properties.

Covalent bonds have the most concentrated electron cloud between atoms so they are much stronger than metallic bonds, or ionic bonds. The center column elements of diamond and silicon possess four covalent bonds; the most before these bonds become metallic with further increases of kissing number. As a result of this high density of covalent bonds, diamond or silicon possesses the most rigid crystal lattice among all structures formed by the same and higher period elements, respectively.

Carbon sits at the top of the middle column elements so its atoms are the smallest. Hence, diamond has the highest bond energy density of all materials. Thus, being located at the crown throne of the periodic table, it is not accidental that diamond is the miracle material with so many extreme properties that pale all other materials.

2.3 The Expansion of Diamond Market

With the availability of diamond in more sizes and shapes, diamond applications increases steadily. The reduction of diamond prices also prompted the growth of existing market. Thus, the diamond sales have been increasing for more than four decades, a feat unmatched by the growth of any other material (Fig. 2.5). The increasing trend of all other industrial commodities will inevitably level off in less than a decade (e.g., steel topped the production at around 800 million tons after world war II). This monotonous increase of industrial diamond sales implies that its market is far from being saturated. It would appear that industrial diamond sales would continue to grow almost without limit due to the constantly new discoveries of novel diamond applications, provided that the price of diamond materials will continue to slide because of the improvements of manufacturing efficiency.

Human civilization is built on the foundation of useful materials available. Thus, woods and stone were the first materials available for prehistoric men. They were used as tools and shelters for hunter and gathers as well as agriculture settlers. Then ancient societies found useful of extractive metals of copper, iron, and aluminum. These metals were made in swords and spears, as well as construction structures. At the dawn of industrial revolution, the metals were superseded by the designed materials of plastics. Hence, utensils and garments made of plastics became commonplace. Modern time the civilization entered the silicon age. Computers and electronics have been parts of our daily life. However, all these materials are used for their convenience. They are not the ultimate material for performance. As diamond has the most extreme properties of all materials, its dominance in functional devices will be inevitable with the continual declining in prices and increasing in availability. Hence, its importance will rise with time (Fig. 2.6).

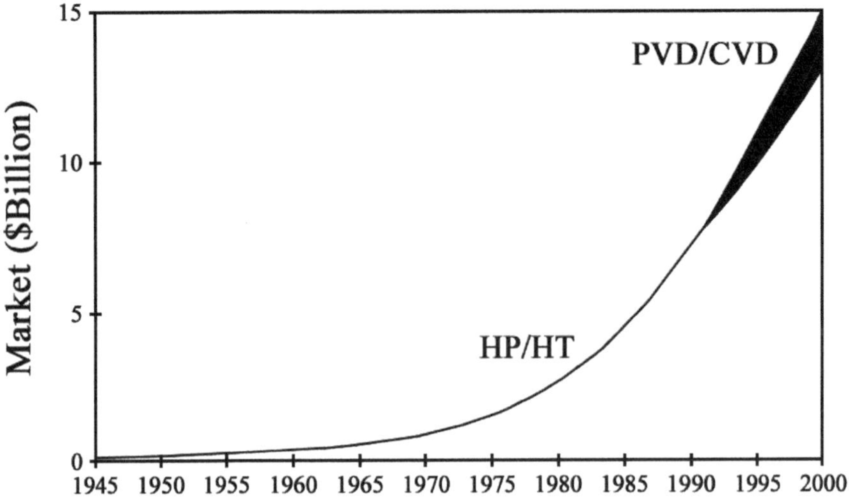

Figure 2.5 The relentless increase of diamond sales. The long-term upward trend has been smoothed out to remove the temporary fluctuations due to the short-term global economic setbacks. The high pressure-high temperature synthesized diamond now accounts for more than 95% of superabrasives sales (<5% is for natural diamond). However, CVD diamond films and PVD diamond-like carbon coating are now taking off. They may serve primarily in the non-mechanical applications in the future.

The rise of diamond material as the dominant building blocks of human civilization is not dependent on its abundance as with the other dominant materials in the past. The contribution of diamond to the value of life lies in its prestige and quality. Diamond's dominance as the key elements for functions is more due to its enabling capability than to its cost effectiveness. For example, diamond's performance in superabrasives, heat spreaders, radiation windows, acoustic filters, and semiconductor circuitry cannot be substituted by competing materials.

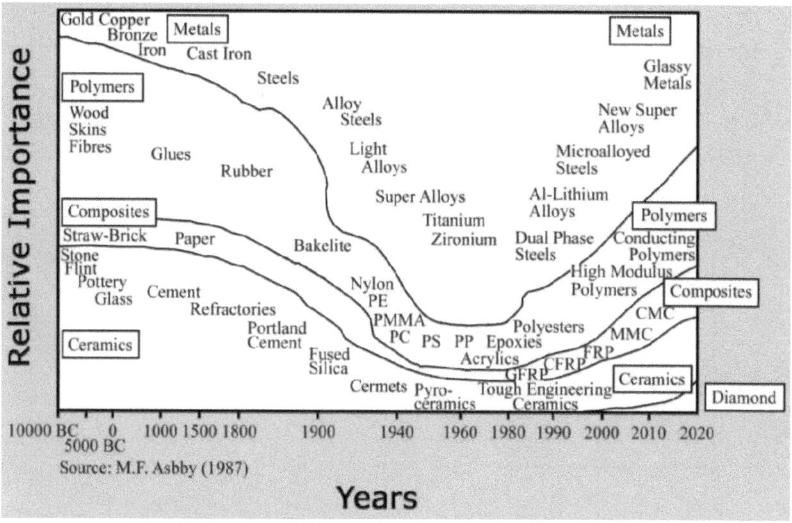

Figure 2.6 The relative importance of materials as the foundation of human civilization. Note the rise of diamond in this century.

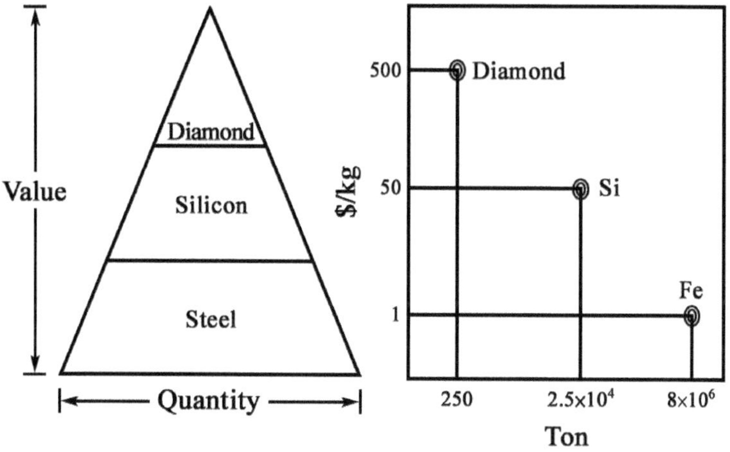

Figure 2.7 The quality-quantity pyramid of standard industrial materials. Interestingly, the three civilization enabling materials that rule in the past, now, and future (iron, silicon, and diamond) lie exactly in a straight line on the logarithmic scale of a time pyramid.

The value and quantity of industrial materials often form an inverse relationship that is representable by a product pyramid. Since diamond is the king of industrial materials, it will become the icing of cake for all product pyramid. Diamond superabrasvies have already been massively produced for various industrial applications, although many non-mecanical applications are just beginning (Fig. 2.7).

2.4 The Ultimate Semiconductor

Among many of diamond's champion properties are its semiconductor capabilities. Diamond structure forms the basis of almost all semiconductors, including the most prevailing material today: silicon. However, with its highest hole mobility and temperature stability of all possible semiconductors, diamond can be the dream CPU chip for future supercomputers (Fig. 2.8). Moreover, by possessing also the unmatched thermal conductivity, diamond will be the only semiconductor that can dissipate heat effectively by itself.

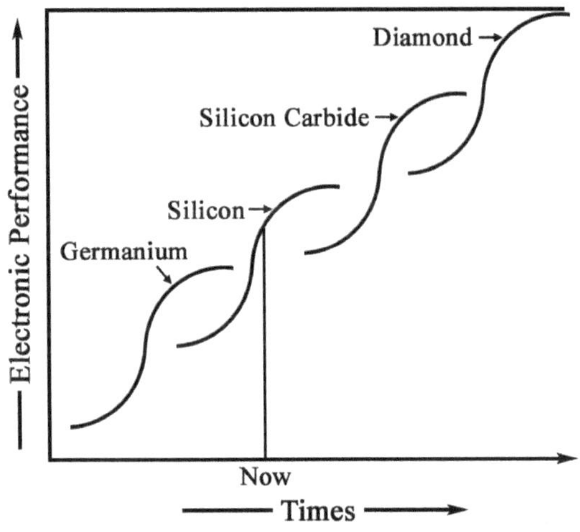

Figure 2.8 The incremental evolution of semiconductor's performance can be achieved by optimizing the circuitry design, but the quantum revolution may only be attained by changing for a more effective material. The history of semiconductor progression is moving upward along the tetrahedrally bonded elements (Group IV) located at the center column of periodic table. It has been progressed from Ge to Si now. It will gradually advance to SiC, the half diamond semiconductor. Eventually, the advancement of semiconductor chips requires the use of full diamond, the terminal material for all semiconductors.

In each of the major industries, computer, communication, energy, there are bottlenecks that can be cleared by using diamond devices; for example, the heat spreader for CPU/GPU/NPU; in the power devices, the IGBT (insulating gate bipolar transistors) chips, etc.

As the computing material, diamond has another unique property-negative electron affinity. Diamond containing material can be the most efficient cold cathode that is capable to emit electrons at extremely low voltage (less than 10 V). With an array of miniature electron guns, diamond can revive the vacuum tube, the antique computing unit that was replaced by silicon transistors in 1950s. Thus, the ultimate computing machines in the future will be matched between diamond transistors and diamond emitters. In either case, diamond will win the race as the champion muzzle of all computing materials.

Although computers today are electronic machines, they may also be run by other mechanisms, such as by chemical potential (e.g., DNA computers) or electromagnetic waves (e.g., optical computers). In particular, photons have been envisaged as a much more efficient carrier of information denoted as bits of 0 or 1. Not only photons can travel at the speed of light that is much faster than electrons moving either along a conductor in the case of diamond semiconductor or through vacuum in the case of diamond emitter, but their motion also consumes less energy and hence, a minimal amount of wasted heat will be generated. Current semiconducting supercomputers can reach the speed of teraflops (trillions floating point operations per second, a float point operation is equivalent to one arithmetic operation). New hypercomputer using diamond electron emitters may boost the speed 1000-fold to petaflops (quadrillions floating point operations per second). Despite this, it may eventually be superseded by photonic ultracomputers that is capable of running at an astonishing speed of 1000 times faster than a future hypercomputer.

Diamond still comes ahead as it is the most efficient optical conduit possible. Moreover, diamond has the smallest openings of all materials; its (111) face covers myriad hexagonal helical openings of only 2.4 angstrom across (Fig. 2.9). These atom-sized openings can filter light like an optical sieve that blocks all wavelengths longer than the X-ray with a half wavelength equal to diamond's opening size. Hence, diamond can be used to construct a photonic ultracomputer with message carried by the highest frequency possible (10^{18} Hz). This capability, coupled with other superior diamond properties such as fast heat dissipation will make diamond the ultimate computing material ever.

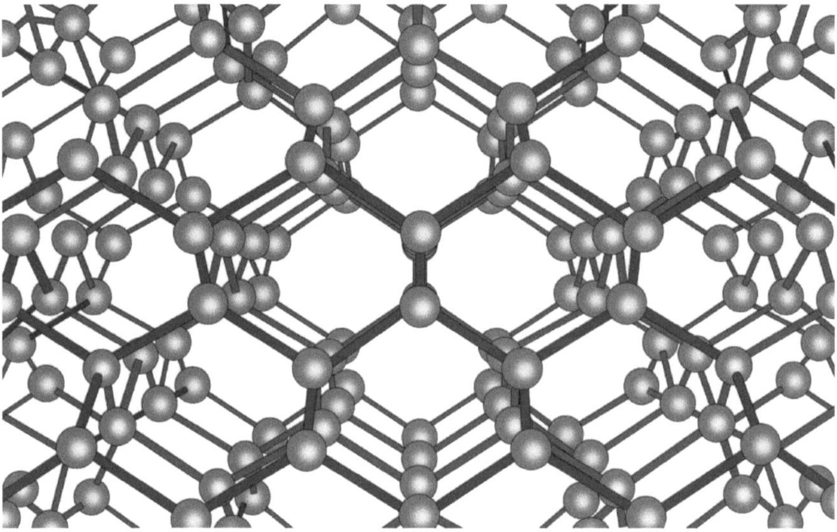

Figure 2.9 The atomic sized optical sieve of diamond when looked through its octahedral (111) surface. The hexagonal helical holes will change to square spiral holes when diamond's cubic (100) face is looked on. Only hard X-rays with a wavelength shorter than twice the opening may go through these atomic sieves.

2.5 The Diamond Family

Diamond and graphite are polymorphs of the same carbon elements. The two structures are inter-related in such a way that by stretching (111) planes a diamond may transform into graphite. In reverse, by puckering the basal planes, graphite may transform into diamond. Hence, we may call diamond a puckered graphite—and graphite, a stretched diamond. The above illustration indicates that the two polymorphs may convert entirely from one phase to the other, and vice versa. In addition, diamond and graphite may transform into each other only partially. In this case, the intermediate structure may contain both diamond's (sp^3) and graphite's (sp^2) bonds. In fact, there is no pure diamond structure as it would require that perfect alignment of the lattice infinitely in all directions, so all diamonds inevitably contain a certain amount of graphitic bonds.

There are several ways that graphitic bonds can be incorporated in a diamond structure. The most common source of graphite bonds is located on the surface of a diamond crystal where at least one of the diamond bonds of the carbon atom is terminated to form an unpaired single electron, i.e., a dangling bond. Because the grain boundary is the source of graphitic bonds so the smaller the grain, the more are the graphitic bonds. In fact, a nanocrystalline diamond with about 1000 atoms (10 atoms on each edge in average) will have about the same number of graphitic atoms located on the surface as diamond atoms located in the body, so this nanoparticle should be renamed graphitic diamond. Another way to introduce graphitic bonds is to drill holes through diamond (e.g., by making an optical sieve) or to induce defects in the crystal. In this case, the situation is reversed, instead of graphitic bonds surrounding diamond bonds; graphitic bonds are enclosed in diamond bonds.

Another way to form graphitic diamond is to mix the two types of bonds in the same volume. This is the structure of amorphous diamond, a diamond-like carbon that contains no foreign element (e.g., hydrogen free).

A third way to mix the two type bonds is to introduce a slight diamond character into each of the atom in graphite. The net result is to curl up the basal planes, and depending on the distribution of the curled atoms, the structure may form either a ball or a tube, the former is known as bucky balls, and the latter, carbon nanotubes. These curled graphitic layers may also be viewed as hollow diamond balls or tubes as they may also be derived by eliminating the interior atoms from a solid diamond.

With the above relationship between graphitic bonds and diamond bonds in mind, DLC, bucky balls, nanotubes, and other structures that contain partial characteristics of sp^3 bond may be considered as diamondoids (partial diamond) materials (Fig. 2.10). This terminology is relevant as the unique shapes and properties of bucky balls and nanotubes are indeed derived from this partial diamond character. Because of this diamond connection, bucky balls and nanotubes have phenomenal properties and they have been envisaged to be the enabling materials for many exotic applications (e.g., hydrogen storage, field emission, drug carrier, molecule scale, nanotransistor, etc.).

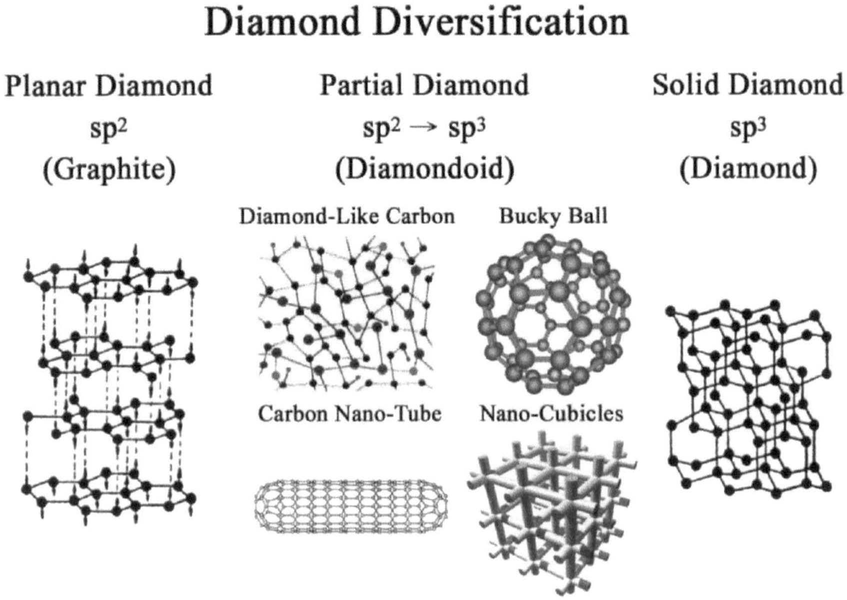

Figure 2.10 The diamond trend of carbon materials. This diagram shows the connection of many carbon materials that exhibits various degrees of diamond characters. The nanocubicles represent a horde of hypothetical structures that may be formed by assembling carbon nanotubes in various forms like using steel beams for constructing houses. This molecular manipulation of carbon nanotubes may be developed in the future.

2.6 Diamond Is Forever

With the trend of diversifying the product folio and the reduction of manufacturing costs, numerous new applications will be made debut. For example, diamond is the best component for microelectromechanical devices (Fig. 2.11).

When diamond products become commonplace, humans' material civilization will enter the last dynasty—the diamond age (Fig. 2.12). As there is no element that is above carbon in the periodic table, diamond will never succumb to any other material in the future. Thus, although diamond is chemically metastable and it is easily shatterable, "diamond is forever!" may not be just a figurative expression of desire, but a literal prediction for the ultimate stage of human civilization.

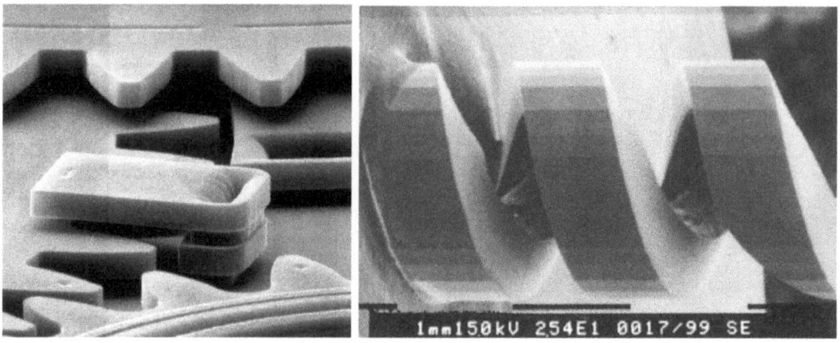

Figure 2.11 The laser-cut diamond microelectromechanical components of perpetual gears (left diagram) and the strongest screw (right diagram) possible. Both of them can last indefinitely even in a highly acidic environment.

The Scottish mathematical physicist James Clerk Maxwell derived the entire theory of electromagnetism in his head in 1861. He was amazed that this pure theoretical theory could explain the nature of ordinary light then unknown to scientists. In fact, the theory actually predicted the speed for the first time. Moreover, this theory predicted the existence of other invisible lights that might transmit with different frequencies. Maxwell envisaged that these then undetectable waves could be used to transmit messages over a great distance without making direct contacts. This foresight prompted Heinrich Rudolf Hertz to discover molecular energy radio waves in 1886, Guglielmo Roentgen to discover atomic energy X-rays in 1895, and Paul Villard to discover nuclear energy gamma radiations in 1900. In 1901, Guglielmo Marconi received the first trans-Atlantic radio message. This feat confirmed Maxwell's vision of acting at a distance without attaching a string.

Maxwell was so amazed that something that is everywhere could be so beautiful and perfect. Legend was that he exclaimed, "nothing can be too wonderful to be true." Indeed, without such a possibility, life would be seriously flawed. Fortunately, we do have diamond, the omnipresent material that is also omnipotent. Light and diamond make our world much more interesting and worthwhile to live. In fact, it is light that sparkles diamond in brilliance. Diamond will also become the ideal optical window that ushers light into numerous wonderful applications in the future.

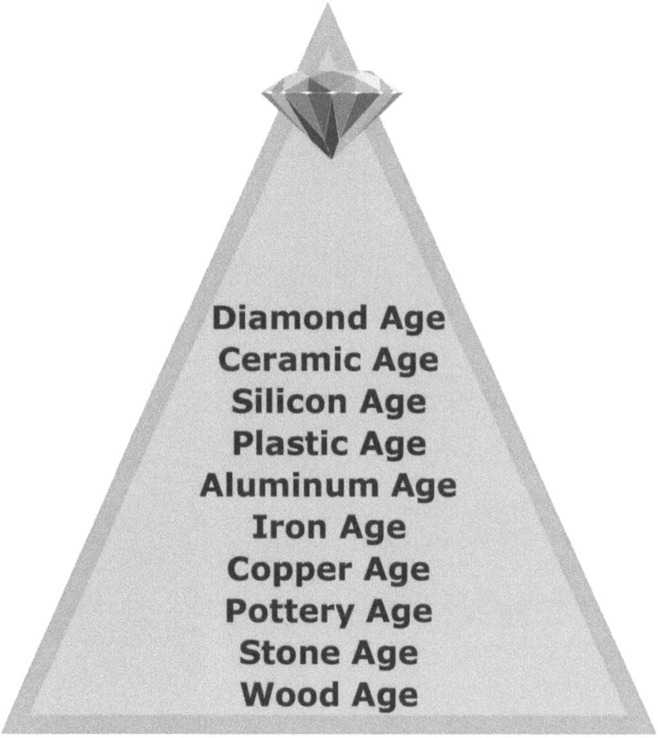

Figure 2.12 Materials civilization is succeeded by implementing more advanced materials from time to time. The conventional materials are not replaced in wholesale; rather, they are supplemented by the new champion material that can better meet with humans craving demand.

Chapter 3

Ultrahard Materials

"The hardness of diamond is like the speed of light that is impenetrable."

—Francis Bundy, 1986

3.1 Superhard Materials

Superhard materials include diamond and those materials that possess more than about half of diamond's hardness, such as cubic boron nitride (cBN), and possibly C_3N_4. Ultrahard materials are those with hardness that surpasses even that of diamond. Although no scientist has come up with a valid idea of what may make a material ultrahard, this chapter will discuss many possibilities that such superdiamond materials could be designed.

The intrinsic hardness of a material measures the weakest link of its atomic lattice. In order to minimize this weakness, each atom must be surrounded by at least four atoms to form the simplest solid—tetrahedron. In this way, the tetrahedral bonds of atoms can extend out to form a three-dimensional network

Handbook of Industrial Diamonds: Diamond Films and Carbon Coatings
James Sung

ISBN 978-981-4310-00-0 (Hardcover), 978-1-003-18045-6 (eBook)
www.jennystanford.com

without very weak cleavage planes. Such a structure is called diamond structure, and diamond is carbon atoms arranged in a tetrahedral network. With the increasing of the surrounded atoms known as kissing number (contacting neighbor or coordination number), the crystal structure will become more symmetrical and hence the possibility of having very weak cleavage planes is further reduced.

A carbon atom has four valance electrons ($2s^2+2p^2$). These four valance electrons can form either $sp^2\pi$ bonds with kissing number of 3, or sp^3 bonds with kissing number of 4 (Fig. 3.1). The former is the low pressure-high temperature allotrope of graphite; and the latter, high-pressure–low-temperature allotrope of diamond. Because graphite atoms cannot form three-dimensional structure, they spread out as hexagonal nets known as basal planes. These basal planes are loosely adhered by very weak van der Waals force. Consequently, graphite becomes one of the softest materials known. However, graphitic $sp^2\pi$ bond (1.45 Å) is covalent bond that is reinforced by metallic bond, so it is actually shorter than diamond's $sp^2\pi$ bond (1.54 Å). Consequently, graphite bonds are stronger than diamond bonds. The theoretical bulk modulus of graphite bond is 588 GPa, which is much higher than diamond's 443 GPa). Thus, although graphite's each basal plane is harder than diamond, the structure is soft because of these planes are separated too far apart (3.35 Å).

A harder lattice must be packed very tight so no weak plane exists. One way to pack atoms tight is to increase the packing efficiency. The packing efficiency increases with the increasing kissing number (Table 3.1). Hence, if everything else being equal, the higher the packing efficiency of a crystal structure is, the harder it can be. However, if the kissing number becomes larger than valance electrons, the bonds become metallic that is significantly weaker than covalent bonds. In this case, although the crystal lattice is symmetrical, the bonds are not strong to make the structure hard. For example, iron (austenite) is made of closest packed atoms with kissing number of 12, but it has only six valence electrons, so its hardness is not very high.

The packing efficiency is limited by the size of atoms. If atoms are smaller, more atoms can be packed in a giving volume than by surrounding each atoms with more contacting neighbors.

This is a much more effective way to increase the atomic concentration. Moreover, by keeping the kissing number low, covalent bonds can be maintained to make the structure strong. Thus, although diamond's packing efficiency is low, its atoms are small, so they can be packed so tight as to attain the highest atomic concentration (176/nm^3) of all structures. Consequently, diamond becomes the hardest material known.

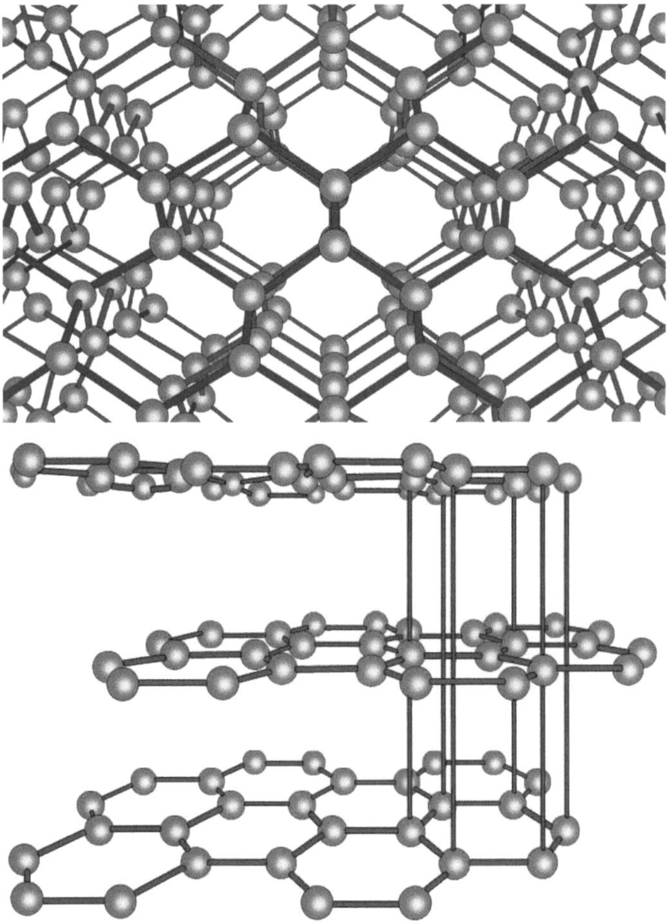

Figure 3.1 The 3D superhard diamond (top diagram) and the 2D ultrahard graphite (bottom diagram). Despite the fact that the latter is harder than the former on the plane, it is extremely soft due to the presence of very weak van der Waals force that holds basal planes together.

Table 3.1 Kissing number and its packing parameters

Kissing number	Crystal structure	Packing efficiency	Size ratio
3(2D)	Triangular	0.91	>0.155
4	Diamond	0.34	>0.225
6	Simple cubic	0.52	>0.414
8	Body center	0.68	>0.732
12	Face center	0.74	1.000

Note:

[1]The size ratio refers to the size of an atom surrounded by size 1 of contacting neighbors.

[2]Face center cubic (cubic closest packing) structure has the same packing efficiency as that of hexagonal closest packing lattice.

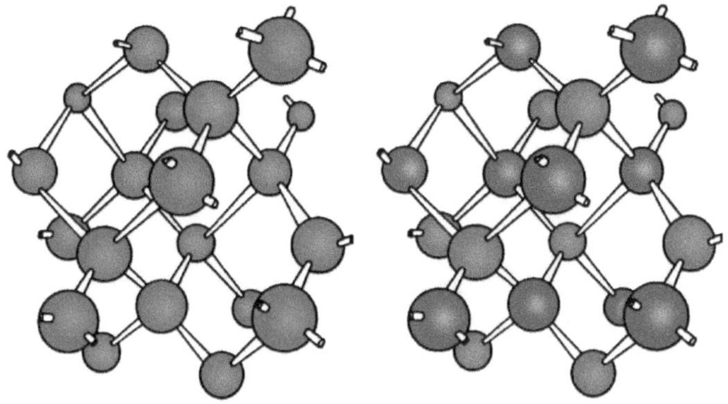

Figure 3.2 The isostructural compounds of diamond and cBN. The latter can be derived from the former by replacing B–N bonds with C–C bonds. The bonding electrons of C–C are located at the middle. However, they shift toward N in B–N. This shift weakens cBN so its hardness is only about half of diamond's.

The substitution of C–C by B–N in diamond will keep the structure intact. This isostructural compound is called cubic boron nitride (cBN) (Fig. 3.2). Because cBN contains exclusively light elements, it becomes the second hardest material known. The B–N bond formed between dissimilar atoms is slightly ionic. Hence, it is weaker than the pure covalent bond of C–C formed between identical atoms. As the consequence, cBN's hardness is only about half that of diamond. Even so, it is still the second hardest material known.

3.2 The Scale of Hardness

A harder material can penetrate a softer material because the former contains tighter packed atoms. Hence, the earliest hardness scale is based on abrasion test on minerals known as Mohs relative hardness (Table 3.2).

Table 3.2 Relative hardness of minerals

Mineral	Composition	Mohs	Knoop, N/mm^2
Talc	$Mg_3Si_4O_{10}(OH)_2$	1	
Gypsum	$CaSO_4 \cdot 2H_2O$	2	320
Calcite	$CaCO_3$	3	1,350
Fluorite	CaF_2	4	1,630
Apatite	$Ca_5F(PO_4)_3$	5	4,300
Orthoclase	$KAlSi_3O_8$	6	5,600
Quartz	SiO_2	7	8,200
Topaz	Al_2SiO_4	8	13,400
Corundum	Al_2O_3	9	21,000
CBN	BN	9.5	45,000
Diamond	C	10	70,000

Note: Wooddell has extended Mohs' scale beyond corundum (9) to 19 for cubic boron nitride, and 42.5 for diamond.

Mohs' scale of hardness is not linear, in particular that diamond (10) is more than 4 times of corundum (9). More quantitative measurements (e.g., Vicker or Knoop scale) of hardness were developed based on the pressure applied to the deformed area using a wedge shaped diamond indenter. Figure 3.3 shows the relationship between "absolute" scale and relative scale of hardness.

As hardness reflects the atomic concentration in a crystal lattice, elements with smaller atoms are generally harder. Since the lighter atoms are smaller, the hardest materials known all contain light atoms at least half of B, C, N or O atoms. In the case of superhard materials such as diamond and cBN, they are made of 100% of light atoms. For other less hard materials, they may include some heavier atoms.

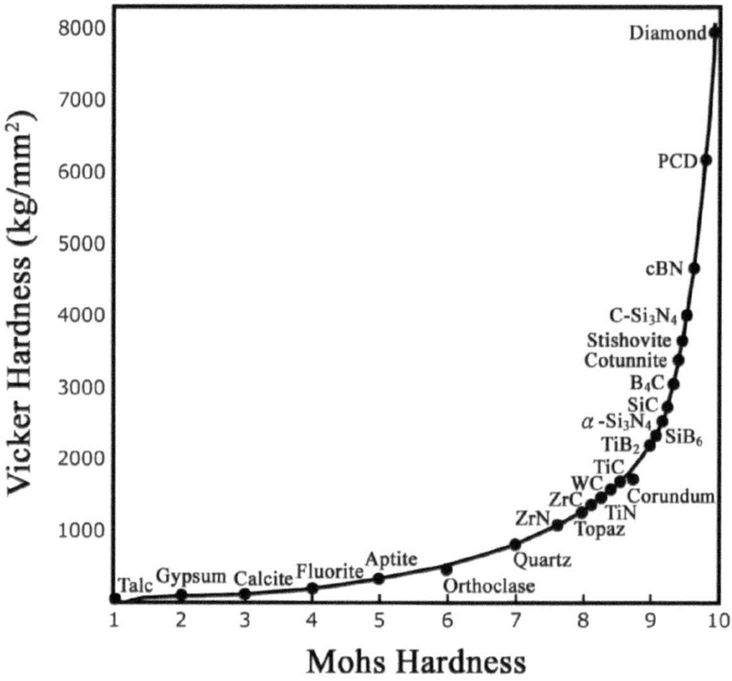

Figure 3.3 The comparison of Vicker hardness measured by indentation (ordinate) and abrasion (abscissa). In the above diagram, the lowercase c refers to cubic lattice. The measured Vicker hardness figures are: spinel-Si_3N_4 = 43 GPa, Cotunnite TiO_2 = 38 GPa, TiB_2, BP, and $(Si,C)_3N_4$ = 33 GPa.

The echelon of the third hardest materials include B_6O in rhombohedral structure, Si_3N_4 in spinel structure, SiO_2 in stishovite structure, and TiO_2 in cotunnite structure (space group Pnma). These phases are all high-pressure allotropes of common ceramics corresponding to boron oxide, α- or β-Si_3N_4, quartz, and rutile, respectively. The high pressure has increased the packing efficiency by raising the kissing number of cations. For example, Si atoms in silicates are normally surrounded by four neighboring atoms, but in the case of spinel phase, the kissing number of 2/3 of their atoms have increased to 6. For stishovite, all Si atoms are bonded to six oxygen atoms. As a consequence of increased atomic concentration, their hardness increase substantially. However, these third echelon materials can never match the hardness of superhard diamond and cBN because they contain atoms larger atoms of heavier elements (e.g., Si, Ti...etc.).

Based on the above discussions, very hard materials contain mostly light elements, superhard materials contain only light elements, and for ultrahard materials that are harder than diamond, not only there should be all light elements, but also more valence electrons must be involved in forming covalent bonds. In this case, the packing efficiency is increased without making the structure metallic. This is the only way to make a crystal lattice harder than diamond.

3.3 What Is Hardness?

The intrinsic hardness of a crystal structure reveals the weakest link of the lattice in all directions. Thus, although graphite may be harder than diamond on basal planes, the three-dimensional structure becomes one of the softest because of the weak van der Waals force that links between basal planes.

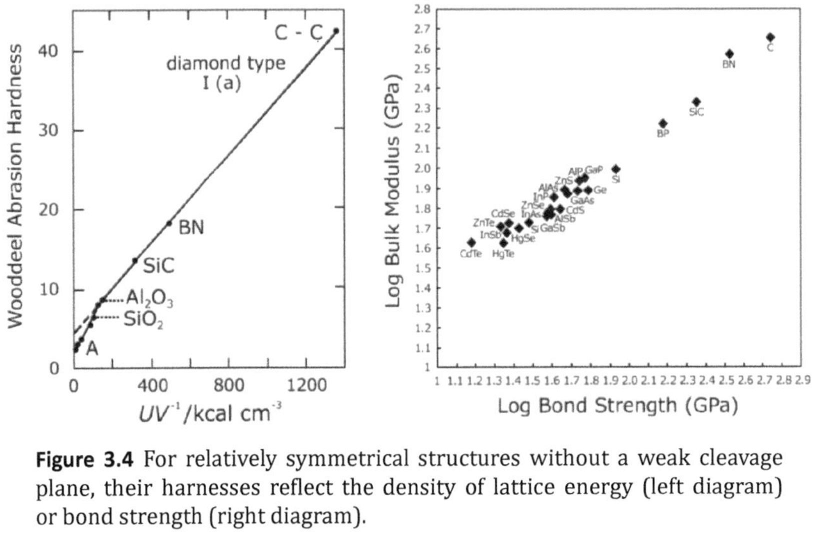

Figure 3.4 For relatively symmetrical structures without a weak cleavage plane, their harnesses reflect the density of lattice energy (left diagram) or bond strength (right diagram).

Hardness is typically expressed in unit of pressure (or stress). However, pressure (F/A, where F is force and A is area) is equivalent to energy density ($F \times L/A \times L = E/V$, where L is distance, E is energy, and V is volume), so hardness measures the energy density that is needed to move a group of atoms away from their equilibrium positions in the lattice. As atoms are anchored by atomic bonds; hence, for a symmetrical structure without obvious weak planes, hardness manifests the concentration of bond energy that holds atoms in place (Fig. 3.4).

3.4 The Limit of Hardness

The supreme hardness of diamond was established millennia ago when it was used to inscribe other gems. From time to time, scientists dreamed to find a superdiamond material that is harder than diamond, but inevitably, these attempts failed. For example, when cubic boron nitride (cBN) was first synthesized by Robert Wentorf of General Electric Company in 1957, for a brief period of time, it was thought to be harder than diamond. But this elation soon vanished, and diamond's royal hardness remained unchallenged until now.

Francis Bundy, who led GE's scientists to synthesize first man-made diamonds under high pressure in 1954, probably spent more time than others experimenting new materials that might be harder than diamond. In 1986, he received the Bridgman Award, the highest honor for a high-pressure scientist to bestow. During the award speech at AIRAPT International High Pressure Conference held in Kiev, Bundy reported his decades-long quest for a superdiamond material. His conclusion was that diamond's hardness might represent a natural barrier analogous to the speed of light that is impossible to penetrate. Indeed, Bundy dismal prediction seems to prevail, as new quests for the superdiamond have so far been futile.

However, just as the case that before diamond was actually synthesized, there were sniffs of possible success, indications are there that diamond's hardness may not be nature's impenetrable boundary after all. For example, it was demonstrated that twinned lamellae in diamond are locally harder than diamond itself. Moreover, "work hardened" diamonds that contain high concentration of dislocations are harder than unstrained diamond. By the same token, nanocrystalline diamond with ample grain boundaries may also be harder than single crystal diamond. The same may be true for supercooled liquid carbon as well.

However, the above examples of possible superdiamonds are all based on the irregularities of diamond structure itself. The higher hardness achieved is not the intrinsic material property, but representing the extra energy required to move carbon atoms across the crystal defects. In order to be a real superdiamond, the composition, and/or the bonding structure must be altered to make the structure more rigid than diamond.

3.5 The Criteria of Superdiamond

There are two approaches to boost this energy concentration: either by increasing the atomic concentration or by strengthening the bond intensity. The atomic concentration (mole's number) can be increased by incorporating smaller atoms in the structure, or by improving the packing efficiency. The smallest atoms belong to Period 2 elements (Period I elements of H and He do not have enough bonding electrons to form a rigid three-dimensional structure), so the superdiamond must be made from atoms of period 2 elements.

The packing efficiency of a crystal structure is reflected in the number of immediate neighbors an atom has, known as kissing number (KN) or coordination number (CN). For a structure to be hard, it must not possess weak planes. Hence, the atoms ought to have coordination in three dimensions. The simplest volumetric solid is a tetrahedron; therefore, KN must be at least four to be able to support the structure in all directions. This is why the same carbon can be as soft as graphite when its KN = 3, and it becomes as hard as diamond when its KN = 4. With further increases of KN, not only atoms are packed tighter, but also the symmetry is increased. The latter trend will further eliminate weak planes. The result is a more isotropic structure, not only in hardness, but also in other properties (e.g., resistivity).

There are three ways to distribute valence electrons around an atom: metallic, covalent, and ionic. If valence electrons are fewer than four, as in the case of Group I to III elements, the electronic repulsion among neighboring atoms are weakened, so each atom can have more atoms around. As a result, the kissing number is high (6, 8, or 12). Because the few valence electrons must be shared by the more neighboring atoms, the bonds become metallic. On the other hand, if valence electrons are four or higher, as in the case of Group IV to VIII elements, the electronic repulsion among neighboring atoms are enhanced. As a result, the kissing number is low (0, 1, 2, 3, or 4). Because the more valence electrons are shared by the few neighboring atoms, the bonds become covalent. If metallic atoms (e.g., Li) form bonds with covalent atoms (e.g., F), these bonds become sim ionic (e.g., LiF).

In order to make the structure hard, the bonds between atoms must be strong. The bond strength is proportional to the electron concentration located between the bonding atom. For metallic bonds, the number of valence electrons is less than the kissing number; hence, the bond strength is greatly reduced by the thinning effect of the shared electron concentration. For covalent bonds, the number of valence electrons is equal or more than the kissing number; hence the bond strength is high. For ionic bonds, the shared electrons are off center, so the shared electron concentration also decreases. Hence, the strength of covalent bond is much higher than either metallic or ionic bond. The van der Waals bond is even weaker as it is not an atomic bond, but formed between polarized molecules.

In order to make the structure superhard, the bonding must be predominantly covalent. Hence, it would eliminate all metallic elements (Group I to III) as superhard materials. Moreover, the structure must be three-dimensional so the kissing number must be at least 4. This requirement would eliminate all covalent elements that form multiple bonds (Group V to VIII). Thus only carbon group (Group IV) remains. Group IV elements are uniquely located at the middle of these two opposite trends: many weak metallic bonds and few strong covalent bonds. As a result, C, Si, Ge atoms possess the highest amount (4) of single covalent bonds that is just enough to support a three-dimensional diamond structure. Hence, these elements possess the highest bond energy per atom so they can form the hardest structure made of the same or the higher Period elements.

As the hardness reflects the concentration of bond energy, the atoms must not be larger than carbon. This requirement would eliminate all Group IV elements except carbon. Thus, carbon is the only superhard element remaining. This would explain why diamond is thought to possess the highest hardness possible for all materials.

The above-discussed criteria for increasing hardness are summarized in Table 3.3.

Table 3.3 Criteria of hardness (bond energy/volume)

Criteria	Approach	Design parameter
Atomic density	Use smaller atoms	Use period 2 elements
	Pack atoms tighter	Increase KN number
Bond intensity	Covalent bonding	Use non-metallic elements
	More bonds	Increase KN number

3.6 The Kissing Number

The crystal structure of a superhard material must possess no weak planes; hence, its symmetry ought to be high. In order to avoid such weak planes, the kissing number must be at least 4. Hence, the kissing numbers for superdiamond structures include 4, 6, 8, and 12. In fact, these structures are interrelated. All of them can be derived from the closest packing such as cubic closest packing (CCP) that has a face centered cubic (FCC) lattice.

A CCP structure has a kissing number of 12. In addition to be surrounded by 12 neighbors, each atom is also surrounded by 6 octahedral void (O), and 12 tetrahedral voids (T). Each O void is surrounded by six atoms, and each T void is surrounded by four atoms. Hence, the ratio of O void to atom (i.e., O/A) is 1, whereas that T/A = 2.

If all O voids are occupied by the same atoms in a CCP structure, the result would be two interpenetrating lattices of CCP, each occupying the O voids of the other. This is a simple cubic structure with a kissing number of 6. If every other T void is occupied by the same atoms in a CCP structure, the final structure contains two interpenetrating lattices of CCP, each occupying the T voids of the other. This is a diamond structure with a kissing number of 4.

In a CCP structure, all T voids form a simple cubic structure, so if both O and T void are filled up, the composite structure contains two interpenetrating lattices of simple cubic structure. This is the body-centered cubic structure (also CsCl, AgMg, AlNi) with a kissing number of 8.

In the case that all O voids are filled up and every other T voids are also occupied, the finished structure is equivalent to leaving O voids empty but filling up all T voids. This is a structure of CaF_2 (also ZrO_2) or its reverse, the structure of Li_2O (also Na_2O). This structure will be used to describe a hypothetical superhard materials C_2Ne (see latter description).

Based on the above scenario, all symmetrical structures can be derived from CCP by filling up various portions of interstitial voids as shown in Table 3.4.

Table 3.4 Symmetrical structures derived from CCP

Old structure	Void filled	New structure	Kissing number
CCP	None	CCP	12
CCP + CCP	O	Simple cubic	6
CCP + CCP	T/2	Diamond	4
CCP + CCP	T	C_2Ne	8 + 4
CCP + CCP + CCP	O + T	Body-centered	8

Alternatively, the above filling process may be depicted by the interpenetrating of two or three identical CCP structures. Thus, simple cubic structure with KS = 6 may be formed by the interpenetrating of two CCP structures at each other's O voids. Diamond structure with KS = 4 may be viewed to be interpenetrating of two CCP structures at each other's half T voids. Body-centered cubic structure with KS = 8 may be constructed by interpenetrating of three CCP structures among there O and T voids (Fig. 3.5).

3.7 The Concentration of Atoms

Diamond's carbon atoms are the smallest among Group IV elements, so it can pack the most atoms per unit volume ($176/nm^3$) in the structure (Fig. 3.6). Thus, although diamond's kissing number is only 4 so its structure is relatively open, but in terms of the number of atoms, or number of bonds, diamond's packing is the densest.

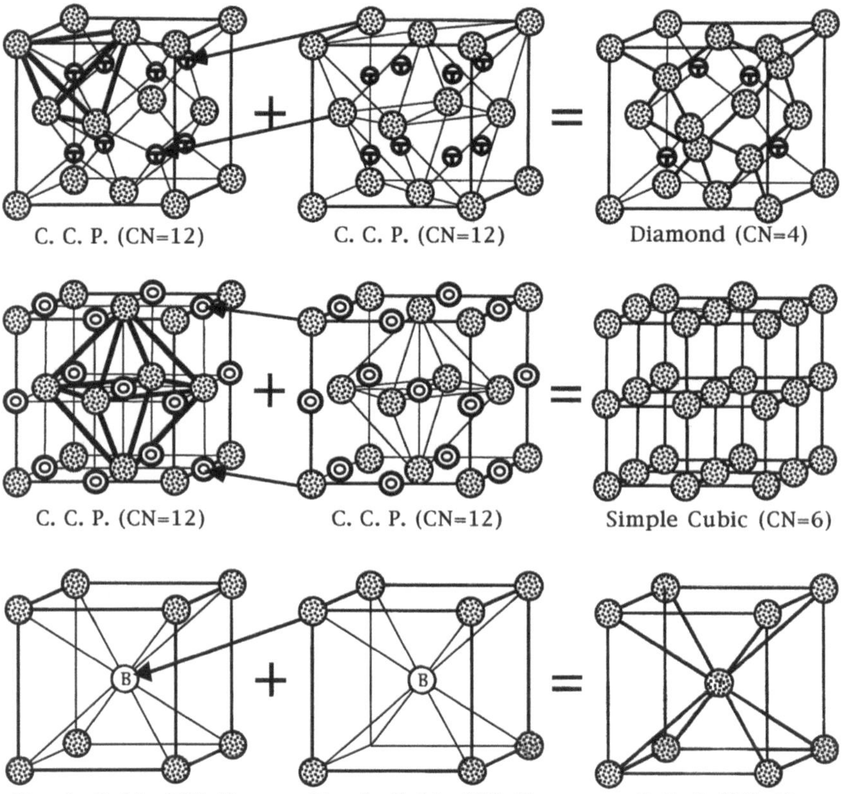

Figure 3.5 Superdiamond structures may be formed by interpenetrating CCP structures at each other's voids by forming single covalent bonds with neighboring kissing atoms. In this case, all valence electrons are used so there will be no dangling bonds or lone pairs to destabilize the structure. The top diagram shows that the mutual penetration of two CCP structures in each other's tetrahedral voids will form a diamond structure of KS = 4. The middle diagram shows that the mutual penetration of two CCP structures in each other's octahedral voids will form a simple cubic structure of KS = 6. The bottom diagrams shows that mutual penetration of two simple cubic structures in each other's body-centers will form a body-centered structure of KS = 8. This structure is equivalent to that formed by mutual penetration of four CCP structures in tetrahedral and octahedral voids of all of them.

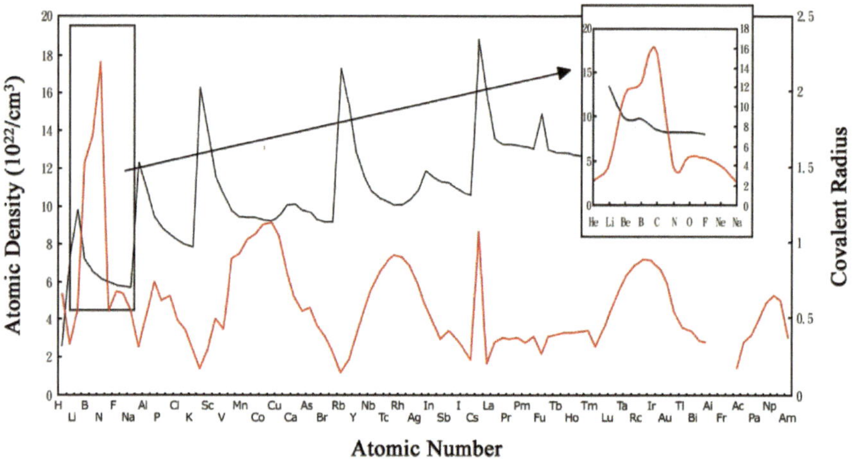

Figure 3.6 Atomic sizes and atomic concentration. Carbon atoms are small but not the smallest. However, the small carbon atoms coupled with four neighbors make diamond the tightest packed structure of all materials.

In general, the material hardness of a covalent compound is proportional to the atomic concentration (Fig. 3.7). However, if the degree of covalence of the bond is compromised, the hardness may reduce substantially. The bond covalence may be lost if the bonding electrons are fewer than the coordinated neighbor, as in the case of metals. The bond covalence may also be reduced if different atoms are incorporated, as in the case for ionic compounds.

Group V to VIII elements possess more than four valence electrons. These electrons tend to form lone pairs that repel neighboring atoms. As a result, they form diatomic molecules (e.g., N_2, O_2, F_2) and single atomic molecule (e.g., Ne). Because of the formation of lone pairs, the number of bonding electrons are reduced. Hence, these elements can form less than four single covalent bonds per atom. The concentration of bond energy of a structure is directly proportional to the number of single covalent bonds each atom has. Diamond's atoms have the most (4) single covalent bonds per atom. As a result, diamond's concentration of bond energy is the highest (Table 3.5), so is its hardness.

According to the above discussion, diamond has been the hardest material because it possesses the largest atomic concentration and strongest atomic bonds. In order to be harder than diamond, the crystal structure must pack more atoms, and/or the atoms are bonded stronger. But according to the above discussions, there seems to be no element that is capable of forming such a tighter structure with a higher concentration of bond energy.

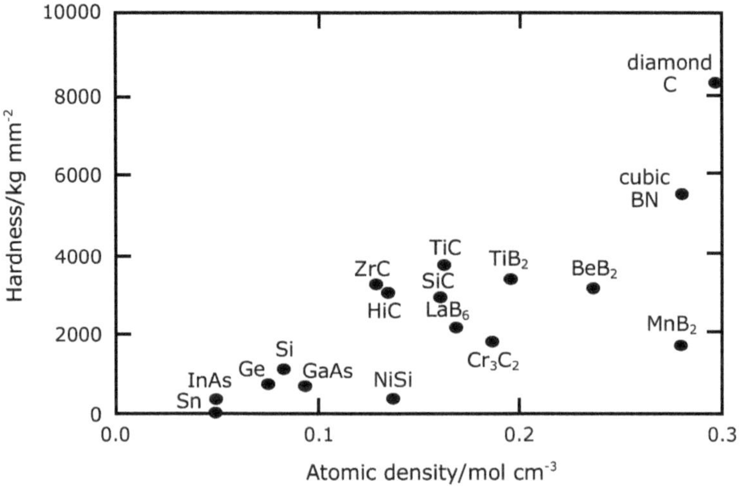

Figure 3.7 The hardness is proportional to the atomic concentration for covalent structures. However, this hardness can be compromised if the bond becomes either metallic or ionic.

Table 3.5 Bond energy (eV)/atom of period 2 elements

Element	Bond energy (eV)	No. of bonds	Total energy (eV)
Li	0.55	1	0.55
Be	0.61	2	1.2
B	3.1	3	9.3
C	3.7	4	14.8
N	4.2	3	12.6
O	4.6	2	9.2
F	4.8	1	4.8

There have been claims that certain structures may be harder than diamond. But these allegations cannot stand to the test of the above criteria. For example, it was surmised that C_{60} (fullerite) could be polymerized under high pressure to form an "ultrahard" structure. Moreover, this ultrahard fullerite may be twice the hardness of diamond when measured by surface scratches using NanoScan (Blank, et al., 1998). However, the ultrahard fullerite is not denser than diamond, so its atomic packing cannot be higher. Moreover, the structure is held primarily by $sp^2\pi$ bonds, so its bond number is not higher than diamond. Thus, the bond energy per unit volume of ultrahard fullerite must be noticeably lower than diamond. However, the arched $sp^2\pi/sp^3$ bonds could exhibit an apparent surface hardness that may be ultrahard. Such a hardness is like that of three-dimensional graphite (bct-4 or H-6, see Fig. 3.5, bottom) with distorted $sp^2\pi/sp^3$ bonds. But the theoretical hardness in terms of bulk modulus for 3D graphite is on par with cubic boron nitride. Hence, we expect that the true volumetric hardness of ultrahard fullerite should be about half of diamond, even though its surface hardness may be twice.

3.8 The Design of Superdiamond

Diamond's hardness comes from its small atoms that each of them forms four covalent bonds. As a result, diamond's atoms can be packed in such a tight structure that it contains the highest atomic concentration of all substances. To make a structure harder than diamond, its atoms must be smaller than carbon, and/or these atoms form at least four covalent bonds. The first consideration would strike off all elements with period number higher than 2. The second criterion would eliminate all elements lighter than carbon. Although these elements may form mono-atomic structures that is packed tighter than diamond, but their bonding electrons are less than kissing number so they form weak metallic structure (e.g., H) or van der Waals solid (e.g., He). With all these elements ruled out, only C, N, O, F, and Ne are possible candidates of superdiamond (Fig. 3.8).

However, in order for N, O, F, and Ne to become a superdiamond, these elements must form mono-atomic structures with kissing number higher than 4. Moreover, no lone pair electrons are allowed, so all their valence electrons must be involved in single covalent bonds. The superdiamond structures so formed contain atoms smaller than diamond's so the crystal structure is tighter. In addition, each atom has at least four covalent bonds to allow them to form symmetrical lattice with no weak cleavage planes.

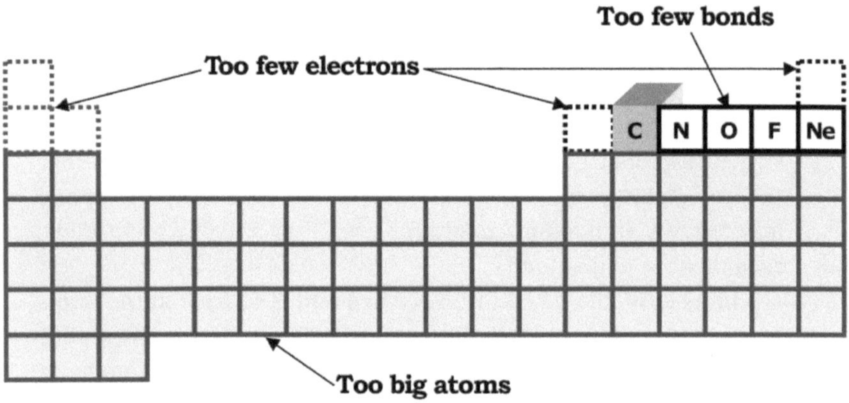

Figure 3.8 Elements with electrons less than carbon's (6) cannot form four covalent bonds that is required to support a rigid structure. On the other hand, elements with electrons more than neon's (10) are too big to pack tight enough to form a superdiamond. Elements with 7–10 electrons contain lone paired electrons, so they form gases. Hence, diamond becomes the hardest material. However, if the lone pairs can form single covalent bonds, then N, O, F, and Ne atoms could pack tighter than diamond so they may become superdiamond.

In order to squeeze more atoms in a given volume, the structure must be mono-atomic. That means, the lone pair of non-bonding electrons must be invoked to form single covalent bonds (Fig. 3.9). Although such structures are not available from these covalent elements, there is no a priori physical law that prohibits them to form. In fact, the second law of thermodynamics would require the formation of covalent bonds at the expense of lone pair electrons because this is the way to reduce the entropy. Under extreme pressure, electronic states tend to become equal; otherwise, the volume cannot be reduced effectively.

Under the extreme pressure, the packing efficiency must increase to reduce both volume and entropy. The packing of single atoms and diatomic molecules with lone pair electrons are very loose due to the weak van der Waals bonds. Hence, if pressure is high enough, these lone pair must break up to form mono-atomic structure with single covalent bonds. Indeed, when diatomic molecules of common gases, such as H_2, N_2, or O_2, were solidified cryogenically, and they were squeezed under extreme pressure using diamond anvils, these diatomic molecules all show the tendency to dissociate and form mono-atomic structures.

As these hypothetical structures may possess superdiamond hardness, their concentration of bond energy would be higher than diamond. So once the atoms are packed in place, these structures could be locked up indefinitely regardless of whether they are stable or not. Hence, after synthesis, such a superdiamond may survive at ambient conditions even if it is highly metastable, as in the case of diamond.

In a sense, the quest for superdiamond is to pack more atoms to a crystal structure than diamond. The increase of atomic concentration to make superhard materials may be compared with the increase of nucleus mass for the synthesis of superheavy elements. Just as an island of stability for superheavy elements (N = 114–184) is predicted to exist beyond uranium (N = 92), the heaviest stable isotope known (the heaviest artificial isotope is now approaching N = 112), so may an island of stability for superdiamond structures be there awaiting discovery.

Once the atoms are packed tighter, their structure would become harder. Diamond can cut other materials because its atoms are packed tighter. If atoms of any element of N, O, F, Ne can be packed tighter than diamond, such element can be used to cut diamond just like diamond cutting other materials (Fig. 3.10).

3.9 The Invention of Carbon Nitride

The simplest way of making a superdiamond is to incorporate smaller atoms into the diamond structure. Of the four superdiamond candidate elements, N is a common impurity of diamond. Most natural diamonds belong to a category called Type Ia that may contain several thousand PPM of nitrogen atoms. However, most nitrogen atoms in natural diamonds form included aggregates that can weaken the structure. On the other hand, diamonds synthesized with molten metal catalysts (e.g., Fe, Ni, Co or their alloys) under high pressure typically belong to Type Ib. Instead of forming inclusions as natural diamond, synthetic diamonds may contain a few hundred PPM of nitrogen atoms that substitute carbon atoms in the diamond lattice.

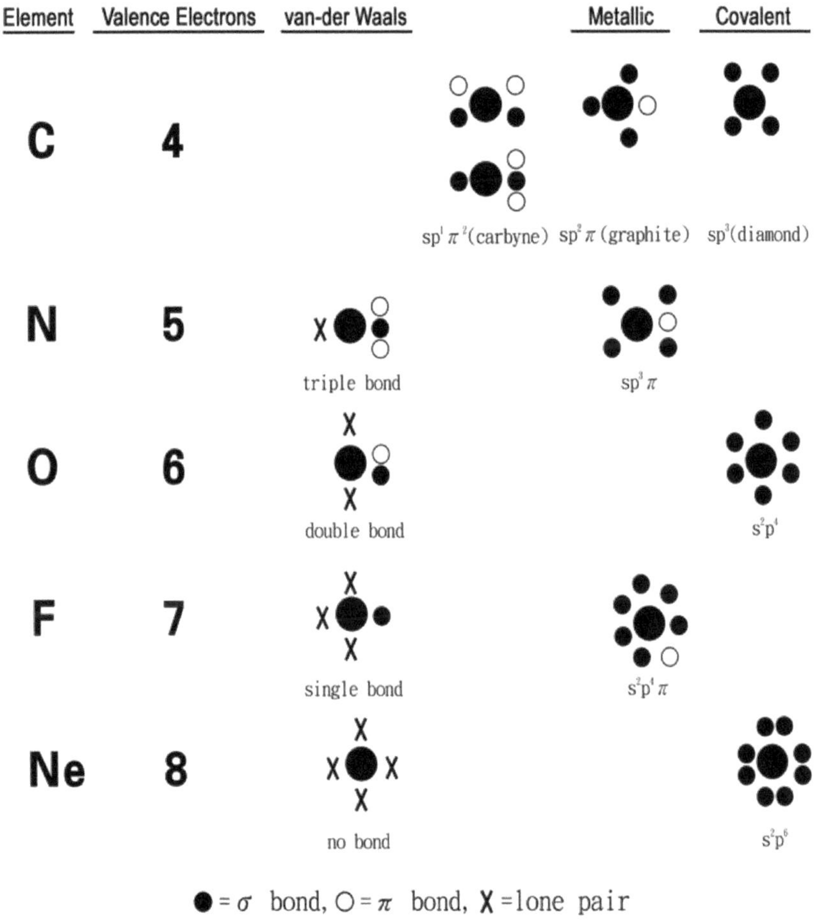

Figure 3.9 Two-dimensional illustration of possible covalent bond types for C, N, O, F, and Ne. Each legend for s bond, π bond, and lone pair contains two electrons.

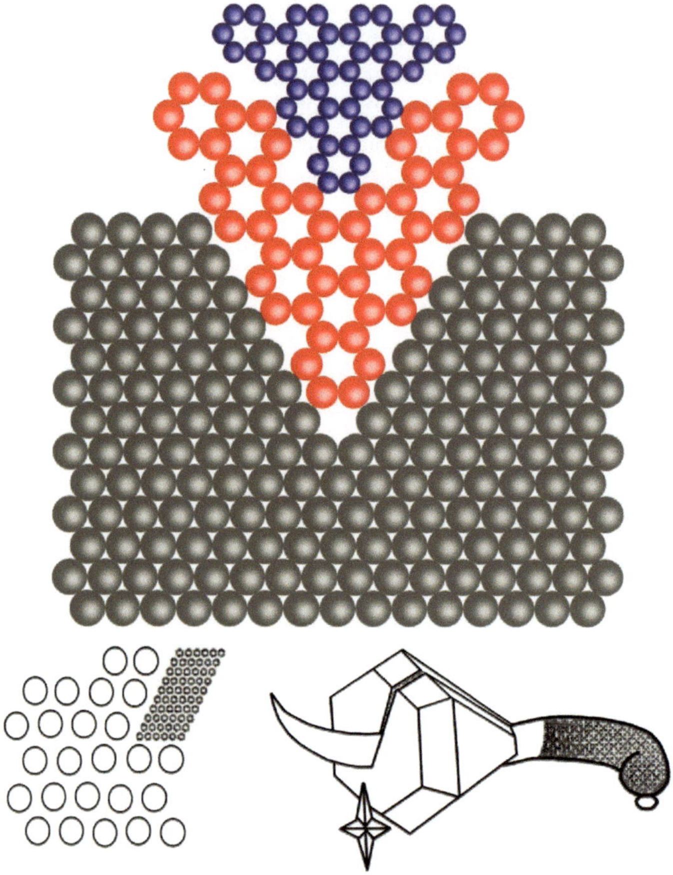

Figure 3.10 Materials with tighter atomic lattice can penetrate into those with looser structures (top diagram). Hence, diamond can cut into other materials (bottom left). However, a superdiamond with even tighter atomic structure can slice diamond just as easy as diamond splitting other materials (bottom right).

A perfect Type lb diamond may be harder than pure diamond (Type lla) because the former has a high atomic concentration due to the presence of smaller nitrogen atoms. However, when the concentration of nitrogen atoms increases, they tend to segregated along (100) faces. In this case, nitrogen atoms may be adjacent to one another. As a consequence, their extra electron will combine to form double or triple bonds. The clustered nitrogen aggregates has made (100) faces of diamond softer.

In order to make diamond harder than its intrinsic lattice, very small amount (PPM levels) of nitrogen atoms must be incorporated as substitutional atoms (Fig. 3.11).

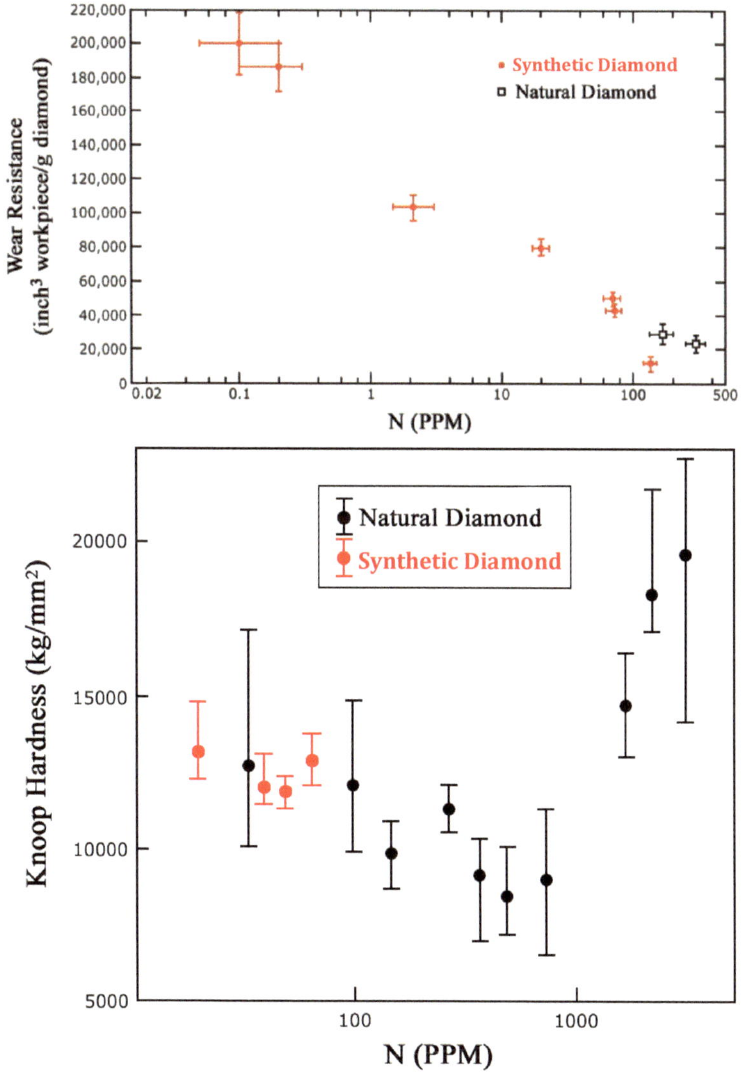

Figure 3.11 The incorporation of nitrogen atoms in platelets in diamond may weaken the structure (top diagram). However, if nitrogen atoms can substitute carbon atoms in the lattice, the hardness of diamond may increase (bottom diagram).

Recently, Dr. Chih Shiue Yan seemed able to incorporate more nitrogen atoms in substitutional positions by high-pressure annealing (e.g., 6 GPa, 2000°C, 10 min) of nitrogen-doped CVD single crystal diamond. The precursor diamond was grown homo epitaxially on (100) face of a Type Ib diamond seed. Contrary to normal Type Ib diamonds that reveal softer (100) faces, this time it became superhard with Vickers hardness much higher than even the purest Type IIa natural diamond (Fig. 3.12).

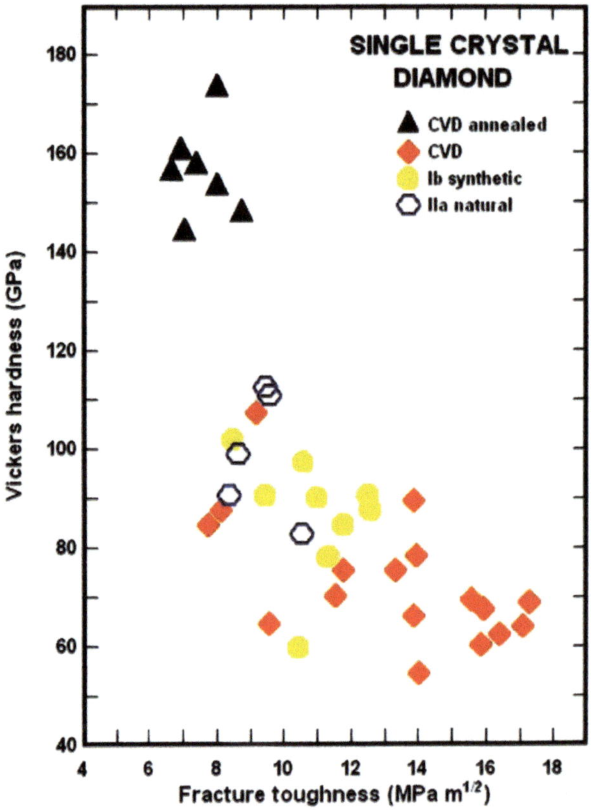

Figure 3.12 Dr. Yan's measurements indicated that nitrogen-doped (100) diamond surface after high-pressure annealing may become much harder than diamond itself. The original diamond was grown by CVD in microwave plasma with a gas mixture held at a pressure up to 220 torrs. The CH_4/H_2 ratio was up to 20%; and N_2/CH_4 ratio, up to 5%. The CVD diamond was homo epitaxially deposited onto a synthetic diamond seed of 1 mm in size to from a much larger single crystal diamond with the same orientation.

If more nitrogen atoms are to be incorporated in substitutional positions, an outlet of their extra electrons must be found. One possible way to tie up nitrogen's dangling electrons is to incorporate hydrogen atoms. As hydrogen atoms are extremely small (covalent size = 0.64 Å), they can hide insterstially by attaching to the side of nitrogen atoms. If this were the case, the superdiamond may have a stoichiometry of C–N–H. However, the single bonding is distinctly different from hydrogen cyanide (H-C≡N) that contains triple bonds. This hypothetical compound of CNH has a much higher atomic concentration; hence, its hardness could be significantly higher than diamond. The compressive hardness of a symmetrical structure may be estimated from its bulk modulus (volume incompressibility). Using a simple empirical model, the author predicted that such a structure may have a bulk modulus of 4.9 GPa, which is 17% higher than 4.3 GPa of diamond (Sung and Sung, 1996).

Most hard materials are either carbides (e.g., SiC) or nitrides (e.g., BN). These hard materials attain their hardness mainly by incorporating the small atoms of carbon or nitrogen. If carbon and nitrogen can combine to form a new structure, it may be superhard. In order to allow such a compound to form, the extra electron of each nitrogen atom is combined with another bonding electron to form a stable pair. In this case, each nitrogen atom can form only three bonds instead of four. However, carbon atoms can still maintain their original tetrahedral coordination like diamond does.

Although multiple bonded graphitic phase of C_3N_4 was mentioned in the literature, the author was not aware of it. In 1984, the author proposed that single bonded C_3N_4 may be synthesized to form three-dimensional structures (Sung, 1984) (Fig. 3.13) that are isostructural to αand β phases of Si_3N_4, the common hard ceramics (Fig. 3.14). Moreover, the two isostructural phases should form solid solution with a continuous change of composition of $(C,Si)_3N_4$. Because nitrogen atoms (covalent size = 1.40 Å t are considerably smaller than carbon atoms (covalent size = 1.54 Å), C_3N_4 structures, if formed, could be harder than diamond. But each nitrogen atom can form only three bonds that is one bond less than the carbon atom, so the tighter packed structure contains only 6/7 of bonds when compared to diamond, so the relative hardness between the two was debatable.

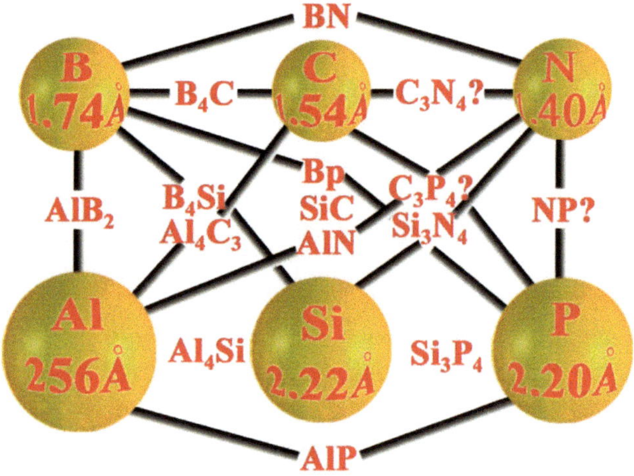

Figure 3.13 Light elements and their covalent compounds. The circles represent the covalent atomic sizes. All compounds shown are very hard, but only those made of at least 50% of C or N are superhard. In 1984, the author proposed that hypothetical compounds with question marks may be synthesized.

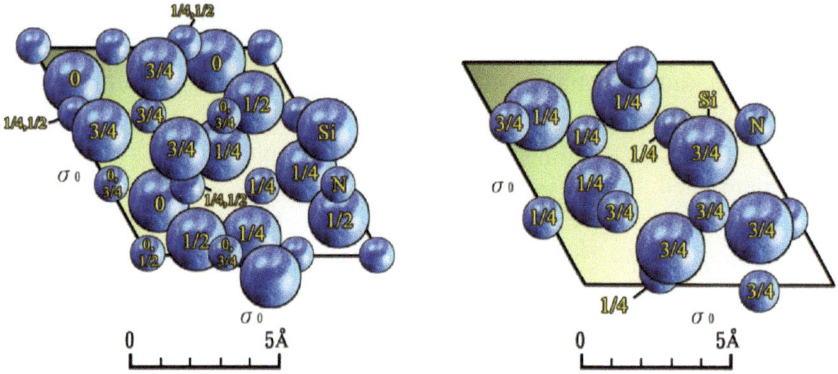

Figure 3.14 The two hexagonal phases of Si_3N_4 that may be emulated for C_3N_4. Because of this isostructural relationship the two compositions may form continuous solid solution of $(C,Si)_3N_4$.

The diagram shows α phase on the left, and the more symmetrical β phase on the right. Large circles represent Si;

and smaller ones, N. The fraction denotes the height of the atom from the basal plane.

In order to estimate the hardness of hypothetical C_3N_4 structures, the author hired Prof. M. L. Cohen to compute their theoretical values. In 1985, Cohen came up with an empirical formula that may be used to calculate the bulk modulus of a diamond-like compound of any composition. His results indicated that the hardness of C_3N_4 structures could indeed rival with diamond. Because C_3N_4 structures were then patentable ideas of the author, Cohen removed their descriptions from the text and published only the empirical formula for estimating bulk moduli (Cohen, 1985). However, he acknowledged that the research was initiated by the author and the paper was published under the sponsorship of Norton Christensen, Inc., where the author was employed.

In 1989, Cohen formally unveiled the possible existence of C_3N_4 compounds in Science magazine (Liu and Cohen, 1989). The possible existence of these superdiamonds created a sensation in the scientific community so many scientists began the attempt to synthesize C_3N_4. After a decade of intensive research worldwide, a record of scientific research may have been achieved with about 1000 scientific papers published on the subject of this elusive material. About 100 scientists claimed the success of synthesizing C_3N_4 crystallites, but the indisputable proofs are still lacking. However, there have been many carbon nitride thin films deposited, some with measured hardness approaching that of diamond. As one of the benefits of this research, CN films are now routinely used as tribological coatings (e.g., for protecting hard drives).

3.10 The Synthesis of Silicon Carbon Nitride

Ceramicists have long been trying to combine the two most important advanced ceramics; SiC and Si_3N_4, without success. The composite is predicted to be tougher than either end member, a property that is highly desirable for brittle materials. However, the quest for C_3N_4 can now provide an alternative synthesis route. The solid solution of $(C, Si)_3N_4$ may prove to join the three merits together, hardness, toughness, and thermal stability (Fig. 3.15).

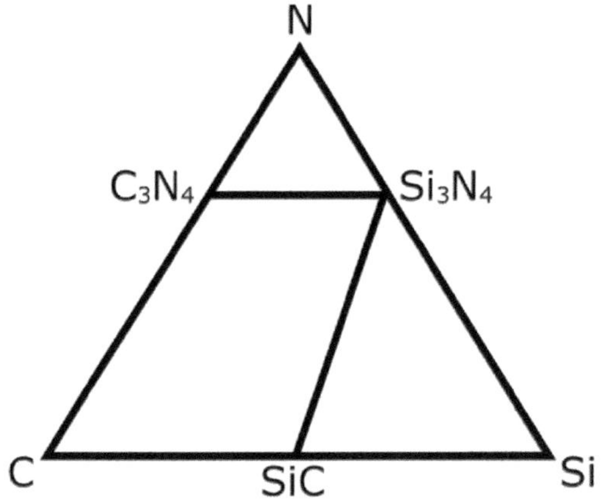

Figure 3.15 The tie line between Si_3N_4 and SiC appears to be an impasse for synthesis. However, the ternary compound of the above two end members, $(Si,C)_3N_4$, can be grown by routing through C_3N_4.

C_3N_4 is intrinsically metastable because carbon atoms are too small to keep nitrogen atoms from approaching each other, as a result, the increased repulsion of their lone pairs electrons tends to destabilize the structure (Fig. 3.16).

In contrast, the large silicon atoms can keep nitrogen atoms far enough to minimize their mutual repulsion. Hence, Si_3N_4 becomes a stable phase. But if Si is replacing part of C atoms in C_3N_4 structure, the solid solution of $(C,Si)_3N_4$ may become stable. Indeed, large (100 micrometers) single crystals of $(C,Si)_3N_4$ solid solutions was synthesized at Taiwan University (Fig. 3.17) in 1998, and also at Pennsylvania State University in 1998. Thus, it confirmed the author's prediction made in 1984 based on the hypothetical isostructural relationship between C_3N_4 and Si_3N_4.

Although the carbon nitride research has been extensive with over 100 papers published a year, all of them are academic works involving supplicate vacuum technologies (e.g., sputtering, laser ablation, etc.). Hence, there have been no crystals available for practical application. In order to mass produce $(C,Si)_3N_4$, a catalyst must be needed. Hence, the author suggested using iron group metal as the catalyst to mass produce $(C,Si)_3N_4$ crystals (Sung, 2001). By decomposing gases of hydrogen and hydrides of C (e.g., methane), Si (e.g., silane), and N (e.g., ammonia) on a molten film of metal catalyst, the CVD process can yield large quantities of $(C,Si)_3N_4$ crystals. This process is analogous to the conventional vapor–liquid–solid (VLS) route to grow carbon or ceramic fibers, and it has also been used to mass produce carbon nanotubes (Fig. 3.18).

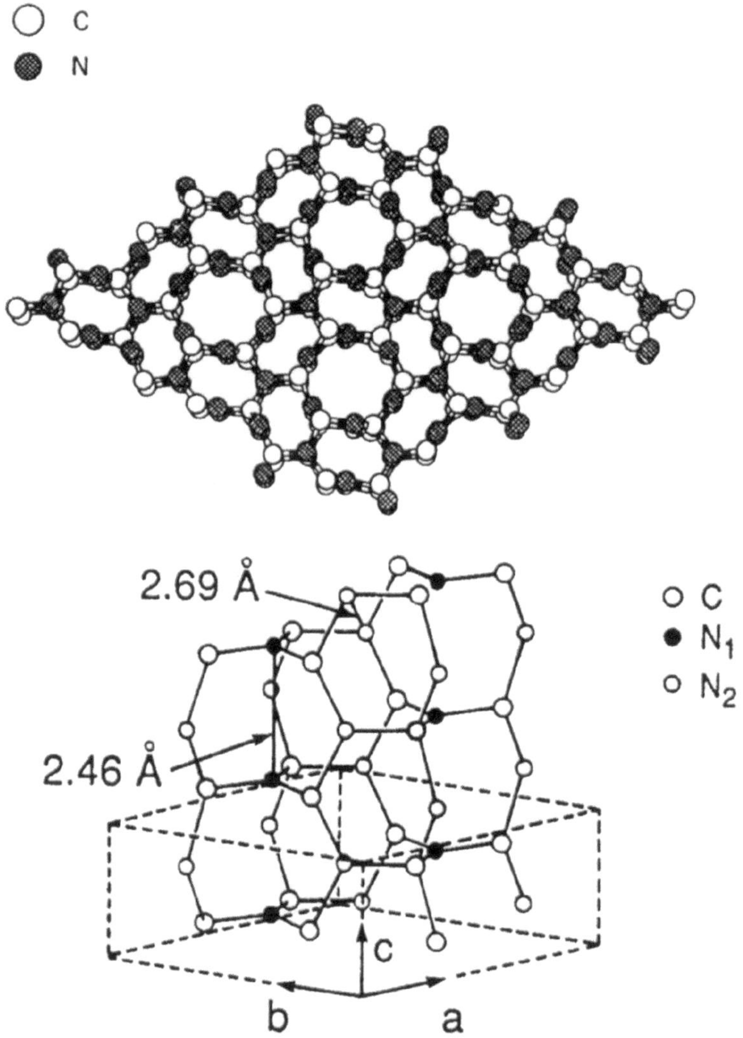

Figure 3.16 Crystal structure of beta-C_3N_4. The nearby N1 atoms may repel each other with their lone-pair electrons. The result is a destabilized structure. The N1–N1 distance can be increased by substituting C atoms with Si or Ge atoms.

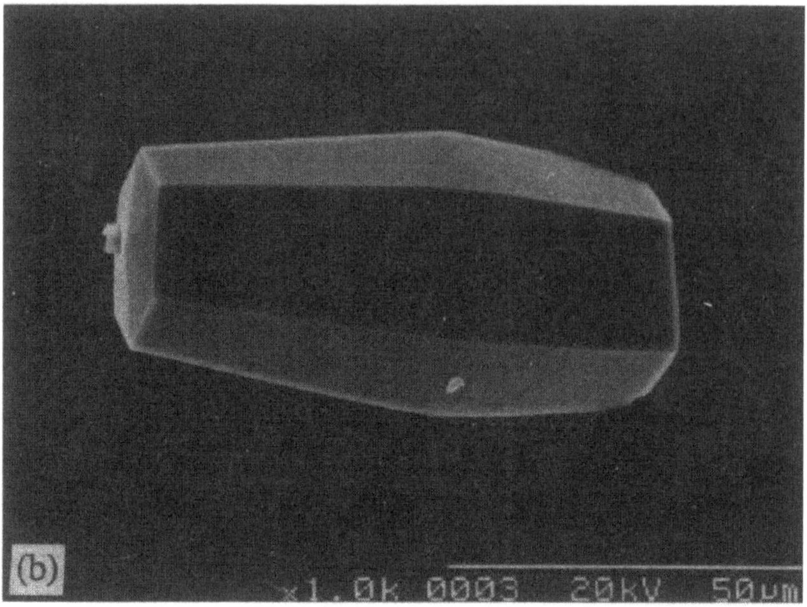

Figure 3.17 The hexagonal prism of $(C,Si)_3N_4$ synthesized at Taiwan University contained a low amount of Si (about 5 atom%). It possesses a direct band gap (3.8–4.7 eV) that is capable of emitting blue light.

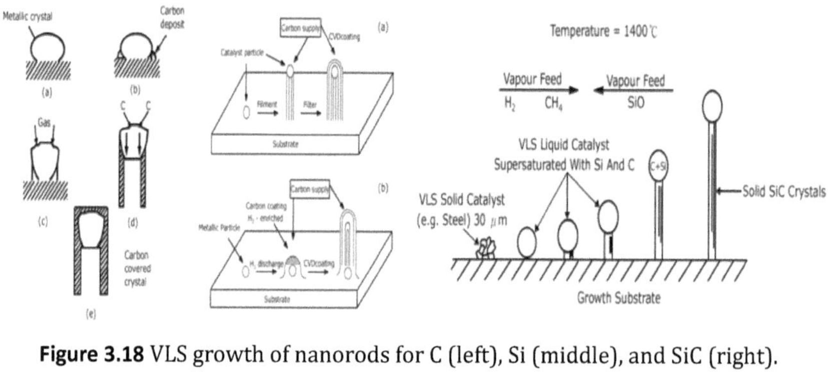

Figure 3.18 VLS growth of nanorods for C (left), Si (middle), and SiC (right).

In order to test the feasibility of this catalytic process of VLS, he author funded a project at Pennsylvania State University. Indeed, tens of carats of $(C,Si)_3N_4$ crystals up to 100 microns have been grown. If this method can be implemented on industrial scale, $(C,Si)_3N_4$ crystals might be produced at a cost as low as about $10/Kg. This cost is comparable to that for the manufacture of conventional premium abrasives, such as seeded gel process for making polycrystalline alumina (e.g., Norton's SG™ or 3M's Cubitron™). In contrast, the superabrasives nowadays is priced at $1/carat. Hence, there is a price saving by more than 1000 times. Such a drastic price reduction could revolutionize the entire abrasive industry.

If $(C,Si)_3N_4$ crystals can be mass produced at low cost, it can replace much expensive superabrasives of diamond and cubic boron nitride (cBN). It has been demonstrated that if the Si content is low in $(C,Si)_3N_4$, the crystal can be much harder than any ceramic material (e.g., SiC or B_4C) used as abrasives. In fact, its hardness may be approaching cBN. Thus, it may be possible to produce $(C,Si)_3N_4$ as the substitute of much more expensive cBN. $(C,Si)_3N_4$, being a nitride, it possesses higher thermal stability and chemical inertness than diamond, so it should be able to machine ferrous alloys effectively just like cBN. Moreover, with its cost thousand times less than diamond, it may have the potential to replace diamond superabrasive as well.

The successful syntheses of Si–C–N ternary compounds also represent a technical triumph as materials scientists have long attempted without success to synthesize a solid solution between common ceramics of SiC and Si_3N_4. Thus, although large C_3N_4 crystals are yet to be produced, the quest for C_3N_4 compounds have already chartered a new frontier of materials science and engineering.

3.11 Three-Dimensional Graphite

In graphite, the four carbon valence electrons form three in-line covalent (sp^2) bonds plus one parallel metallic π bond. The reinforcement of these two types of bonds makes graphite bonds (length 1.45 Å) shorter than diamond's (sp^3) bond (length 1.54 Å). Hence, graphite's basal planes are literally harder than diamond. However, its superdiamond hardness is restricted to the two-dimensional plane, so the three-dimensional structure is in fact extremely soft.

If sp^2 bonds can be twisted to form a three-dimensional structure, the weak planes are removed, so the structure could be superhard. There are two ways to twist graphite bonds; by 60° (H-6) or 90° (bct-4) as illustrated in Fig. 3.19.

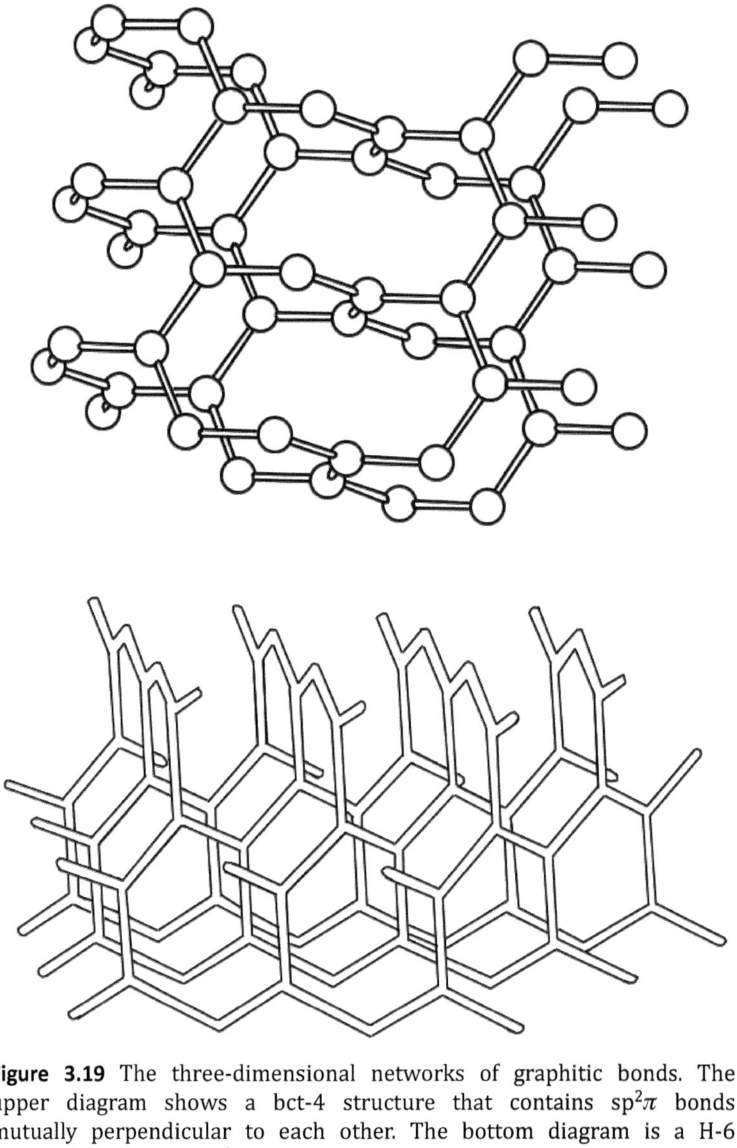

Figure 3.19 The three-dimensional networks of graphitic bonds. The upper diagram shows a bct-4 structure that contains $sp^2\pi$ bonds mutually perpendicular to each other. The bottom diagram is a H-6 structure that allows $sp^2\pi$ bonds to twist 60° successively.

In either case, the twisting has extended the bond length from 1.42 Å of planar configuration to 1.45–1.47 Å for twisted versions. But even so, they are still much shorter than diamond's bond length of 1.54 Å. Hence, the twisting of $sp^2\pi$ bonds only weakens the bond strength slightly. As a result, the three-dimensional networks of graphite are indeed superhard. If hardness is expressed in terms of bulk modulus, H-6 has a value of 372 Gpa; and bct-4, 362 Gpa. These figures are on a par with cubic boron nitride's 367 Gpa. These three-dimensional structures of graphite are still softer than diamond (bulk modulus 443 Gpa) because their kissing numbers is 3/4 of diamond's. In other words, if more carbon atoms can be packed into the structure, the 3D graphite could become a genuine superdiamond.

3.12 Diamond-Like Nitrogen

Although $sp^2\pi$ bonds of graphite cannot be packed tighter because of the three way coordination, $sp^3\pi$ bonds can allow the tetrahedral coordination but with shorter bonds than diamond's. In other words, if diamond's carbon atoms can acquire one more valence electron, they could form a diamond with graphitic bonds! Such a structure would be a superdiamond. The diamond structure with graphitic bonds can be assembled with N atoms that carry five valence electrons instead of four. If the additional valence electron of the N atom can form a reinforced π bond, as in the case of forming graphitic basal planes, then diamond-like nitrogen would be a superdiamond. On the other hand, if extra electrons repel each other, as in the case of separating graphitic basal planes, then the hypothetical structure may not be that hard.

A total energy calculation (Wu, 1996) has revealed that in a hypothetical structure of diamond-like nitrogen the length of $sp^3\pi$ bond is 1.77 Å, that is much longer than diamond's 1.54 Å. As a result, the bulk modulus of such a hypothetical structure is only 1.6 GPa that is significantly lower than diamond's 4.3 GPa. This implies that diamond-like nitrogen is relatively hard (about that of WC), but much softer than diamond.

The above calculation seems to indicate that the extra electrons of nitrogen atoms tend to destabilize the structure of diamond-like nitrogen. However, this is not necessarily so. For example, the extra electrons in benzene and carbonate are resonating among bonds and this resonance are in fact strengthening the structure. If nitrogen atoms could form resonant double bonds, the interatomic distance of the diamond-like nitrogen may be further shortened to 1.2 Å. Such a compact structure would make diamond-like nitrogen much harder than diamond. As pointed out above, twisted $sp^2\pi$ bonds can form superhard structures of 3D graphite. Hence, it is not unlikely that twisted $sp^3\pi$ bonds may make diamond-like nitrogen a superdiamond.

On the other hand, if dangling bonds are indeed a destabilizing factor, they may be tied up by tucking in with hydrogen atoms that fill the crystal lattice interstitially (Fig. 3.20). In this case, the extra electrons are no longer free to move, nor are they repulsive to one another. The interatomic distance in the diamond-like NH crystal should be in the order of 1.4 Å that is significantly shorter than diamond's 1.5 Å. Hence, this hypothetical structure could be a superdiamond.

Diamond and diamond-like nitrogen may also form mixed crystals or a compound with the composition of CN. In this case, the extra electrons are associated only with N atoms. Again, hydrogen atoms may be tucked in to tie up the dangling bond. Hence, a composition of CNH may be formed. Although this composition is identical to hydrogen cyanide (HCN), they have distinctly different structures.

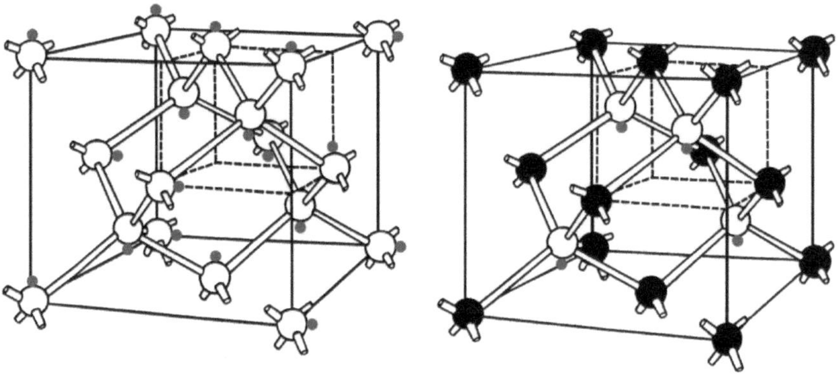

Figure 3.20 Diamond-like N or NH (left diagram) and its compound with diamond, diamond-like CN or CNH (right diagram). NH may be viewed as diamond hydride; and CNH, the compound of diamond and diamond hydride.

In diamond-like nitrogen, the N–N bond might also be substituted by C–O bond. In this case, the extra valence electrons are contributed only by oxygen atoms, as a consequence, the C–O bond becomes ionic. Hence, diamond-like CO would not be as hard as diamond-like N. The relationship between diamond-like N and diamond-like CO is analogue to that between diamond and cBN. The dangling bonds in diamond-like CO may also by plugged by hydrogen atoms. Hence, a structure with the composition of COH_2 may be formed.

3.13 Diamond-Like Light Atoms

Figure 3.21 shows an example of electron concentration maps of diamond-like B, C, N, and O (Wu, 1996). It would appear that diamond-like structures of B and C are highly covalent; hence, they are electrically insulating. On the other hand, diamond-like O may be electrically conducting. But diamond-like nitrogen is somewhere intermediate, so it is likely semiconducting. Indeed, ultrahigh-pressure experimentation of diamond anvils performed at Geophysical Labs of Carnegie Institution of Washington confirmed that nitrogen is a semiconductor at very high pressures.

Electrons are bonded too tightly to atoms in an insulator, so they cannot flow. In this case, the electrical resistivity is extremely high. In contrast, electrons are bonded too loosely to atoms in a metal, so they can flow. But electrons must move around individual atoms by overcoming the drag force of atoms. Consequently, there is a moderate electrical resistivity in metal. However, in diamond-like nitrogen, the extra electrons are perched on the balance between covalent insulator and metallic conductor. In this case, electrons may move in a straight line like the propagation of a crack through a brittle material. As the drag forces of neighboring atoms are canceling out, electrons in diamond-like nitrogen may move without resistance. Hence, diamond-like nitrogen may be superconducting, similar to that of many superconducting nitrides, such as NbN. But in diamond-like nitrogen, electrons do not have to go through atoms of different types, so its critical temperature of superconductivity would be much higher than that for all superconducting nitrides. Moreover, diamond-like nitrogen is much harder than any nitride that contains ionic bonds held between different atoms, so its superconductivity are likely not be disturbed by the rise of temperature. This means diamond-like nitrogen may be a room temperature superconductor. This is the Holy Grail that many physicists have quested for in vain. Current records of superconductivity are held by oxide ceramics that contain layered structures. The critical temperatures still lie in cryogenic regime of lower than −250°C. Furthermore, diamond-like nitrogen has another dream property if it is superconducting. As it is an isotropic solid in three-dimensions, so its superconductivity would not be blocked in any direction. Such a homogeneous superconductor is much more desirable than two-dimensional superconductors of oxide ceramics.

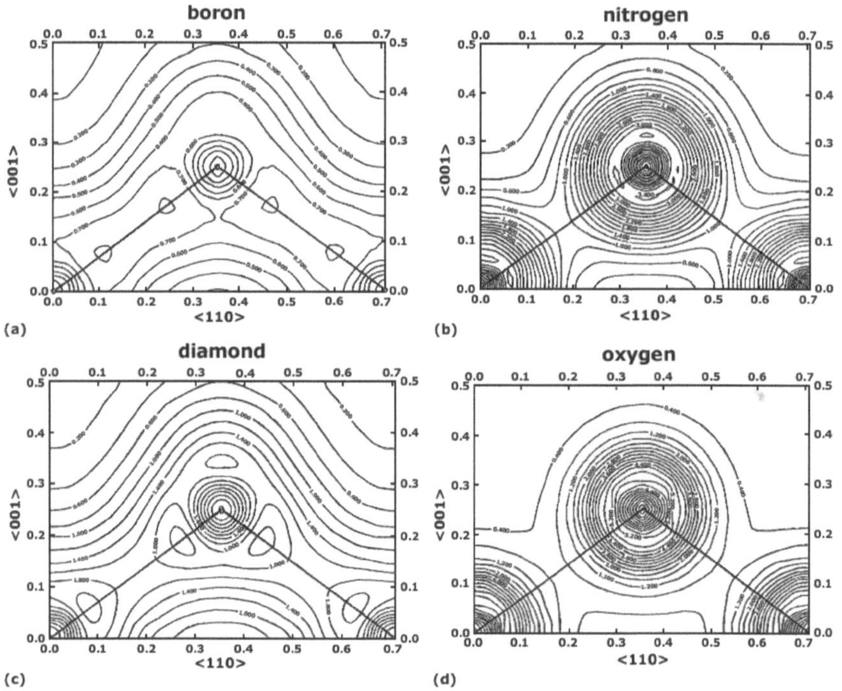

Figure 3.21 Electron concentration maps of diamond-like B, C, N, O. Note that B and C possess covalent bonds; and O, metallic bonds. N's bonds are intermediate, so diamond-like nitrogen is likely a semiconductor.

Diamond-like nitrogen may have many superb characteristics that exceed all other materials including diamond. Among them are hardness, thermal conductivity, and chemical inertness. It may be possible to synthesize diamond-like nitrogen by dissociating ammonia molecules to form a plasma that contains nitrogen ion. The nitrogen ions can then be collimated along the tetrahedral angle, and they are accelerated toward a common center. At the point of collision, these nitrogen ions may combine to form the mono-atomic structure of diamond-like nitrogen. Moreover, the properties of such diamond-like nitrogen clusters, if formed, may be studied with the newly discovered femtochemistry as invented by Dr. Ahmed Zewail, a Nobel laureate. Thus, a laser strobe light tuned to femtoseconds (10^{-15} sec) may be used to monitor the structure taking shape in situ; and the properties of the forming structure, such as its hardness may be measured instantaneously. If diamond-like nitrogen can be synthesized, they may be stabilized by quenching and maintained as a metastable structure just as in the case of diamond.

3.14 Simple Cubic Oxygen

With increasing pressure, the crystal structure must become more compact and eventually its KN will have to increase. Thus, it is predicted (Yin and Cohen, 1980) that at a pressure of about 2300 GPa, diamond would transform into a simple cubic lattice, similar to the structure of NaCl, MgO, or TiC (Fig. 3.22). However, if a distorted tetrahedral structure called BC-8 is lower in energy than diamond, diamond would convert to BC-8 first at a much lower pressure, and the latter would then transform into simple cubic structure at a much higher pressure (Fig. 3.23).

If carbon atoms can form six covalent bonds, each shares two electrons with the neighbor, the simple cubic structure would be much more concentrated in bond energy, and therefore it may be much harder than diamond.

However, the energies of two 1s electrons in carbon atoms are at least seven times lower than that of the spread for four valence electrons of 2s/2p; hence, the core electrons will not participate in forming the bond. The bond of simple cubic carbon is therefore weakened due to the dilution effect of four valence electrons shared by six neighboring atoms. The bond strength is further reduced by the increasing metallic character due to the enhanced overlap of valence bands. As a result, the bond length of simple cubic carbon has increased to 1.7 Å, and its bulk modulus plummeted to 3.4 GPa (Wu, 1996). Thus, although simple cubic carbon may be as hard as cubic boron nitride, it is not a superdiamond.

The six valence electrons in oxygen atoms have an energy spread that is about twice that of the four valence electrons in carbon (Fig. 3.24). If oxygen atoms could form a simple cubic structure, their six valence electrons may form six single covalent bonds. In this case, the simple cubic oxygen would have a much higher concentration of covalent bonds, so it could be much harder than diamond (Fig. 3.25).

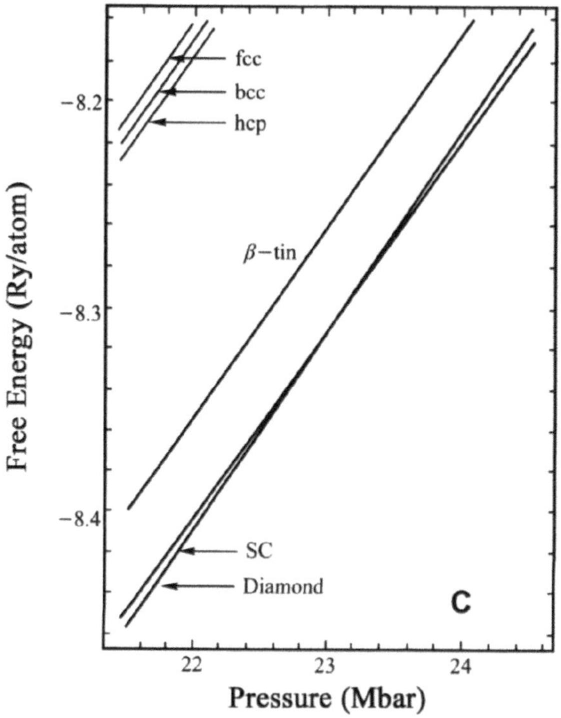

Figure 3.22 The rise of total energy for different structures of carbon. Note that diamond with KN = 4 will become unstable above a pressure of 23 Mb, and it may transform into simple cubic structure with KN = 6.

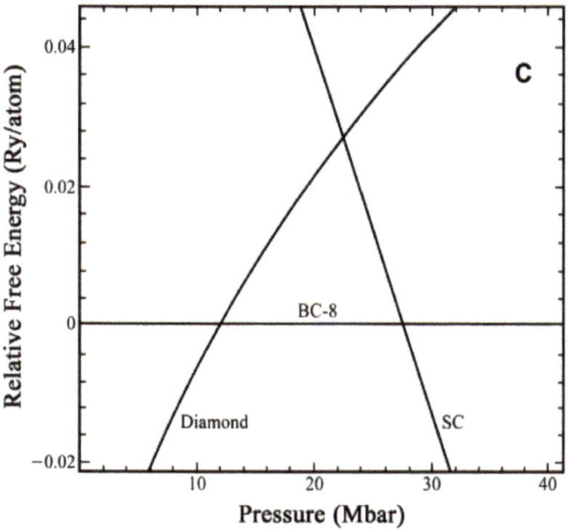

Figure 3.23 If the distorted tetrahedral bonding of BC-8 structure can be formed, diamond will first transform into BC-8 structure at a lower pressure, and then the latter converts into simple cubic structure at a much higher pressure.

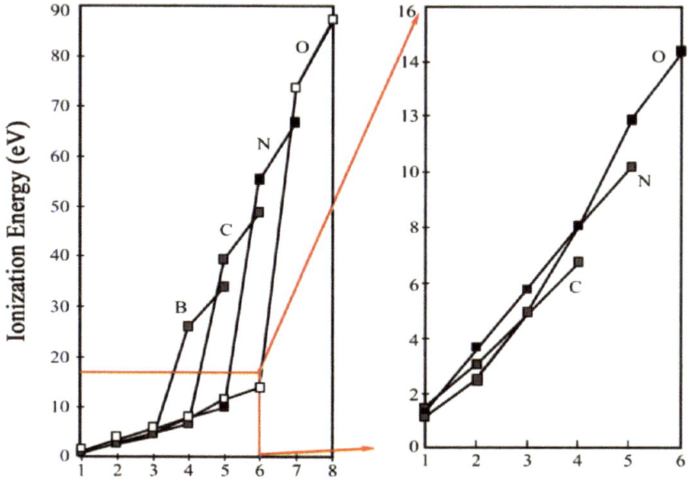

Figure 3.24 The ionization energy increases with the number of electrons removed away from an atom. Note that the core electrons are too difficult to be involved in bonding. However, the valence electrons are capable to form single covalent bonds if pressure is high enough to provide the required energy to dissociate diatomic molecules into mono-atomic structure.

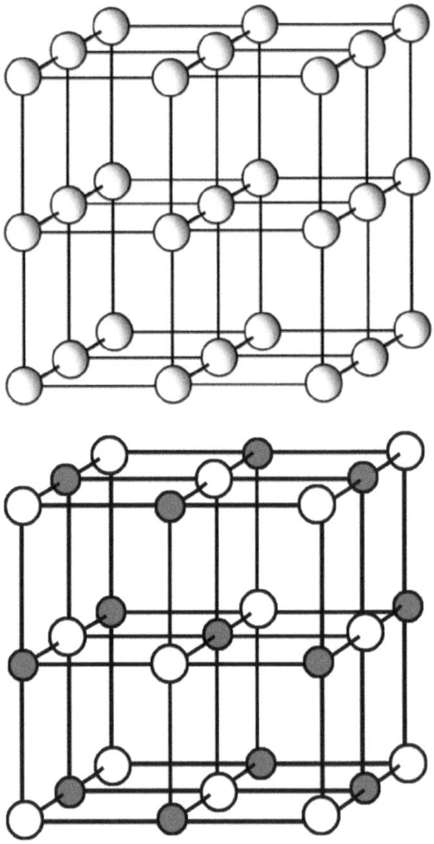

Figure 3.25 Simple cubic oxygen may be a superdiamond, if all six valence electrons of its atoms can form six distinct single covalent bonds (left diagram). Simple cubic NF may also be ultrahard, but because its bonds are ionic, it would not be as hard as simple cubic oxygen (right diagram).

Oxygen atoms do form mono-atomic solid such as MgO and Al_2Q_3 where their atoms are arranged in the closest packing. However, these oxygen atoms are ions so their bonding would not be strong. Even so, these oxides are among the hardest ceramic materials known.

One way to estimate the possible hardness of simple cubic oxygen is based on the isostructural compounds of NaCl, MgO, and TiC. All of them have the simple cubic structure if their atoms are the same. These three structures are with increasing degree of covalence so their hardness increase substantially. Based on a simple inverse cube (actually 3.5) law of the interatomic distance, the hardness of simple cubic oxygen can be calculated to be much harder than diamond.

Alternatively, each O–O bond in simple cubic oxygen may be replaced by a N–F bond. This simple cubic NF compound may also be ultrahard. However, because the bonding electrons are shifted from the middle position toward F, so the bond becomes ionic. As a consequence, simple cubic NF is likely less hard than simple cubic oxygen. The relationship between simple cubic oxygen and simple cubic NF is analogous to that between diamond and cBN.

The oxygen group element sulfur is known to form octahedrally coordinated S_6 molecules that constitutes the basic unit of a simple cubic structure. Thus, it may be possible that oxygen could form the simple cubic structure by linking O_6 clusters. High-pressure experiments have demonstrated that at a pressure of about 10 Gpa, diatomic oxygen molecules did regroup to form squares made of four atoms. With further increases of the pressure, it is conceivable that oxygen squares will reconfigure into O_6 clusters that eventually link together and become a simple cubic structure. If indeed such a covalent structure is formed, the interatomic distance would be only 1.46 Å. The shorter bonds, coupled with the higher KS, would make simple cubic oxygen much harder than diamond (Sung and Sung, 1996).

A total energy calculation (Wu, 1996) indicated that simple cubic oxygen has an interatomic distance of 1.9 Å. As a result, the bulk modulus is only 2.1 GPa, on a par with SiC. The long bond and low hardness imply that only four of oxygen's six valence electrons are involved in bonding. The dilution of four valence electrons by six kissing neighbors would make the structure metallic, and hence, less hard.

If the two additional s electrons can be hybridized to form s^2p^4 molecular bonds, the simple cubic oxygen would be much more compact in structure and its hardness much higher than diamond. Such a hybridization may occur metastably at very high pressure. If so, it could be quenched and preserved metastably.

It may be possible to synthesize simple cubic oxygen by colliding oxygen ions at the common center from six cubical directions. For example, water molecules can be dissociated by a DC arc to form mono-atomic ions of oxygen. These ions can be collimated by magnets to form ion jets. These jets can then be accelerated toward the negative biased target by an electric field. The collision effects of oxygen ions can be monitored by laser strobe light in femtoseconds.

3.15 Simple Cubic Fluorine

If oxygen can form simple cubic structure, so may fluorine and neon. In the case of F, if seven valence electrons are involved in bonding, there is one additional electron that may repulse each other. The dangling electrons could combine to form metallic π bonds that may further strengthen the structure to form a superdiamond.

Moreover, the π bonds are electrically conducting, or perhaps, superconducting. In this case, the structure would be a superconducting superdiamond. A simple cubic structure contains atoms aligned in orthogonal lines; hence, if it is superconducting, the electrical current is not confined to a particular direction. Furthermore, the straight electrical paths are less susceptible to thermal agitation so the critical temperature of this superconducting superdiamond may be quite high and it may become a room temperature superconductor. If this were the case, this superconducting superdiamond of simple cubic fluorine would be an ideal material for industrial applications.

Alternatively, the dangling bonds from adjacent atoms in simple cubic fluorine may combine to form a hybridized metallic π bond that may resonate among neighboring atoms. In this case, the structure would not be electrically conducting. However, the bond strength of the simple cubic fluorine would be further enhanced by the addition of π bond. As a result, the hardness of this superdiamond would be even higher.

Each of the unpaired electrons in simple cubic fluorine may also be tied up by plugging with a hydrogen atom. The simple FH, so formed may be ultrahard.

Simple cubic oxygen may also form a compound with simple cubic fluorine with a composition OF (Fig. 3.26). In this case, the structure will become ionic due to the shift of bonding electrons toward F atoms. Hence, simple cubic OF is likely not as hard as simple cubic F, but it may still be an ultrahard material. Again, this ultrahard structure may be impregnated with hydrogen atoms so the unbonded electrons of F atoms are passivated. The composition of this superdiamond is OFH.

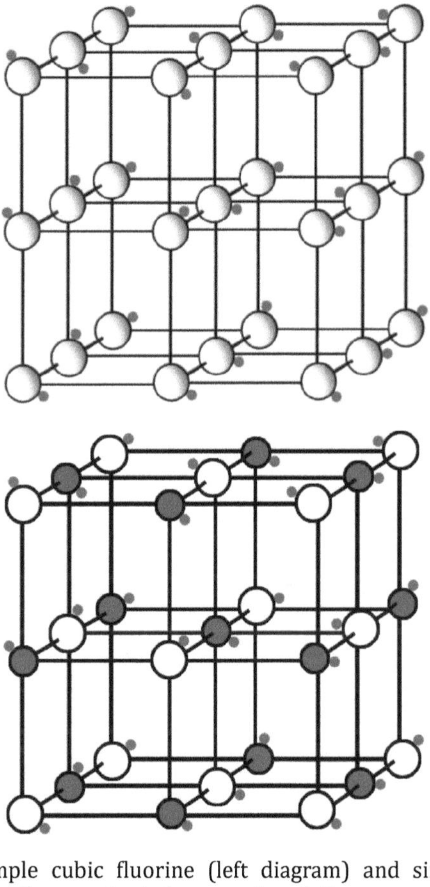

Figure 3.26 Simple cubic fluorine (left diagram) and simple cubic OF (right diagram). The unpaired electron of each F atom may be bonded to a hydrogen atom in either structure.

Another variation of the similar structure may be to replace each F–F bond in simple cubic fluorine with an O–Ne bond. Again, the covalent bond between two different atoms would become ionic. Hence, simple cubic O–Ne is not as hard as simple cubic fluorine, but it may still be ultrahard.

3.16 Body-Centered Neon

Neon could also form a simple cubic structure with its six p electrons involved in covalent bonding. In this case, it might also be a superdiamond, much like the non-conducting simple cubic oxygen. If the two additional 2s electrons are also involved in forming covalent bonds, neon atoms may configure to form a body-centered structure with KS = 8 (Fig. 3.27). In this case, the packing density of atoms is greatly increased; hence, its hardness could be significantly higher with the potential to make it the hardest structure possible. However, the increased coordination of BCC neon would draw atoms closer so their electron orbitals may overlap. If so, BCC neon may become a superhard metal. In this case, the hardness could be compromised due to the loss of the covalent character in bonding.

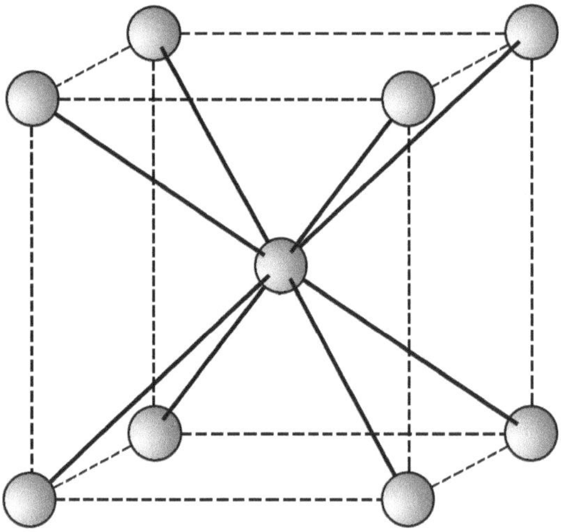

Figure 3.27 Neon with body-centered cubic structure could be a superdiamond.

As neon is a noble element, it would be extremely inert toward any form of chemical reaction. Hence, if BCC neon may be synthesized, it may become the most efficient cutting tool, and in essence, a super superabrasive.

BCC neon may be synthesized by directing collimated neon ions from the eight corner directions of a cube toward the body center. Again, femtochemistry may be applied to monitor the status of the synthesis.

BCC neon may also team up with diamond in such a way that they form an interpenetrating superlattice between cubic closest packing of Ne atoms and simple cubic structure of C atoms. Neon atoms are located at all body centers of the simple cubic structure, and carbon atoms are occupying all tetrahedral voids of the ccp (fcc) lattice. In this case, the compound has a stoicheometry of NeC_2 with each Ne atom encircled by eight C atoms, and every C atom surrounded by four Ne atoms. Such a superlattice is essentially the anti-fluorite structure of CaF_2 with Ne taking Ca's position, and C standing in for F (Fig. 3.28).

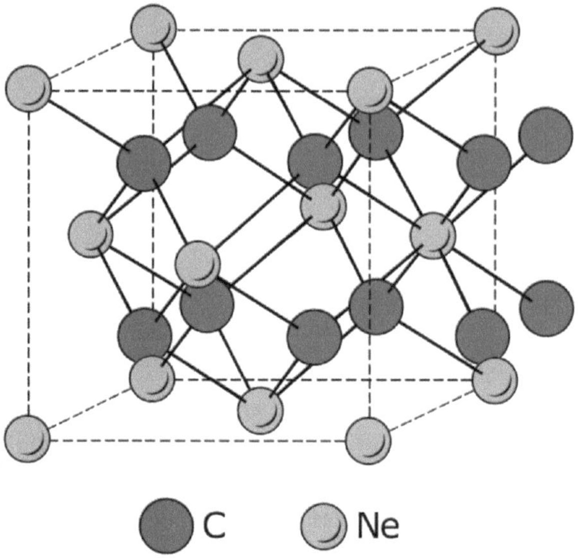

Figure 3.28 NeC_2 may assume fluorite's (CaF_2) structure or anti-fluorite's (Li_2O) structure with the potential to become a superdiamond.

Although there are two elements involved so the bond between Ne and C are slightly ionic, the high packing density of atoms may make up for this deficiency and render NeC_2 the status of superdiamond.

3.17 Conclusion

Hardness is the manifestation of the concentration of bond energy. Hence, to make a superdiamond, either the atoms must be smaller than carbon, and/or it should have more than four covalent bonds. The first requirement would be to eliminate all periodical elements with the period higher than 2. The second requirement would be to eliminate all elements lighter than carbon that contain fewer than four valence electrons. These two criteria would leave only five elements left: C, N, O, F, and Ne. However, these elements cannot form covalent bonds more than four per atom, unless they can reconstruct the structure as mono-atomic lattice with kissing number higher than 4. In this case, their double and triple covalent bonds can degenerate to become multiple single covalent bonds. Moreover, their unbonded lone pairs of electrons can also be split up to form additional single covalent bonds. If indeed these elements can reconfigure their structures this way, they stand the possibility to become superdiamonds.

The atoms of simple cubic carbon must share four valence electrons with six kissing neighbors; hence, their bonds are weakened by becoming metallic, so it would become a subdiamond with hardness possibly rivaling with cubic boron nitride. The structures that may be superdiamond are diamond-like nitrogen, simple cubic oxygen, fluorine, neon, and BCC neon. If all the valence electrons of the atoms can form single covalent bonds, the structure so configured could indeed be a superdiamond. On the other hand, if the bonding electrons are fewer than the kissing atoms, the structure could be weakened by the metallic character. In this case, diamond's hardness could be as insurmountable as the speed of light.

The above hypothetical superdiamond structures may be synthesizable by directing collimated ions from the appropriate angles to hit at the common center. Such a technique was developed by Nobel laureate Y. T. Lee decades ago. The possible instantaneous formation of these exotic structures may be monitored by laser strobe light flashed at femtoseconds (10^{-15} sec). Such femtochemistry has already been invented by Dr. Ahmed Zewail, another Nobel laureate.

Although the above-described hypothetical superdiamond structures may be highly metastable, because they are trapped by an insurmountable potential well of bond energy, once they are formed and quenched, they may be preserved at ambient conditions indefinitely like metastable diamond is.

Chapter 4

CVD of Diamond Films

Diamond deposited from gas can be larger and purer than that made in nature.

4.1 CVD Diamond Formation

Although graphite is stable at low pressures, it refers to large structures of more than hundreds of atoms. If the clusters of atoms are less than several hundred, diamond, bucky balls, or carbon nanotubes may still be more stable. This is because the edges of two-dimensional graphene planes are more energetic than three-dimensional diamond surfaces, bucky balls with no ends, and rolled graphene planes with less ends. Hence, diamond, bucky balls, and carbon nanotubes may form in smaller clusters of carbon atoms. Thus, in terms of thermodynamics, small-sized diamond is still more stable than small graphite even at low pressures (Fig. 4.1).

In addition, the atomic distance (1.54 Å) in diamond lattice is much smaller than inter planar distance (3.35 Å) between graphene planes; hence, if the decomposed carbon atoms have a long mean free path, it can attach to a diamond nucleus much

Handbook of Industrial Diamonds: Diamond Films and Carbon Coatings
James Sung

ISBN 978-981-4310-00-0 (Hardcover), 978-1-003-18045-6 (eBook)
www.jennystanford.com

quicker than starting a new graphene plane. Consequently, diamond may grow prior the formation of graphite. Thus, in terms of kinetics, diamond can grow faster than graphite.

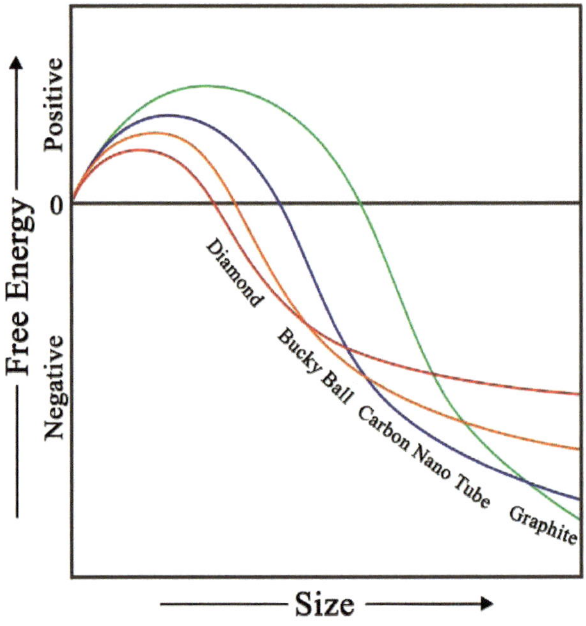

Figure 4.1 The nucleation of diamond is the easiest because of its three-dimensional structure. The nucleation of bucky ball (C_{60}) is the second easiest because it is endless. The nucleation of carbon nanotube is still easier than graphite with further increases of size. The nucleation of graphite is the most difficult because the edge atoms are highly unstable with the presence of more than one dangling electrons. But with larger sizes of graphene planes, graphite becomes the most stable at low pressures. In this diagram, the zero free energy refers to decomposed carbon atoms from a gas phase.

Based on the above consideration, diamond may be preferentially grown from energetic carbon atoms formed from gas if the deposition rate is kept so slow that graphite is difficult to form. In 1952, William Eversole of Union Carbide was attempting to grow diamond this way by pyrolysis of methane at high temperature. In order to promote the diamond formation, he used seeded diamond as the precursor. Eversole might have found the window of diamond growth by decomposing methane gas at very slow rate. If so, he would have made diamond about a year before ASEA's first artificial diamond made under high pressure; and it was two years before the widely publicized GE's synthesis of diamond by converting graphite with metal catalyst.

However, Eversole's pyrolysis process could grow diamond only a few nanometers per hour, as any further rush of the process would yield deposition of only amorphous carbon. The characterization of nano-sized diamond was not possible in 1950s; particularly, it was a thin coating that covered a diamond seed, so Eversole was unable to confirm that he made any diamond at all.

4.2 The Role of Hydrogen Atoms

In 1968, Henry Hibshman suggested that hydrogen may be used to stabilize diamond growth during the pyrolysis of carbonaceous gases. In 1971, John Angus began to duplicate Eversole's process using a hot filament process. This time he tried adding hydrogen gas to speed up the diamond growth. In 1976, Derjaquin started experimenting the CVD growth of diamond by glow discharge. Again, he found that hydrogen addition was beneficial to his process. In 1981, Setaka first introduced microwave plasma for diamond growth using a high proportion of hydrogen gas. In 1988, Hirose found that acetylene flame could grow diamond film in air. This was the first time that vacuum chamber was not used to grow CVD diamond. In 1988, Norton acquired Technion's arc jet technology and began to depositing thick diamond films (the author was responsible for this acquisition).

The modern CVD processes for growing diamond films are dependent on using hydrogen atom as catalyst. Hydrogen atoms can be obtained by dissociation of hydrogen gas at high temperature. Because the dissociation efficiency is low, most CVD processes rely on diluting methane gas (or other carbonaceous gas) by adding an overwhelming proportion of hydrogen gas (e.g., 99%).

The roles of hydrogen atoms are twofold. First, they can maintain the four valence electrons of carbon atoms in pseudo sp^3 configurations so when they join together, they will form tetrahedral bonds of diamond. Second, they can gasify graphite to form methane if graphite is formed. In other words, hydrogen atoms can promote diamond formation and at the same time they may prevent graphite from being formed (Fig. 4.2).

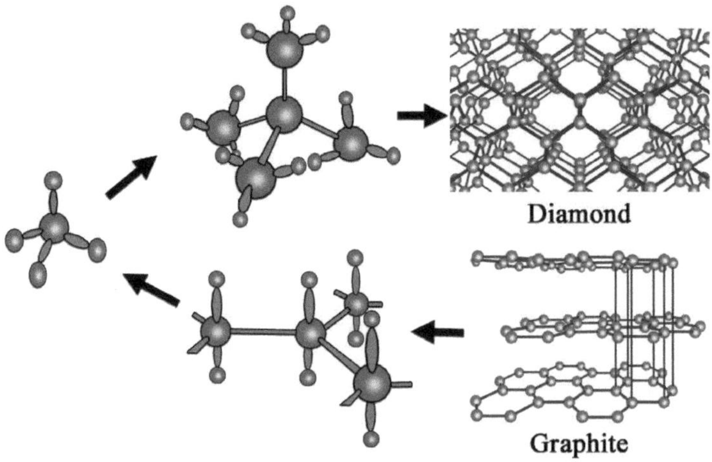

Figure 4.2 The carbon atoms formed by decomposing sp^3 configured methane may be coerced to maintain such configuration if abundant hydrogen atoms can be nearby to act like touch-and-go. When sp^3 configured carbon atoms encounter one another, they will become part of diamond lattice. On the other hand, if graphite is formed, then hydrogen atoms can gasify it by combined graphite atoms to form methane again.

The ability for hydrogen atoms to catalyze the diamond formation lies in the same principle of "touch and go" as described above for the case of high-pressure synthesis. Just as carbon atoms must loosely attach to molten metal atoms so they can freely regroup to form diamond, carbon atoms must loosely attach to hydrogen atoms so they can readily link to deposit diamond. In either case, carbon atoms can maintain their sp^3 configuration and attach easily to the growing diamond because of the constant arouse of metal solvent atoms (e.g., high-pressure synthesis) or hydrogen solvent atoms (CVD growth).

Hydrogen is more effective to catalyze carbon to form diamond than molten transition metals; hence, the threshold temperature of diamond growth for the CVD process is lower than that for the high-pressure method. For example, CVD diamond is typically deposited at a temperature of about 900°C, whereas high-pressure diamond is synthesized around 1300°C. This difference in synthesis temperatures reflects the difference in activation energies. The activation energies for hydrogen assisted CVD processes are about 0.09 eV, and for metal catalyzed processes, 0.13 eV.

In the 1980s, Japanese invented several CVD processes that may grow metastable diamond at high deposition rates. Specifically, hydrogen molecules are decomposed by various means that heat reaction gases to high temperatures. The heating could be achieved by hot filament, microwave discharge, electric arc, or acetylene flame. It was found that the growth rate of diamond films is proportional to the square of the concentration of hydrogen atoms (Goodwin, 1993). As the concentration of hydrogen atoms tends to increase with increasing temperature of the gaseous phase, so is the growth rate of the diamond deposition and the quality of film formed.

During a CVD process, the concentration of hydrogen atoms determines the relative growth rate between diamond and graphite (Fig. 4.3).

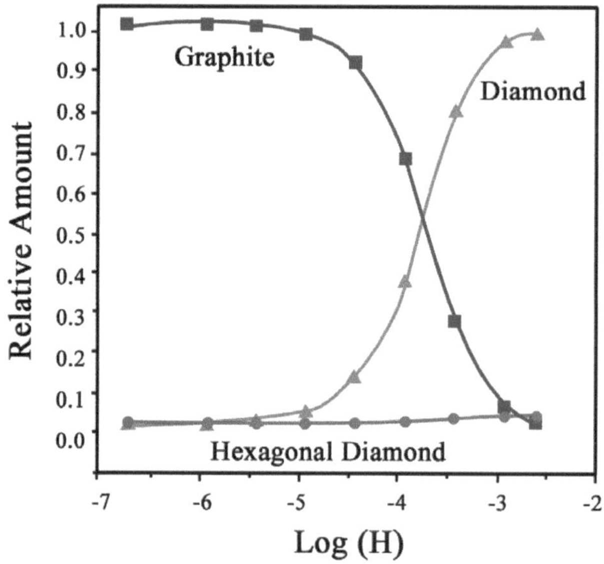

Figure 4.3 The relative deposition rate of diamond and graphite is controlled by the concentration of hydrogen atoms in the gas mixture.

Based on the above description, hydrogen atoms play a key role by keeping a dynamic balance between methane and decomposed carbon atoms. With the touch-and-go relationship, carbon atoms are fooled to maintain its pseudo sp^3 configuration with nearby hydrogen atoms until they join together to form diamond. In order to maintain this dynamic balance, it is critical that the carbon atoms must hit hydrogen atoms all the time and in no cases that carbon atoms should hit themselves, in particular, they must not run into one another consecutively. Hence, the mean free path of hydrogen atoms must be much shorter than the collision distance of carbon–carbon atoms. Consequently, for growing diamond films, there is a maximum carbon concentration that can be supported by the presence of hydrogen atoms, in other words, there is an optimal ratio of CH_4/H_2 in the original gas mixture (Fig. 4.4).

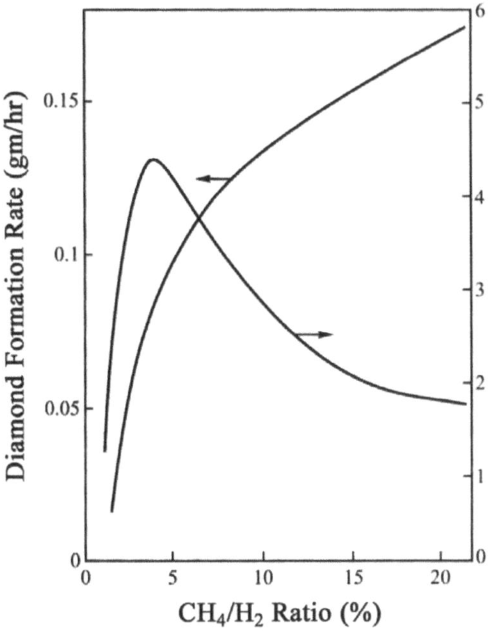

Figure 4.4 Diamond formation rate and its conversion rate as a function of CH_4/H_2 ratio in the original gas mixture.

Thus, the higher the concentration of hydrogen atoms, the faster the diamond film can grow. Alternatively, if the deposition rate is maintained at optimal, then the higher the concentration of hydrogen atoms, the better the quality of the deposited film. The dependence of diamond's growth rate (G) and the film's defect concentration (D) on the concentration $[H]$ of hydrogen atoms may be expressed by:

$$G/D = K[H],$$

where K is dependent on the deposition conditions other than the concentration of hydrogen atoms. Thus, for a given concentration of hydrogen atoms in the reaction chamber, an optical-grade diamond film must be grown very slowly, whereas a tool-grade diamond film can be deposited at a higher rate.

As both the growth rate and the quality of diamond film are dependent on the concentration of hydrogen atoms; hence, all CVD processes for growing diamond are designed to enhance the dissociation of hydrogen molecules in the gas mixture to form hydrogen atoms. The dissociation of hydrogen molecules is highly dependent on temperature and pressure (Table 4.1); hence, most CVD processes are aimed to heat the gas mixture more effectively at a higher pressure.

Table 4.1 The concentration of hydrogen atoms as a function of temperature

Temperature (°C)	$[H]$ (kg mol/m^3) $P = 20$ torr	$[H]$ (kg mol/m^3) $P = 200$ torr
1600	7.62×10^{-8}	2.41×10^{-7}
1800	4.43×10^{-7}	1.40×10^{-6}
2000	1.78×10^{-6}	5.66×10^{-6}
2200	5.47×10^{-6}	1.75×10^{-5}
2400	1.35×10^{-5}	4.43×10^{-5}
2600	2.75×10^{-5}	9.50×10^{-5}
2800	4.69×10^{-5}	1.77×10^{-4}
3000	6.65×10^{-5}	2.92×10^{-4}

A common gas mixture for CVD growth of diamond film contains CH_4 that is fully diluted in H_2. The dissociation of this gas mixture will produce various species of C and H radicals, all of them are temperature dependent (Fig. 4.5).

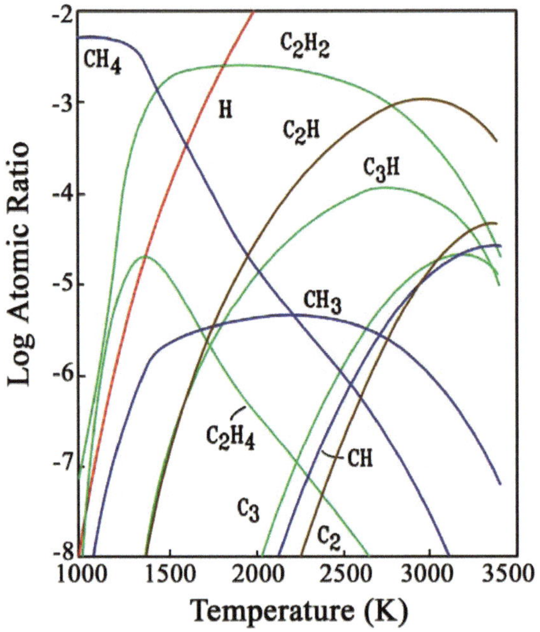

Figure 4.5 The gas species in an original mixture of CH_4 (0.5%) and H_2. The plot is based on the gas pressure at 20 torr.

4.3 The Morphology of Diamond Crystals

Due to its isometric lattice, diamond is crystallized with a very simple habit, the crystal shape ranges from cube with six (100) faces to octahedron with eight (111) faces. The relative dominance of the cubo-octahedral shape is inversely related to the growth rate perpendicular to the particular face (Fig. 4.6).

The crystal habit of diamond is dependent on the probability of attaching carbon atoms along various crystallographic direction of the diamond lattice. In the diamond lattice, there are two dangling electrons in the direction of <110> (i.e., the direction in perpendicular to (110) face), but there is only one in either <100> or <111> direction. Hence, the deposited carbon atoms are most likely joined to (110) face. This means that (110) face will soon disappear because atoms tend to pile up along its perpendicular direction. This is why diamond seldom exhibit (110) face of all origins, including both natural diamond that was formed under ground by converting carbon directly, and synthetic diamond synthesized under pressure by converting graphite using metal catalyst.

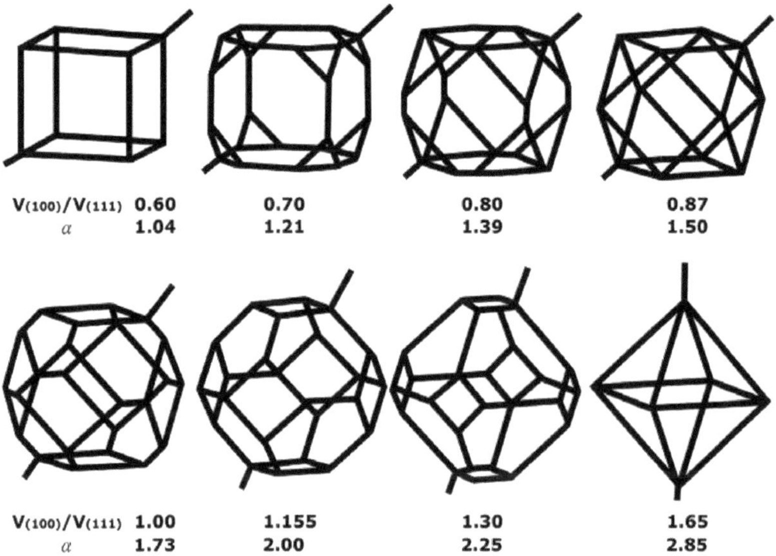

Figure 4.6 The crystal habit of diamond as determined by the relative growth rates (V) perpendicular to (100) and (111) faces. The crystal habit indicator $\alpha = \sqrt{3}\frac{V_{100}}{V_{111}}$. The arrow points to the direction of the fastest growth.

The relative dominance of (100) and (111) faces are dependent on the rate of carbon deposition. These two faces are contrast in atomic density of lattice planes and their separation. (100) faces have the lowest atomic density among major crystallographic faces and their spacing is the closest. In contrast, (111) has the highest atomic density and their spacing separates the farthest. When the deposition of carbon atoms is slow, these carbon atoms can be dumped to <100> direction easier, consequently, cubic habit will be grown out. As the growth rate increases with the increasing carbon concentration (i.e., CH_4 concentration), therefore, the higher the methane content is, the more likely the crystal will be octahedral.

On the other hand, the higher the temperature is, the easier carbon can nuclear a new layer on the top of the octahedral face;

hence, the more likely the diamond crystal may become cubic (Fig. 4.7). This is an opposite trend of crystal habit developed for high-pressure synthesis. In the latter case, the higher the temperature is, the more likely it develops octahedral habit.

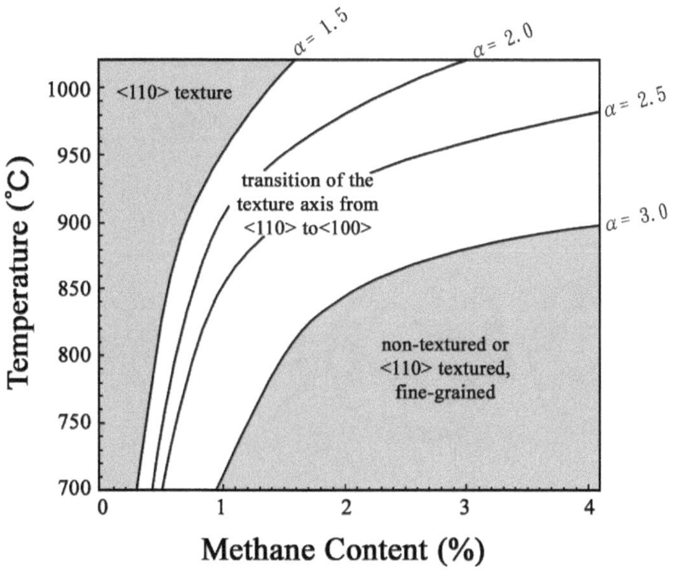

Figure 4.7 The crystal habit of diamond as determined by methane content and deposition temperature.

4.4 The Microstructure of Diamond Films

The sequence of events of CVD diamond growth is to form nuclei on the substrate first. Subsequently, these nuclei will grow, but very soon, they will run into one another to compete for the space of growth. However, the coarser grains will grow faster than the smaller ones. The result will be the expansion of larger grains at the expense of smaller grains. Such a coarsening effect will result in a columnar structure of crystals (Fig. 4.8) that is actually typical for most CVD processes (e.g., CVD deposition of tungsten film).

The columnar structure may continue with increasing thickness of the diamond film, thus, the thicker the diamond film, the coarser the diamond grains on the top (Fig. 4.9).

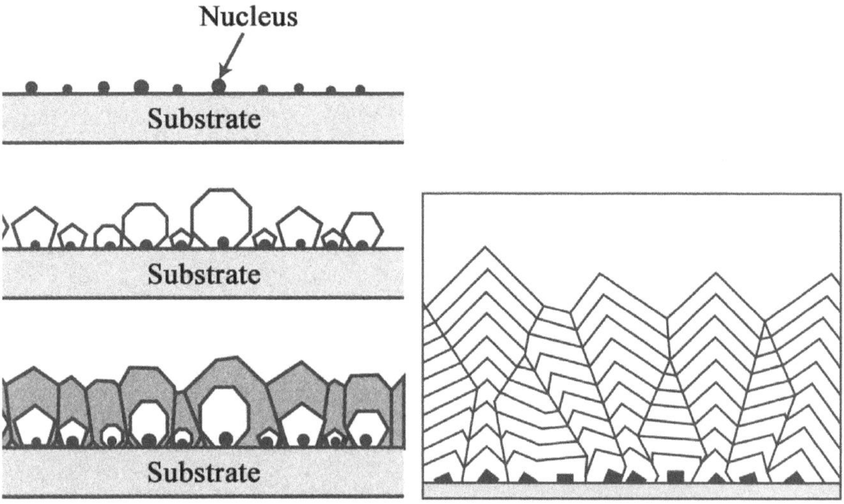

Figure 4.8 Nucleation and growth of diamond film in successive stages.

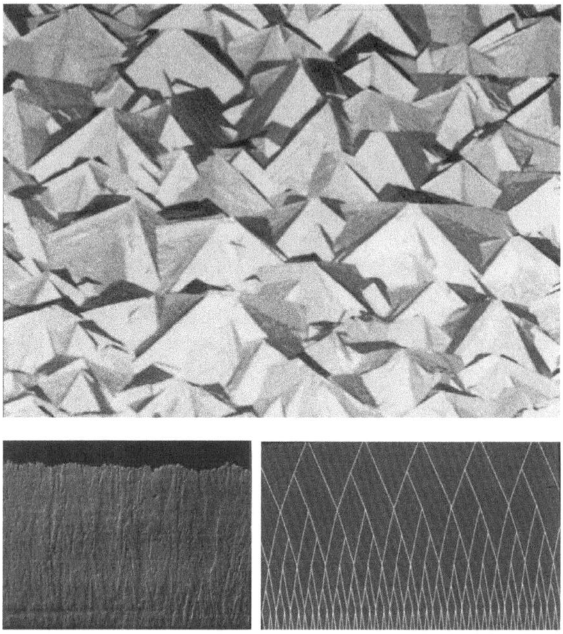

Figure 4.9 The microstructure of diamond film that shows coarse grains on top view (top) and the columnar cross section (bottom left). The columnar structure is formed by pinching off smaller diamond grains due to the coarsening of large grains.

Although the columnar structures are common for CVD growth of diamond films, there are subtleties in the microstructure and crystal morphology due to the difference of growth conditions (e.g., methane content, gas pressure, substrate temperature); hence, the final appearance of diamond films may differ (Fig. 4.10).

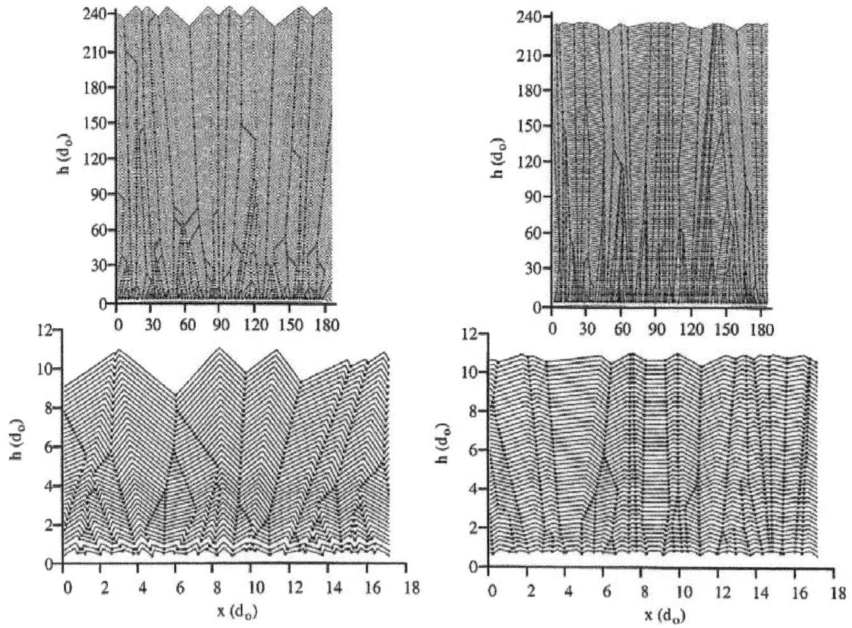

Figure 4.10 The computer modeled grain growth in diamond film and the development of microstructure at different stages. In these diagrams, d_0 refers to the pitch of the original nuclei. α $(=\sqrt{3}\frac{V_{100}}{V_{111}})$ is the crystal morphology indicator. The left diagrams show two stages of diamond film growth with $\alpha = 2.0$; the right diagrams, with $\alpha = 1.5$.

Because atoms on diamond's (110) face has two dangling electrons, whereas they have only one on (100) and (111) faces, so more carbon atoms will attach to (110) face than other ones. Consequently, the <110> direction (35° away from <111> direction) will grow the fastest and most diamond film will soon grow out of (110) face. This is particularly true for optical-grade diamond films that grow at a very slow pace (Fig. 4.11).

Figure 4.11 The plan view (top) and the cross section (bottom) of an optical-grade diamond film.

The preferred orientation of this optical-grade diamond film is confirmed by the center peak of (220) reflection of X-ray diffraction (Fig. 4.12).

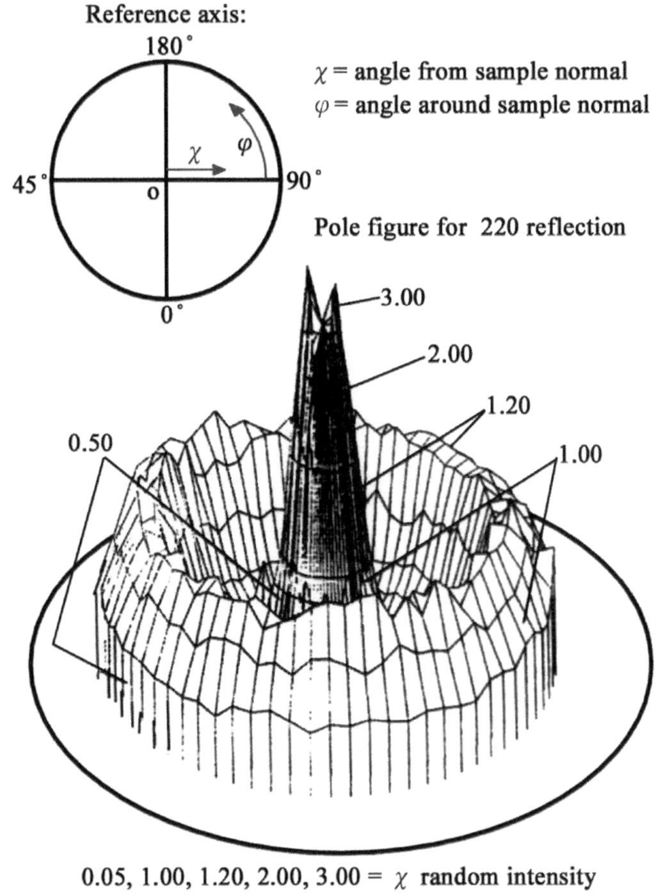

Figure 4.12 The polar diagram of a CVD diamond film of optical grade. Note the strong orientation along the <110> direction for most crystals.

The perfection of the optical-grade diamond film is reflected in its very sharp peak of (220) reflection. The full width half maximum (FWHM) is only 0.142° that is much lower than 0.228° for tool-grade diamond films. The low FWHM confirms that the optical-grade diamond film has a much lower internal stress than tool grade. Because the crystal lattice is virtually stress free so the two sub peaks of copper target's K_α reflection are fully resolved. In fact, even the two sub peaks of (311) reflection are also clearly discernable (Fig. 4.13).

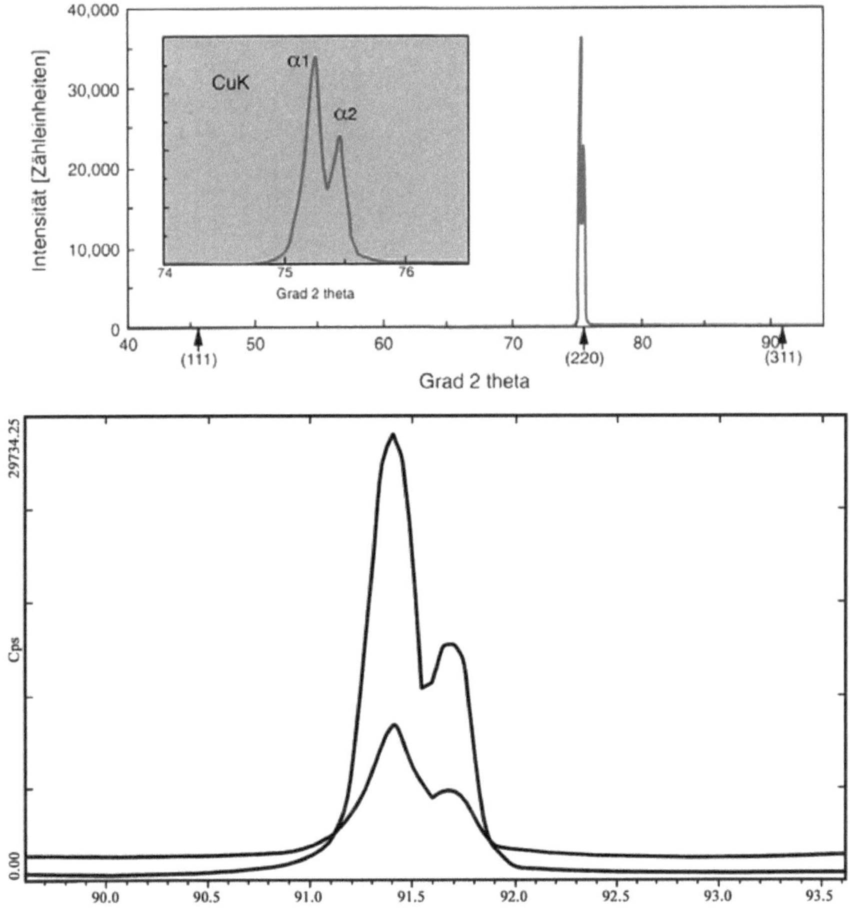

Figure 4.13 The X-ray diffraction pattern with copper target of an optical-grade diamond film showing the full resolution of the two sub peaks of K_{α} at (220) reflection (top, left inset), and also at (311) reflection (bottom). (Element Six brochure).

The diamond bond type (e.g., sp^3) may be deduced from the Raman spectra of the diamond film. The FWHM at the major peak at 1332 cm^{-1} may reflect the presence of non-diamond bonds (e.g., sp^2). The FWHMs of optical-grade diamond films are now comparable to that of natural gem diamond (Type IIa) (Fig. 4.14).

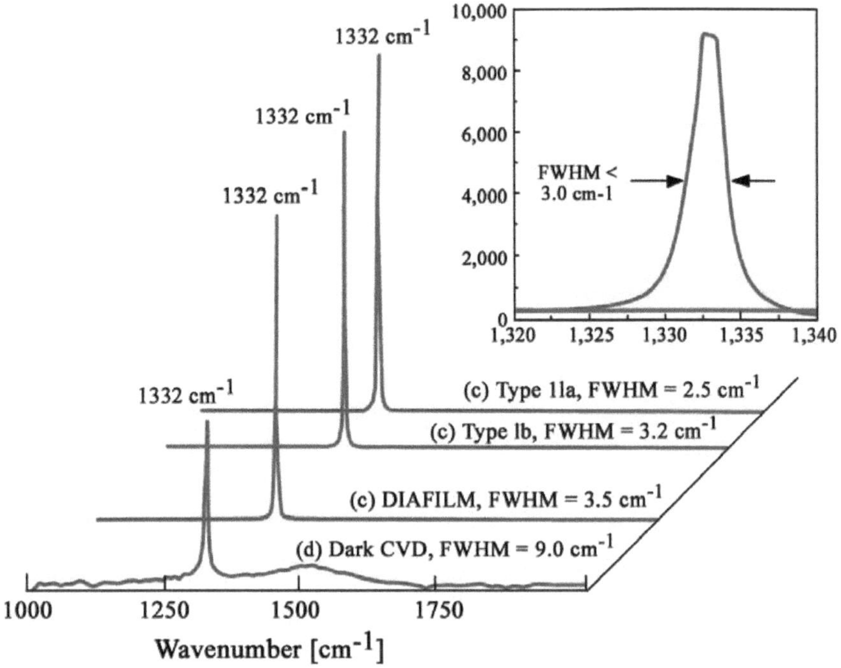

Figure 4.14 The comparison of Raman spectra of various types of diamond. Note that the optical-grade CVD diamond has a quality that rivals the best natural gem diamond.

Normally, the optical grade has a diamond grain size of about 50 microns on the nucleation side, and the grain may grow up to 250 microns on the growth side. But such coarse grains are, although well suited for optical applications, fragile for mechanical applications.

The diamond grain size may control many properties of a diamond film. In general, the small grains can stop or deflect the propagation of cracks so they are tougher. Hence, smaller diamond grains are more suitable for mechanical applications, such as by coating on cWC inserts.

On the other hand, the coarser grains can conduct heat faster due to the less grain boundaries that will slow down the phonons (Fig. 4.16). Consequently, the diamond grain size must be controlled to suit specific applications.

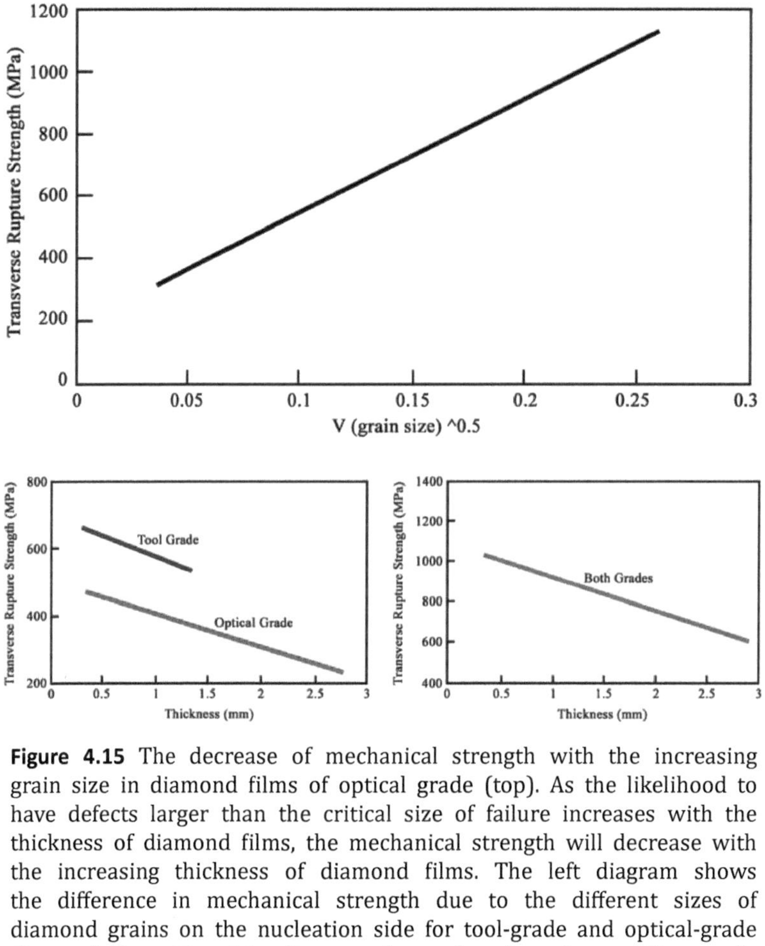

Figure 4.15 The decrease of mechanical strength with the increasing grain size in diamond films of optical grade (top). As the likelihood to have defects larger than the critical size of failure increases with the thickness of diamond films, the mechanical strength will decrease with the increasing thickness of diamond films. The left diagram shows the difference in mechanical strength due to the different sizes of diamond grains on the nucleation side for tool-grade and optical-grade diamond films. The right diagram shows that there is no difference in diamond grain sizes for both grades.

An easy way to control the size of diamond grains is by manipulating nucleation. In general, the more the nuclei on the substrate, the finer will be the grain size (Fig. 4.17). Hence, if coarser diamond grains are desirable, the substrate must be flat and homogeneous. Moreover, a passive element may be used (e.g., Cu) that is difficult to form diamond nuclei. On the other hand, by introducing a lot of nuclei to the substrate, the diamond grains will be much smaller (Fig. 4.17).

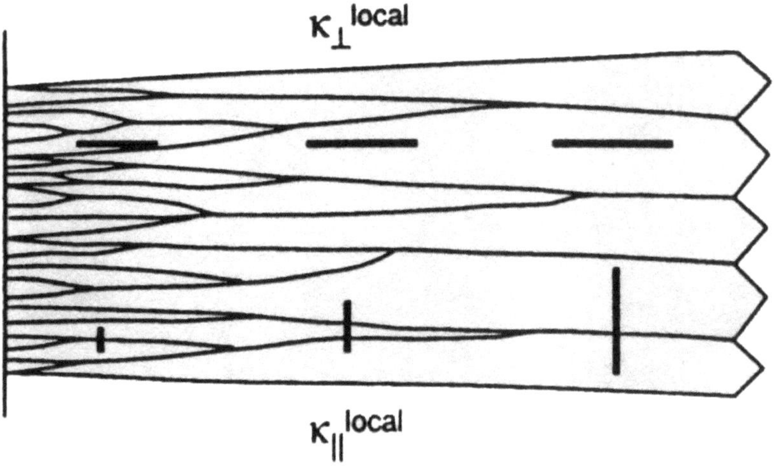

Figure 4.16 Thermal conductivity at various locations of a diamond film. The relative thermal conductivity is shown by the bar length. Note the higher thermal conductivity along the columnar gain, and trans the larger grains.

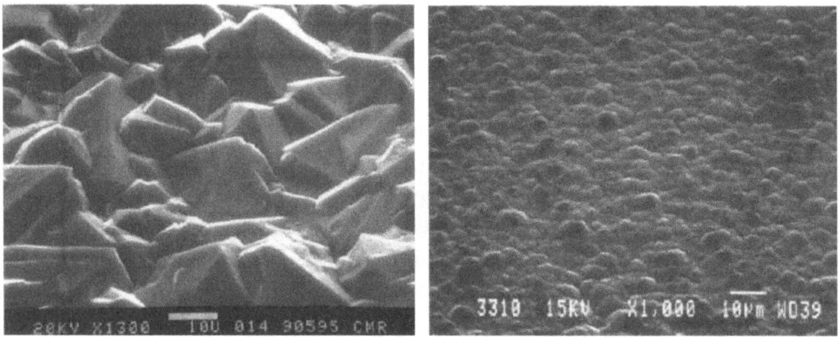

Figure 4.17 Coarse diamond grains (left) can be grown with low concentration of diamond nuclei on the substrate, and fine diamond grains (right) may be formed with extensive nucleation on the substrate.

If ultrafine diamond grains are desired, it may be attained by applying a negative bias to the substrate. Nucleation rate can be increased by a million fold so the grain size will be substantially finer than that deposited without applying such bias to the substrate. An even more dramatic way to increase the nuclei density is to use existing diamond of nanometers in size. For example, by dispensing nano-sized (e.g., 4–6 nm) diamond particles (e.g., produced by detonation of dynamite) in an acetone and ultrasonic vibrate the substrate (e.g., cWC) immersed in the bath, billions of nuclei will be covered on each square centimeter of the substrate. The diamond film so grown will contain extremely fine grains with very high toughness.

In addition to control the nucleation and grain size, diamond grains may also be deposited with preferred orientation. One way to align the crystal orientation is to grow diamond film epitaxially on silicon wafer with special orientation (Fig. 4.18). In addition to control the nucleation surface of the crystal orientation, the growth conditions may also be adjusted to favor the development of one face at the expense of the other. For example, when the growth rate is very slow, carbon atoms will be attached to the densest surface of octahedral face preferentially because the nucleation of an extra layer is difficult. Such a condition will favor the development of large octahedral faces. For example, the decomposition of more diluted methane gas (e.g., 0.5%) and deposit diamond film on a substrate maintained at a lower temperature (e.g., 800°C) will favor the growth of (111) face at the expense of the (100) face.

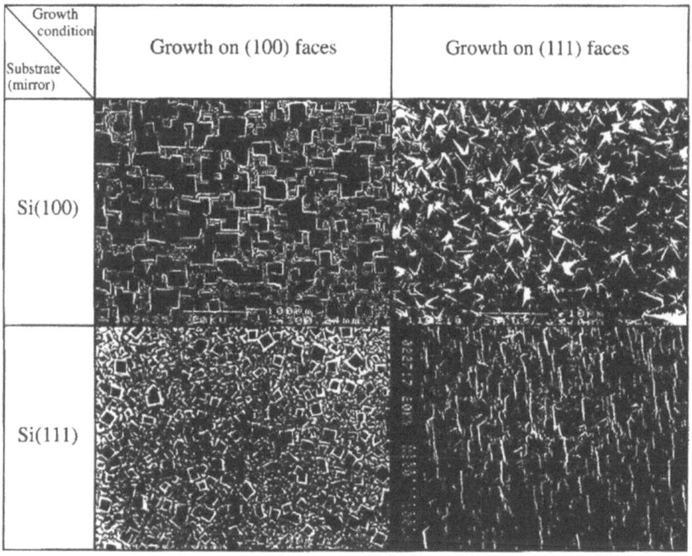

Figure 4.18 The diamond orientation of a CVD diamond film may be controlled by the substrate orientation and growth conditions (top) (data of J. Y. Wu). The (100) textured diamond film on a (100) silicon substrate (bottom) (data of J. T. Glass, et al.).

In addition to grow CVD diamond on planar surfaces, the diamond film may also be formed on the curved substrate or a tubular substrate. By subsequent dissolution of the substrate, freestanding diamond bowls or diamond tubes may be obtained (Fig. 4.19). Such exotic diamond object may have important applications. For example, diamond bowls may be used as a tweeter diaphragms or missile radomes; diamond tubules could be used as water jet nozzle or mini pipette.

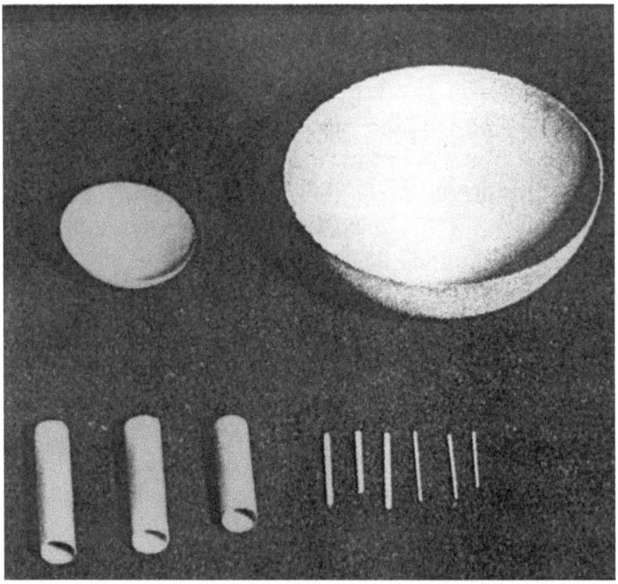

Figure 4.19 Freestanding diamond bowls and diamond tubules.

4.5 The Composition Dependence of CVD Processes

The optimal growth rate of a CVD process for depositing diamond film is dependent on the concentration of hydrogen atoms, and that is in turn controlled by the temperature of the gas mixture. Hence, the higher the temperature of the gas mixture is, the faster of its growth rate. Moreover, the faster a diamond film can grow, the more concentrated of methane gas it needs in the gas mixture (Fig. 4.20).

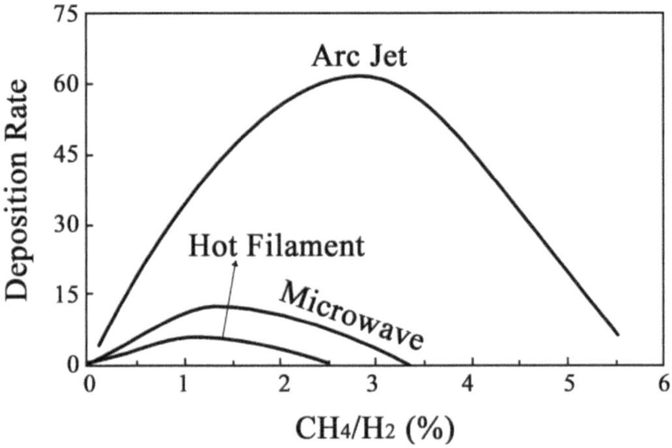

Figure 5.20 The deposition rate of diamond film as a function of CH_4/H_2 ratio. Note that this ratio increases with the increasing temperature of the gas mixture.

Although higher temperatures favor higher growth rates by allowing the use of a higher CH_4/H_2 ratio in the gas mixture, a slower growth rate is still required to grow diamond films of the highest quality. Thus, the CH_4/H_2 ratio may still have to be kept low in order to minimize the concentration of defects (Fig. 4.21).

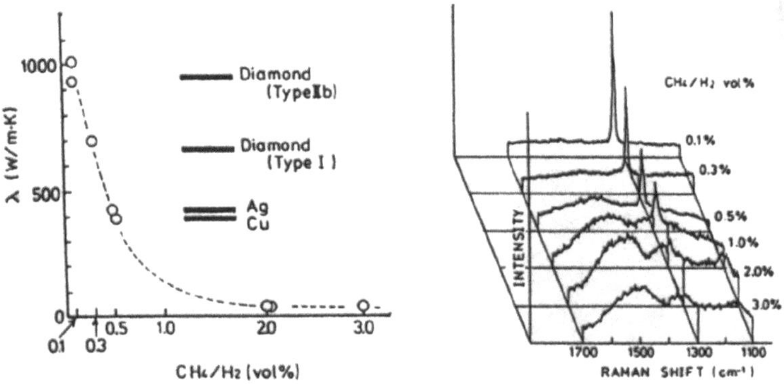

Figure 4.21 The thermal conductivity (left) and the Raman spectrum (right) of diamond films as a function of the CH_4/H_2 ratio in the gas mixture. Note that the slower growth rate can grow diamond films with coarser grains that contain less amounts of grain boundaries and defects. The result is a diamond film with a higher thermal conductivity.

Although hydrogen atoms are effective in removing non-diamond solids from the deposited film, oxygen molecules or atoms are even more powerful to do so. Hence, during diamond deposition, it may be advantageous to incorporate oxygen to purify the diamond formed. However, if there is too much oxygen, the diamond formed may also be oxidized. Therefore, the diamond stabilizing hydrogen and diamond destroying oxygen must be balanced, and this balance point is roughly about the equal amount of hydrogen and oxygen (Fig. 4.22).

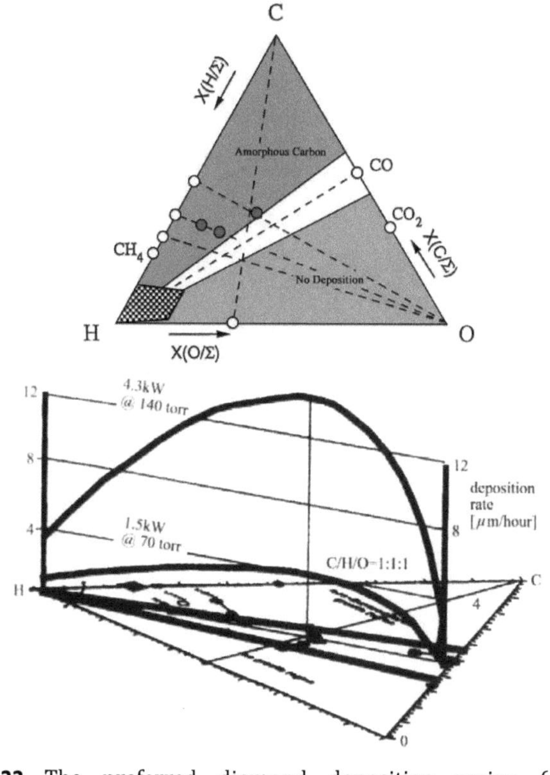

Figure 4.22 The preferred diamond deposition region (top) and the deposition rate (bottom) in the ternary system of C–H–O. Note that graphite will be deposited if carbon concentration is too high; diamond will be oxidized if oxygen abundance is too much. The optimal diamond growth condition appears to be at the equal concentration of C:H:O = 1:1:1. However, this is to assume that hydrogen molecules are fully dissociated at very high temperature. In most cases, the temperature is not high enough, so additional hydrogen molecules must be added to dominate the gas mixture.

4.6 Hot Filament CVD of Diamond Films

There are several ways to heat hydrogen gas. One simple way is to heat the gas with hot filaments (e.g., W, Ta) (Fig. 4.23). However, the temperature of hot filament is relatively low (e.g., 2200°C) so the concentration of dissociated hydrogen atoms is low. Consequently, the growth rate of diamond film is slow (e.g., 1 μm/h).

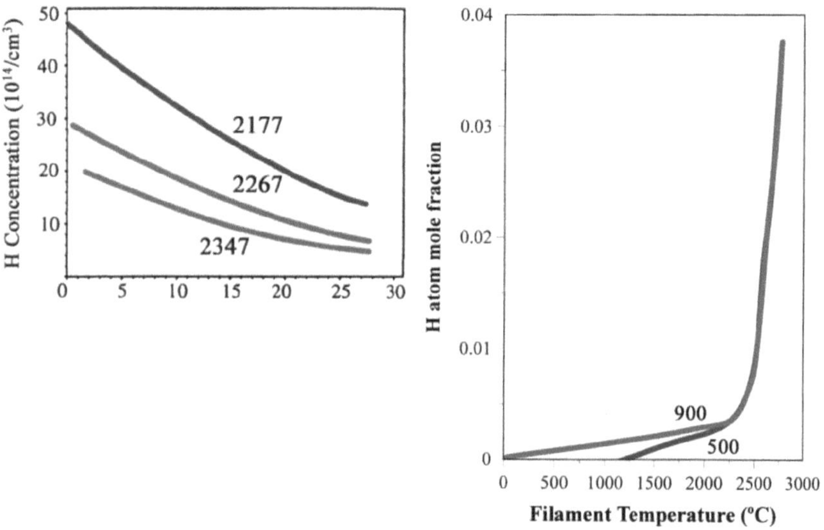

Figure 4.23 The concentration of hydrogen atoms as a function of the distance away from the heating filaments (left) and the filament temperature. The numbers labeled refer to filament temperature (°C).

Hot filaments may increase their lengths at high temperature due to the reaction with the gas mixture to form carbide and hydride. In order to keep them straight, they need to be subjected to a tensile force (Fig. 4.24).

Hot filaments systems have the advantage of spreading out the hot zone to a large area (up to 1 m^2), so they can grow large diamond films with uniform thickness (Fig. 4.25). As the uniformity of diamond film often determines the usefulness of the product, hot filaments systems have been used widely as a means to manufacture diamond film products, particularly that for making cutting tools, dressers, and mini wire drawing dies.

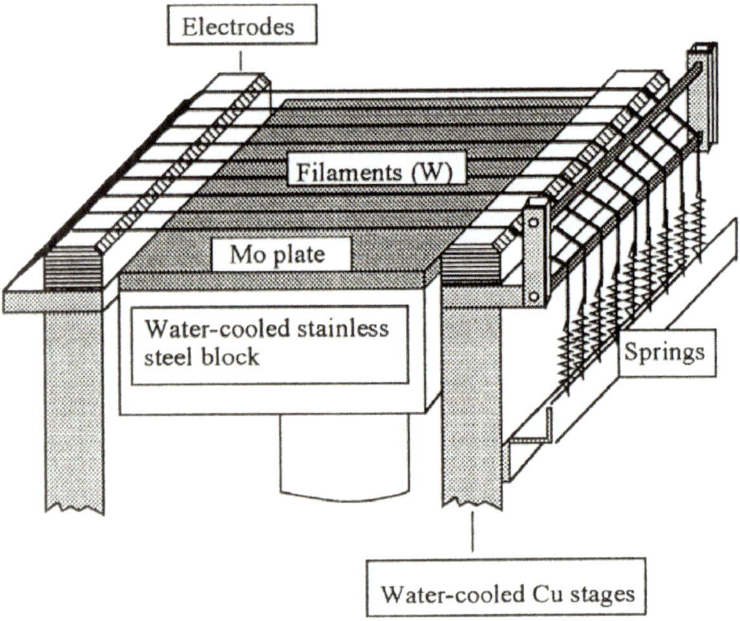

Figure 4.24 A hot filaments system may suspend filaments vertically, or it may stretch them horizontally as shown in the figure. The former system requires special fixtures to fasten the samples. The latter system would need a mechanism to straighten the filaments.

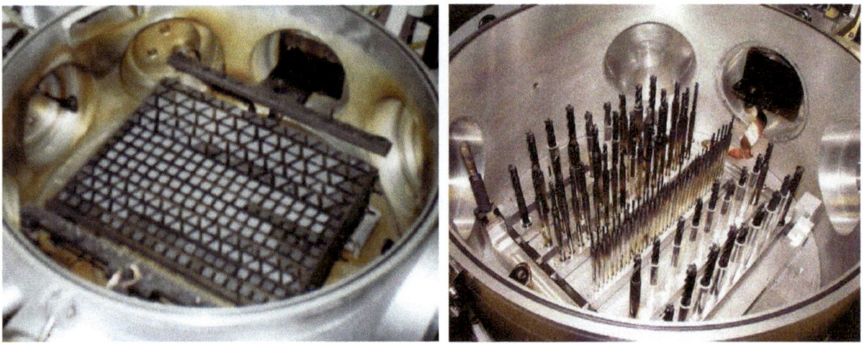

Figure 4.25 Two large hot filament systems used at Kinik Company for coating inserts (top) and round tools (bottom). The deposited area has a size of 30 cm × 40 cm.

Hot filament systems can produce very large diamond wafers (Fig. 4.26), some as large as 1 m^2.

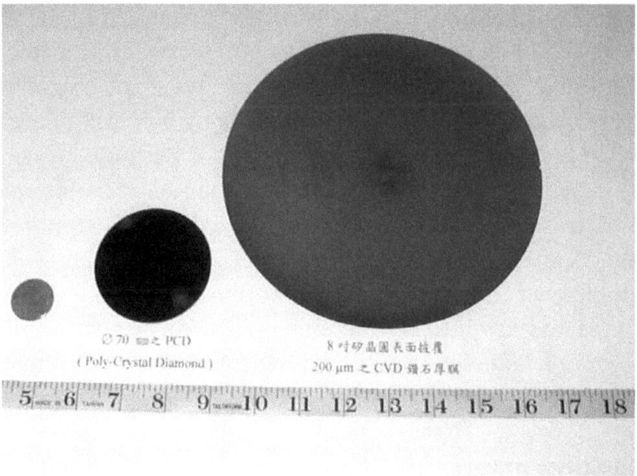

Figure 4.26 An 8 inch silicon wafer was deposited with 200 microns thick CVD diamond (right). For a size comparison, the largest (<3 inches) polycrystalline diamond (center) sintered under high pressure (Element Six product) and a common coin (left) are shown.

The hot filament CVD process may produce large diamond films with uniform thickness (Fig. 4.27).

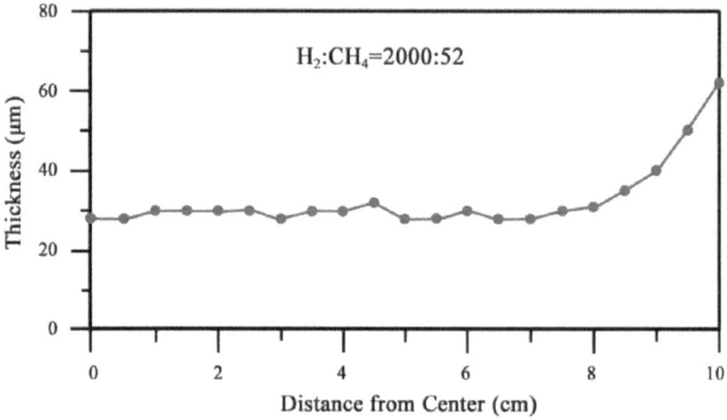

Figure 4.27 Hot filament CVD of diamond film can be uniform in thickness with variation of less than 5%. However, in the above demonstration, the thickness increases dramatically toward the edge of an 8 inch diamond film because the methane gas was fed from periphery. Such thickening effect can be eliminated by dispersing the methane gas more uniformly or by shielding the periphery of the deposition area on the substrate.

Because diamond film is a brittle material, the smaller the grain size, the higher the impact strength; hence, for the mechanical applications, it is preferable not to grow oversized grains. One way to keep diamond grains small is by increase of nucleation density. There are several ways to do so. For example, the substrate may be polished with micron or even nano-sized diamond (e.g., in an ultrasonic bath). The polishing will leave numerous tiny diamond nuclei on the surface. All these nuclei will grow simultaneously in a CVD reactor. Because there is no room left for crystal to grow into, all diamond grains will be very small.

A more effective way to further increase the nucleation density is by applying a negative bias at the substrate. The bias tends to attract positive carbon cations generated at high temperature toward the substrate where they impinge against one another to form diamond nuclei. This method may increase the density of diamond nuclei a million folds (Fig. 4.28). The simultaneous growth of such dense nuclei can form a strong microstructure that is well suited for mechanical applications.

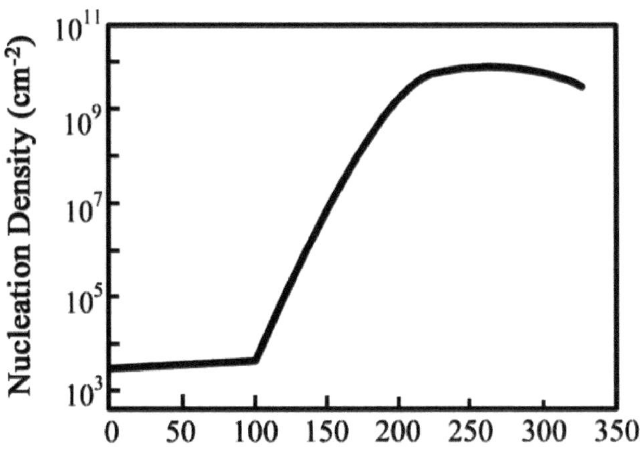

Figure 4.28 The densification of diamond nuclei by applying a negative bias at substrate. Note the million-fold increase of nucleation density achieved.

During the deposition of diamond films, the samples must be placed close enough to the hot filaments so they can be exposed to a higher concentration of hydrogen atoms, yet they need to be far enough from the hot filament to avoid contacting the filament that may be vibrating or drooping at high temperature.

Although installation of hot filaments is a convenient way to grow diamond, this process has its limitations, however. In addition to be slow in growing diamond, it is also contaminated by the vapor of filament material (e.g., W, Ta). Consequently, the deposited film may contain trace amount of metal, and therefore, it is not suitable for making optical-grade diamond films.

One way to avoid using hot filaments is to use a long DC electrical arc with high current (e.g., >100 A) to deliver the energy. Such an arc can be generated in a suitable ionized gas (e.g., by using Ar as ion source) maintained at low pressure. Balzers in Liechienstein was using such a system to coat round tools with a thin (e.g., 6 microns) layer of CVD diamond at a slow rate (e.g., 0.6 micron/h) (Fig. 4.29).

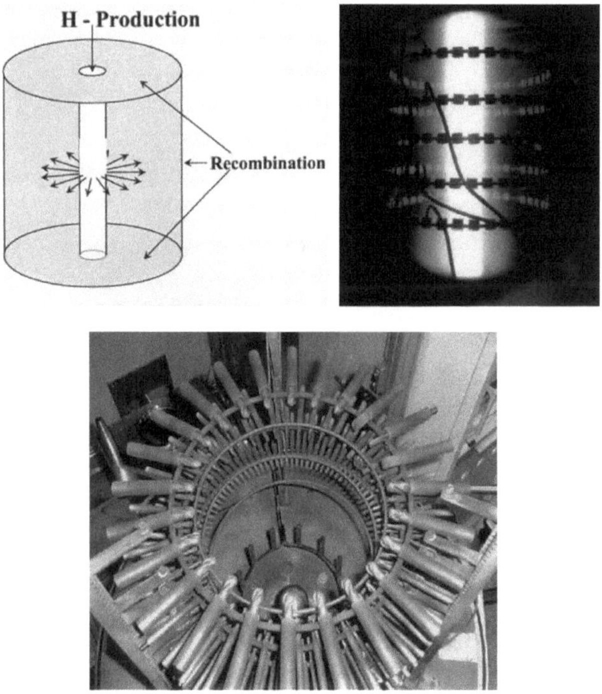

Figure 4.29 The long DC electrical arc system used by Balzers had a diameter of 73 cm and a height of 60 cm (top). The round tools are radiated around so only the cutting tips were coated (bottom).

The tools coated show rough surface and with a larger radius (Fig. 4.30). They may have to be polished before use. Otherwise, such tools can only be used for rough machining work.

Figure 4.30 The insert corner coated by diamond film as deposited from the long arc of high electrical current (top) and its outer appearance (bottom left) and magnified diamond surface (bottom right).

4.7 Microwave Plasma Systems

A more clean CVD system heats the reaction gas mixture by introducing microwave (Fig. 4.31). The energy of microwave will accelerate electrons in gas atoms so that a plasma can be generated. The plasma may reach a temperature exceeding 3000°C, which is much hotter than hot filaments. Hence, a higher concentration of hydrogen atoms may be available in the plasma. As a result, diamond may grow faster than that in a hot filament reactor, particularly if the higher pressure of plasma can be maintained.

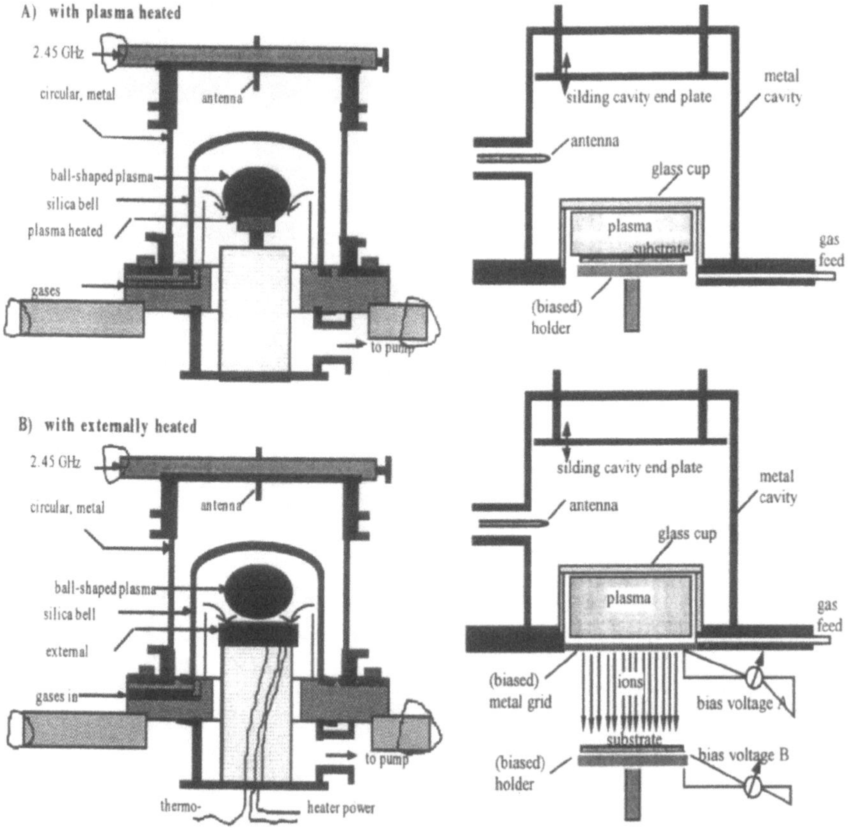

Figure 4.31 Two microwave plasma CVD systems used to grow diamond films. The bell jar system adapted by former Astex (left); the disk reactor offered by former Wavemat (right).

The high power of microwave coupled with a higher gas pressure may boost the deposition rate up to 10 microns/h (Fig. 4.32).

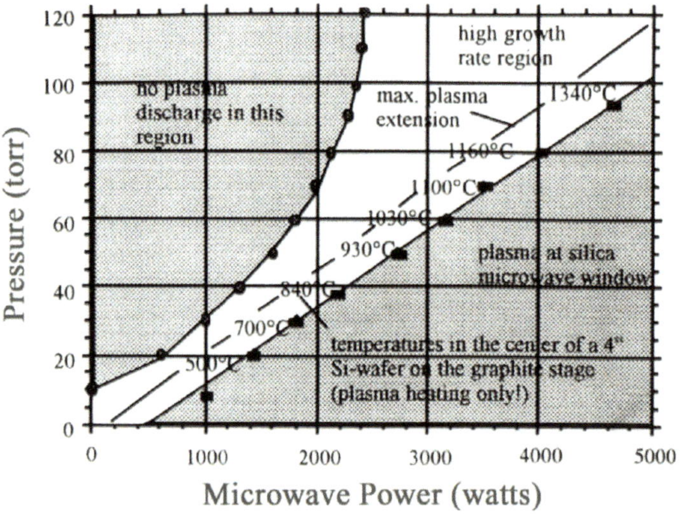

Figure 4.32 The higher microwave power can allow a higher gas pressure to achieve a higher growth rate of diamond.

Figure 4.33 The Wavemat microwave reactor used by the former Norton Diamond Film (left) and its produced coating on cWC inserts (right). The author was responsible for acquiring this technology for Norton in 1988.

The microwave generated is typically at a frequency of 2.45 GHz. However, by reducing the frequency down to 915 MHz, considerable increase of deposition area (e.g., 30 cm in diameter) can be achieved (Fig. 4.33).

The microwave system has the advantage of avoiding contaminants, so it can be used to grow optical-grade diamond films of high quality. However, its deposition area is much smaller than that of hot elements system with the similar power; hence, the cost of diamond deposition is much higher.

4.8 Large Microwave Plasma Systems

Conventional microwave has a fixed frequency, typically 2.45 GHz. However, with decreasing of frequency, larger deposition areas may be covered with less energetic plasma. Hence, a new system is developed that can tune microwave to the frequency desired (Fig. 4.34).

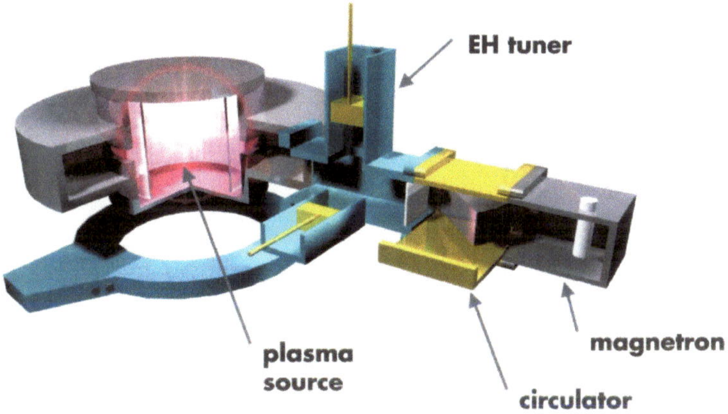

Figure 4.34 The tunable microwave system developed by Iplas.

Another unique system has been developed by Aixtron. It utilizes a metal ellipsoid to focus microwave to the substrate of deposition that is enclosed in a glass bell jar (Fig. 4.35). The ellipsoid reflector design can avoid the uncontrollable shifts of plasma inside conventional cylindrical container.

With such a focused plasma, transparent diamond films can bc dcpositcd (Fig. 4.36).

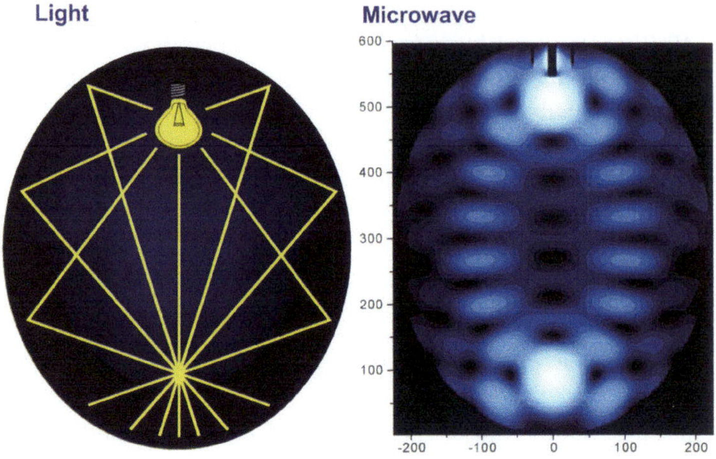

Figure 4.35 The ellipsoid reflector of microwave designed by Aixtron.

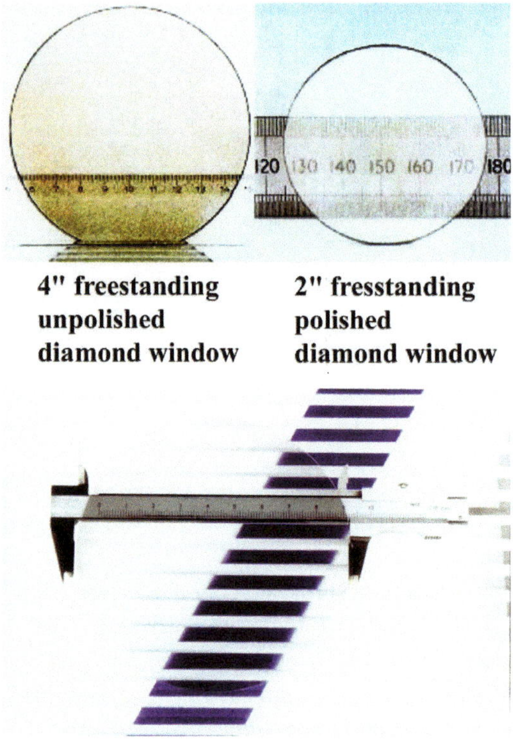

Figure 4.36 Transparent diamond films of various sizes deposited by Aixtron.

Large blanks of thermal grade CVD diamond films can also be deposited in the same system (Fig. 4.37).

Figure 4.37 Diamond wafers of various sizes deposited by Aixtron.

4.9 Flame CVD of Diamond Film

Diamond may also be grown by pressing an acetylene torch against a substrate. The flame end of such a torch is fully oxidized and it may reach the maximum temperature of up to 5000°C. However, at the center of the flame where the combustion is not complete, its temperature is much lower. In this region, decomposed carbon is surrounded by dissociated hydrogen atoms so the conditions are suitable to grow CVD diamond.

The flame method has the attractive conditions of not needing a vacuum chamber and the entire process can be conducted in air. However, due to its large temperature gradient, the diamond film grown is not uniform. Moreover, the flame consumes a high rate of acetylene and oxygen and the conversion rate to diamond is very low. Hence, the cost of depositing diamond is actually higher than other processes. Consequently, such processes have not been harnessed for commercial use.

Figure 4.38 An acetylene flame may be used to grow diamond at the reducing atmosphere at the inside flame (left). Such a flame may also contain carbon cations, so it can be manipulated by applying an external electrical field or magnetic field. For example, the flame may be turned around to spread out a larger area. Alternatively, a negative bias may be applied to attract carbon cations and hence to increase the nucleation density. The result will be a stronger diamond film that contains smaller grains. A flame grown thick diamond film (right).

A more effective way to heat gas mixture to a very high temperature is to bombard it with a high dose of electrons, such as by passing the gas mixture through a pair of electrodes. In order to allow electrons to pass through a path of ionized gas, argon is used to provide the source of ions. Hence, the gas mixture may contain Ar, H_2, and CH_4. Alternatively, CH_4 may be fed independently into the stream of hot gas of Ar and H_2 mixture that passes through the electrodes (Fig. 4.39).

A DC arc nozzle normally contains a cathode (e.g., made of W) located at the center that is surrounded by an anode (e.g., made of Cu). The gas jet is spayed out very fast (e.g., supersonic speed) in order to prevent the electrode from overheating. The plasma formed can reach a temperature higher than that (e.g., 6000°C) at the surface of the Sun. The result is a high degree of hydrogen dissociation. The high concentration of hydrogen atoms may allow a much faster diamond deposition rate (e.g., 50 microns/h) without resulting in the precipitation of graphite or amorphous carbon (Fig. 4.40).

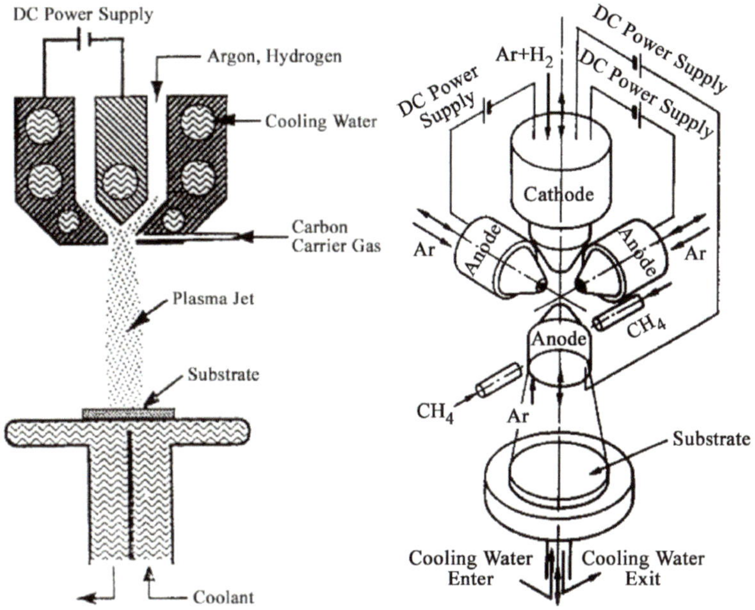

Figure 4.39 The cross section of a typical DC arc system for growing CVD diamond film (left). The employment of multiple anodes of a DC arc system (right) to spread out the deposition area.

Figure 4.40 The DC arc system in action at former Norton Diamond Film (top) and its deposited freestanding thick (e.g., 0.5 mm) diamond films (bottom). The system may consume 60 Kw of electrical power. The diamond wafer formed may reach a size of 18 cm. The throughput achieved was about 6 gm/h. The author was responsible for the acquisition of this arc jet technology from Technion for Norton in 1988.

Although a DC arc system may grow diamond at high speed, the film so deposited tends to cover a small area. In order to spread out the deposition area, a rotating substrate is used to allow an interrupted growth of the diamond film. Alternatively, the plasma generated in the arc jet can be bent and rotated by applying an external electrical or magnetic field so a larger deposition area may be covered. However, even so, the deposition area is still much smaller than that may be achieved by hot filaments methods.

Multiple anodes may also be employed to draw the intense arc jet away from its center. Such a complicated system has not been implemented commercially, however.

The speed of diamond deposition in an arc jet system may depend on the original ratio of CH_4/H_2 in the gas mixture. If this ratio is too low (<0.5%), carbon concentration is too low to sustain a continuous growth of diamond. On the other hand, if this ratio is too high (>5%), carbon deposition is too fast, so graphite or amorphous carbon will be deposited. If the ratio is moderate, then diamond may grow at a relatively high speed (e.g., 30 microns/h) (Fig. 4.41).

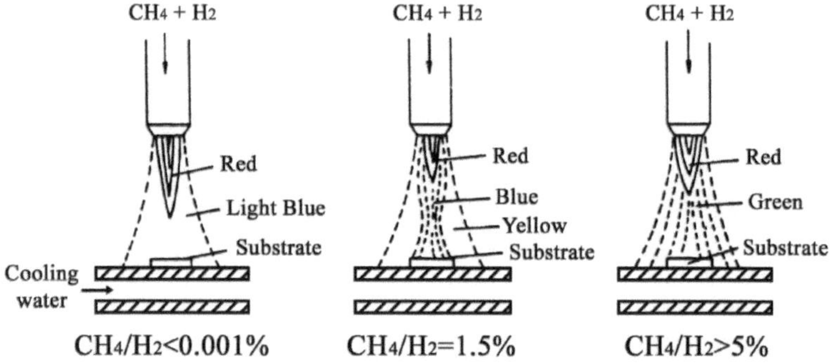

Figure 4.41 The growth of diamond film by a DC arc jet system with different ratios of CH_4/H_2 in Ar. Note the color changes reflecting the difference in gas composition. (*Journal of Shanghai Jiaotong University*, 1997, 31–3, 101–105).

4.10 Comparisons of CVD Methods

As the deposition rate of a CVD diamond film is limited by the availability of hydrogen atoms, which is in turn controlled by the temperature of the plasma, the higher the temperature of the plasma, the faster the diamond film can grow. The higher growth rate is achieved by allowing a higher concentration in the gas mixture so more carbon radicals are also present in the plasma (Fig. 4.42).

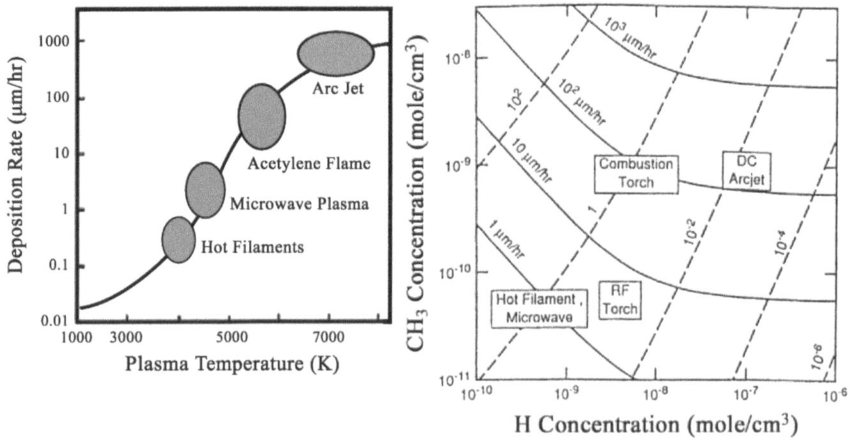

Figure 4.42 The growth rate of CVD diamond deposition as a function of plasma temperature (top). The dependence of CH_3 radical on the hydrogen atoms concentration (bottom). In the diagram, the horizontal grid shows the growth rate of diamond films, and vertical grid marks the gas pressure in mole/cm^3.

Although there are different ways to dissociate hydrogen molecules, the amount of hydrogen atoms formed is directly proportional to the energy provided. Hence, various heating methods can only achieve different distribution of the same amount of hydrogen atoms in space. Moreover, the diamond deposition will follow the same distribution of hydrogen atoms. Thus, hot filament method trades off its growth rate with a large size; on the other hand, DC arc boosts the growth rate at the expense of the deposited area. Microwave plasma and acetylene flame meet in the middle grounds, also compromising their vertical growth rates with horizontal spreading areas. The quantitative relationship between growth rate and deposition size of various CVD diamond deposition reactors for commercial production is listed in Table 4.2.

Table 4.2 Production system of CVD of diamond films

Method	Maker	Km	D (cm)	A (cm^2)	μm/H	g/H	g/KwH
DC Arc	Norton	60	15	180	15	1.0	0.02
DC Arc	Iljin	25	10	80	15	0.4	0.02
DC Arc	Balzers	60	50 × 30	4700	0.6	1.0	0.02
Microwave	Astex	60	15	180	10	0.6	0.01
Hot filament	sp^3	30	35	1200	1	0.4	0.01

Note: The Norton arc was generated with two electrodes close by to achieve a plasma with high temperature for fast growth. Iljin uses seven smaller arcs to spread out the deposition area. Balzers used a long arc between two electrodes to extend the length of the coating chamber.

Table 4.2 confirms that there is a systematic trade-off between the growth rate of diamond films and their deposited sizes for a given rate of energy consumption (Fig. 4.43).

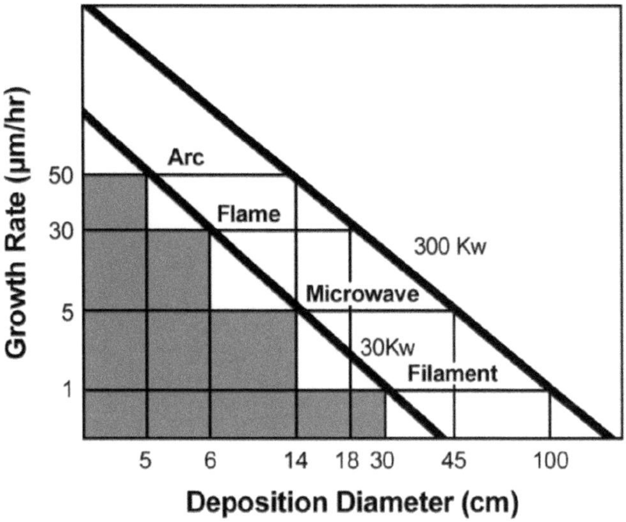

Figure 4.43 Compromise between growth rate and deposition area for various commercial CVD processes of diamond deposition. In this diagram, the energy used in oxyacetylene flame is converted to electrical power for comparison. The capability of depositing diamond for all CVD processes is 1 c.c. (3.5 g) per 100 kWh; hence, the 30 kW deposition line may be expressed by $G = \sqrt{\frac{1000}{D}}$; and 300 kW line, $G = \sqrt{\frac{1000}{D}}$, where G is the growth rate (μm/h); and D, the deposition diameter (cm).

The amount of diamond formed per unit time for all established CVD processes is directly proportional to the energy consumption rate (Fig. 4.44).

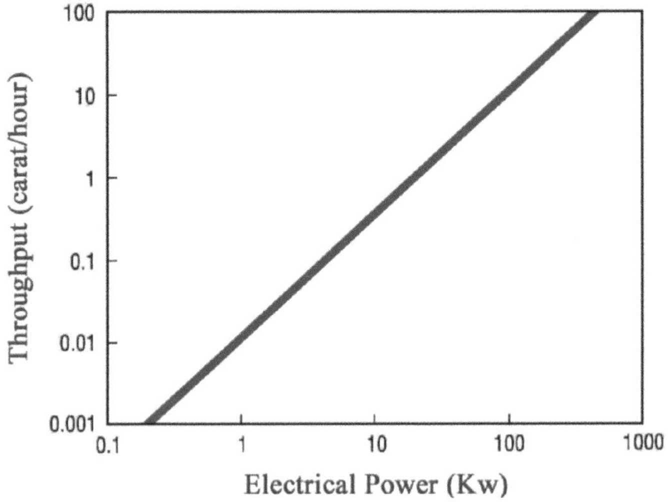

Figure 4.44 The amount of diamond formed per unit time is directly proportional to the energy consumption rate.

For all CVD diamond film deposition processes, the diamond quantity produced can be expressed as energy consumed as follows:

1 carat of CVD diamond = 10 KwH electricity

Although acetylene flame does not involve the use of electricity, the relationship between the diamond formed and energy consumed is the same. So this equation has the general applicability to all CVD diamond deposition processes.

However, there have been claims that much faster diamond growth rate may have been achieved by CVD methods. For example, when flame synthesis of diamond film was first announced, nearly 1 mm/h growth rate was claimed. Similarly, DC arc was said to achieve a diamond growth rate of over 100 microns per hour. Again, QQC process alleged even faster growth rate based on multiple laser bombardment on CO_2 and N_2 gas, this time the deposition was achieved without using a vacuum chamber. However, although very high growth rate of diamond films may be demonstrated in laboratories, none of these fast growth rates has been sustained on manufacturing conditions.

In recent years, P1 Diamond filed a patent indicating diamond could be deposited by soft microwaves (e.g., 150 or 400 MHz) with an anti-electrode forcing the plasma to be very close (e.g., <1 mm) to the deposition substrate. In this way, hydrogen molecules could be dissociated with the produced atoms immediately used in catalyzing the diamond formation. It was said that this process could not only increase the growth rate, but also it could greatly reduce the amount of reaction gas (e.g., methane, hydrogen) required. So ultimately, diamond films can be deposited with much lower cost. In addition, claims are also made by using nitrogen addition to the reaction gas to boost the diamond growth rate during the deposition of (100) face of single crystals of diamond using another single crystal diamond as seed. However, even the growth rate is indeed hundreds microns per hour, it is still much slower than that (e.g., 1 mm/h) could be sustained by the high-pressure growth of diamond.

In contrast to CVD diamond film growth, the energy efficiency of high-pressure diamond synthesis is about 150 carats diamond = 10 KwH. Hence, in term of energy efficiency, CVD diamond deposition is about 0.7% of that for high-pressure diamond synthesis. This ratio is also reflected in the difference of growth rates between CVD diamond deposition (average about 7 μm/h) and high-pressure diamond synthesis (about 1000 μm/h).

The slow CVD diamond deposition versus the fast high-pressure diamond synthesis is ultimately determined by their difference in the mean free path (collision distance) of carbon nutrients, the carbon solute atoms in the molten catalyst are separated by about 0.7% of the distance of the carbon molecules in the reactive gas. It is interesting to note that the price of CVD diamond ($5/carat) for tool grade is also about 150 times higher than low-grade diamond synthesized under high pressure ($0.3/carat).

When various CVD diamond processes were developed in late 1980s, scientists shared the optimism that these processes would phase out high-pressure synthesis of diamond because they did not require ultrahigh pressure for production. But the intrinsic slow growth rate has made CVD diamond films much more expensive. Moreover, it was realized that the large reactors for depositing CVD diamond films cost just as much as a high tonnage of press used for high-pressure diamond synthesis.

Moreover, not only the energy is much more wasteful in depositing diamond from gas, but also the conversion rate (<5%) of carbon to diamond is much less than that (>40%) for high-pressure diamond synthesis. In addition, the utilization of non-carbon gas (e.g., hydrogen) is even poorer, especially for such processes that require fast movement of gases (e.g., acetylene flame, arc jet). Even the gas may be recycled, but the benefit is offset by the additional investment in installation and maintenance of recycling systems. With all these intrinsic limitations, the throughput of diamond and cost of production are trailing behind that for high-pressure diamond synthesis by about 100-fold (Fig. 4.45).

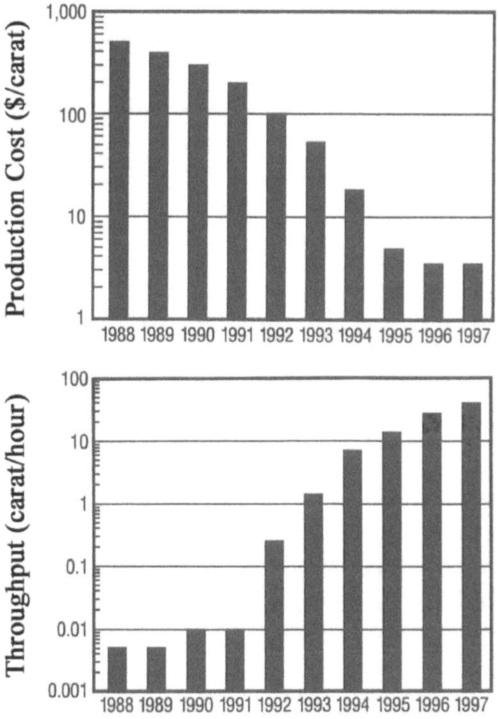

Figure 4.45 The production cost and diamond throughput for commercial CVD deposition of diamond films of tool grade. Note the leveling of cost and throughput in late 1990. The cost and throughput are not competitive compared to that of diamond synthesis under high pressure.

The cost of making CVD diamond films will increase if higher-grade products are needed (Fig. 4.46).

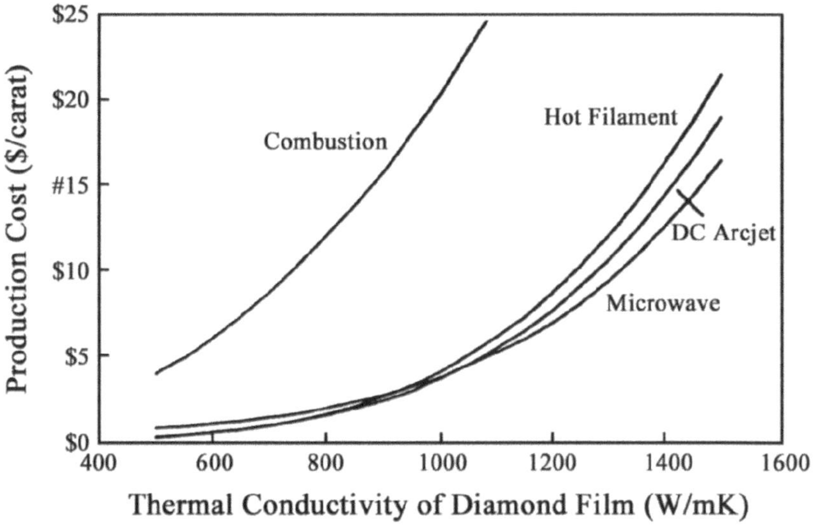

Figure 4.46 The cost trends for making CVD diamond films with decreasing defect density.

The intrinsic limitation of metastable CVD diamond growth has ruled out diamond films as a low cost source of diamond. Hence, diamond films are best used in sizes and shapes that are beyond the reach of high-pressure diamond synthesis. For example, surface acoustic wave filters and boron doped diamond electrodes cannot be produced by high-pressure synthesized diamond, so they may be served by using CVD diamond films.

In the history of commercialization of diamond films, many companies went under because of the poor judgment of project priorities. Some examples include Crystallume, the first commercialized diamond film company; Astex, the company that sold most reactors; and Crystalline Materials, the company that used most governmental funds. But the most dramatic case is Norton Diamond Film; the author was responsible for establishing its technology in late 1980s (Fig. 4.47).

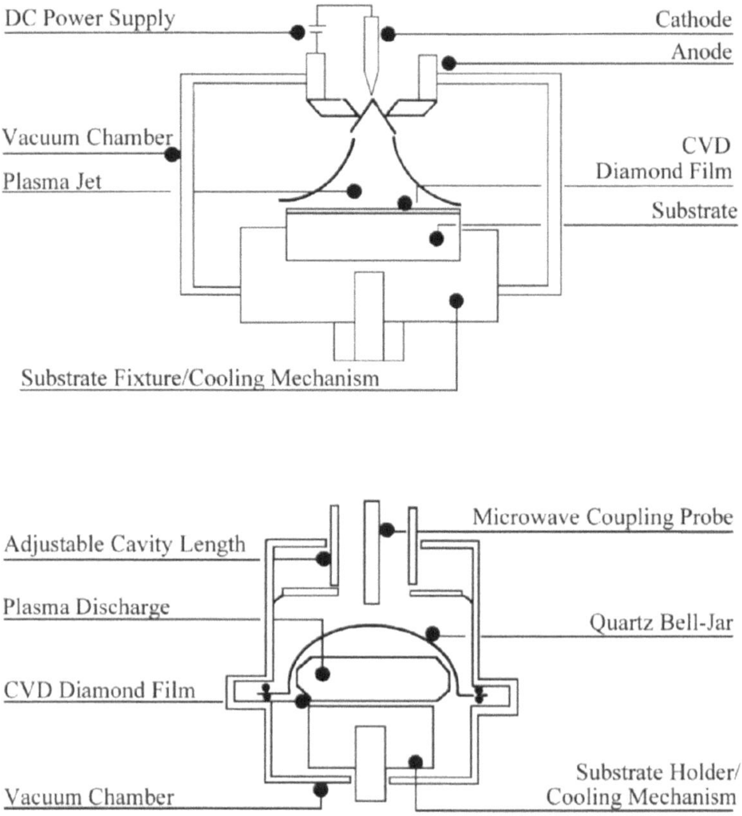

Figure 4.47 The two leading technologies of former Norton Diamond Film that the author was responsible to acquire; Technion's DC arc jet system (top) capable of growing the best thick diamond film, and Wavemat's microwave plasma system (bottom) capable of coating the finest thin diamond film.

Norton Diamond Film was the largest diamond film company in the world in the 1990s. In 1989, Norton projected that it could sell $300 million worth of diamond films by 1993. This optimism boosted its stocks considerably before Norton was acquired by Saint-Gobain a couple years later. However, with tens of millions invested in Norton Diamond Film in 1990, the company's revenue was only a few million dollars by the turn of the century. Eventually, it was closed down by the

French parent company. However, Norton Diamond Film's thick diamond film cutting tools have been the best cutting tool available in the industry. Thick diamond film cutting tools were developed in late 1980s under the author's supervision. It was a highly successful product that generated most revenue for Norton Diamond Film before it was shut down. Had Norton Diamond Film focused more on this product, the company could have been continued to grow, as the industry still lacks a second supplier of thick diamond film cutting tools.

4.11 CVD Diamond Growth without Hydrogen Atoms

CVD diamond may be deposited without the catalytic assistance of hydrogen atoms. For example, by mixing methane (e.g., 1%) with an over abundance of argon and subject the gas to microwave agitation, diamond films may be deposited at a slower rate (e.g., 1 μm/h) than normal CVD processes and at a lower temperature (e.g., 700°C). The diamond films so formed are very thin, and they contain nanocrystalline grains.

Although the concentration of hydrogen atoms formed by dissociating methane is too low to provide much catalytic effect, the bombardment of argon atoms and cations on dissociated carbon atoms forces them to form transient dimers (C_2). These dimers can combine to become a part of nuclei. However, because there is no hydrogen atom to allow further growth of the deposited nuclei, basically this hydrogen-free CVD process can only nucleate diamond without its growth. For this reason, diamond films are deposited with nanom crystals without larger grains. Such nanom diamond films are very smooth on the growing side as well as on the substrate side, so they can be used without polishing. This simplicity can reduce significantly the cost of making a diamond film product such as surface acoustic wave (SAW) filters.

Nanocrystalline diamond may also be formed by dissociating enriched methane (e.g., 1/3) in hydrogen gas at low temperatures (e.g., <600°C) by microwave agitation. The low temperature can slow down the decomposition rate of methane considerably so the deposition of graphite may be avoided. However, the low temperature also slows down carbon transport so diamond grains may not grow much. Moreover, the high methane concentration may promote faster nucleation. The slow growth of large amount of nuclei has prompted the formation of nano-sized diamond crystallites. Nanocrystalline diamond grown this way contains a lot of graphitic (sp^2) bonds in lattice. In contrast, nanocrystals grown by using argon gas as the solvent are essentially diamond.

CVD diamond may also be formed by substituting hydrogen with halogen elements. For example, by subjecting the gas mixture of CCl_4 and Cl_2 gas to high temperature (e.g., in microwave plasma) diamond may be formed. Because the carbon halogen bonds are weaker than that between carbon and hydrogen, lower temperatures are required to grow diamond films. But because of the corrosive nature of the reactive gas, the halogenated CVD processes are seldom used.

Another way of growing CVD diamond without using hydrogen gas is by laser ablation. This method was developed by Pravin Mistry in 1990 accidentally when he bombarded CO_2 and N_2 gases simultaneously with several lasers. He called such a process QQC (French words quel-que-chose).

During the laser bombardment, three types of lasers (excimer, Nd:YAG, Co_2) are used together to irradiate the gas mixture that covers a substrate (e.g., steel). The diamond may form continually as the substrate advances (Fig. 4.48). This process is appealing in many ways; for example, it can coat a long substrate with a continuous diamond film and the entire process can be run without a vacuum chamber. Moreover, the substrate does not have to be a refractory carbide former as most other CVD processes may require. In fact, diamond film may even be coated on even a steel shank that normally cannot be coated normally because steel can back-convert diamond at high temperatures. In addition, the diamond formed can be flattened in situ by the same laser that is used to deposit the diamond film. This step reduces or eliminates subsequent polishing processes that are both difficult and costly. Furthermore, the gas mixture (CO_2 and N_2) is also a lot of less expensive than that (CH_4 and H_2) for conventional CVD processes. Despite of all these promises, the process has not been commercialized. One of the obstacles may be that these lasers are very expensive as multimillion dollar investment was solicited to support this process even after 10 years of its discovery (e.g., the author was asked by Rustum Roy, its vigorous promoter at Penn State University, if money could be raised in Taiwan).

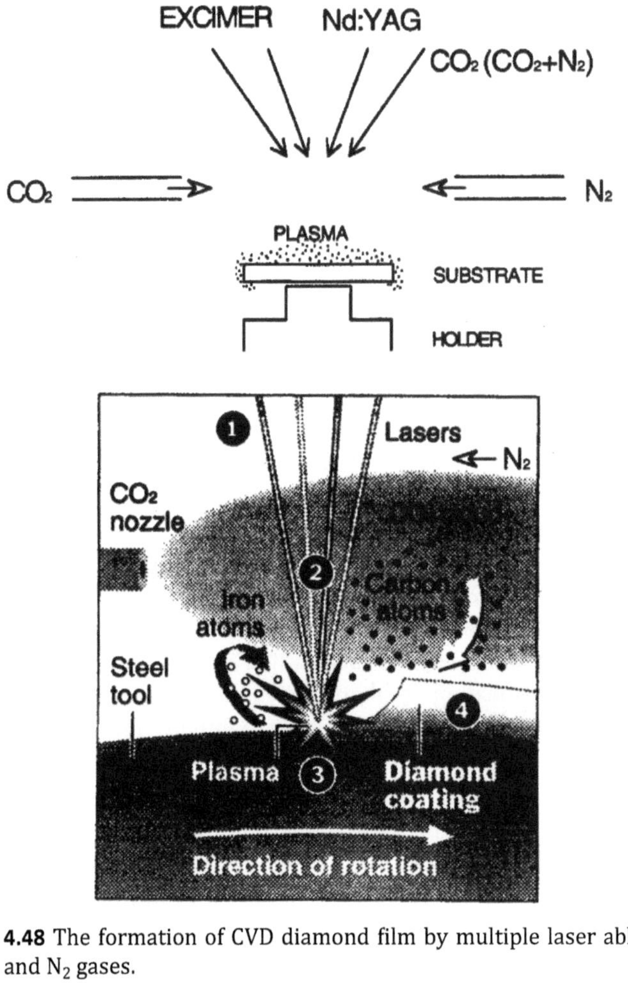

Figure 4.48 The formation of CVD diamond film by multiple laser ablation of CO_2 and N_2 gases.

Although the cause of CVD diamond growth by laser ablation is not fully understood, it is related to the chain reaction of laser bombardments. Specifically, the infrared beam of CO_2 laser can heat up the CO_2 molecules rapidly. The near-infrared beam of Nd:YAG laser will dissociate these CO_2 molecules to form carbon atoms. The ultraviolet beam of excimer laser will ionize carbon atoms and it may also dissociate the tightly bounded N_2 molecules. Together, the energy of these lasers will accelerate the movement of carbon atoms and cations to collide with one another to form diamond under local high pressure.

While the gas molecules are dissociated and accelerated, the atoms in underlying substrate (e.g., W, Co is cWC is used) are also vaporized and mixed in with the deposited diamond. Because the temperature of diamond deposition is very high, it was claimed that the growth rate may reach 1 μm/sec (3.6 mm/h), which is much higher than stable diamond growth under high pressure. However, such an outrageous claim has not been verified by independent researchers.

4.12 Low Pressure versus High-Pressure Growth of Diamond

When various low-pressure CVD processes for growing diamond were unveiled by Japanese in late 1980s, many shared the optimism that these new recipes could produce large amount of diamond at low cost. For one thing, the low-pressure reactors may be much less expensive than high-pressure equipment used in industry for making diamond superabrasives. Moreover, the high-pressure reaction cell is constrained by the assembly of pressure medium, so its production size (e.g., 60 cm in diameter) or volume (e.g., 200 cm^3) is severely limited. CVD reactors do not have this constraint, so the sample size can be much bigger, as many had believed to happen.

However, it now becomes clear that CVD processes cannot compete with high-pressure methods in both throughput and cost of diamond production. Thus, in 2002 more than 400 tons of diamond were produced worldwide (China made more than half of the quantity but one-tenth of the value; Sung, 2003) using the conventional high-pressure method, but less than 1 ton was deposited by CVD methods. Moreover, the price for the premium diamond (e.g., saw diamond) synthesized by high-pressure method was less than \$1 per carat; but it was more than \$5 per carat (about \$10/cm^2 for a diamond film of more than

1 mm in thickness) for a cutting-grade diamond film deposited by a CVD method.

The unexpected disadvantages of CVD deposition of diamond are clearly reflected in the much lower efficiency of their energy consumption when it is compared with high-pressure synthesis of diamond. In a large production system of high-pressure belt apparatus, this energy productivity may be expressed by: g/H = 4 Kw, which is 200 times higher than current state-of-the-art CVD processes. An efficient belt apparatus nowadays can produce more than 100 grams of high-quality diamond (e.g., saw grit) per hour and it may consume no more than 25 kW. In contrast, the most advanced CVD reactor can produce no more than 10 grams of cutting-grade diamond film, but it may require half a megawatt electric power.

The sharp contrast in energy consumption between CVD deposition and high-pressure synthesis is due to two intrinsic differences in the kinetics of diamond formation. CVD diamond deposition is a metastable process that cannot be rushed; otherwise the stable phase of graphite will be deposited. Consequently, the growth rate must be controlled to be no more than a few tens of microns per hour. On the other hand, the high-pressure diamond growth is limited only by the supply of carbon nutrient so the growth rate may attain more than 1000 microns (1 mm) per hour.

Another limitation of CVD diamond deposition is that it is a two-dimensional process, so the energy input can only be used to increase the flux of dissociated carbon. The deposition rate is then limited by the impingement of this flux on the substrate surface. However, the high-pressure assembly utilizes numerous layers (e.g., 100 alternative layers of graphite/catalyst) of interface for growth. Hence, diamond can be formed throughout the reaction volume.

Besides being energy inefficient, CVD processes are also material wasting. The rate for converting carbon to diamond is only a few percent. In contrast, high-pressure methods can convert more than 30% of graphite into transparent diamond (over 50% for growing dark diamond). Moreover, the bulk (>95%) of gas used by CVD processes is hydrogen. Less than 10% of hydrogen molecules are dissociated to catalyze the diamond formation, and almost all hydrogen is discharged into air without use.

Although CVD gases can be recycled, but the process comes with a significant capital investment as penalty.

In addition, it turns out that a large CVD reactor is not necessarily less expensive than a comparable high-pressure system (e.g., $1 million for a 5000 ton production press). Hence, the cost of making CVD diamond may be intrinsically cost prohibitive. For example, the cost of a 300 Kw CVD system may be similar to that of a 5000-ton press. The former can produce only about 6 grams of diamond per hour, but the latter, more than 100 grams per hour.

Although CVD diamond films may be intrinsically costlier to grow than high-pressure diamond grits, their area properties cannot be matched by that of grits. Hence, diamond films can be used in premium applications, such as high-frequency filters, optical window, heat spreaders, to name just a few. Therefore, CVD and high-pressure processes of diamond making are not competing, but they are complimentary in applications. They can work in tandem to increase the overall market size of industrial diamonds.

Chapter 5

Polishing of Diamond Films

It costs more to polish a diamond film than to grow it.

5.1 Mechanical Polishing of Diamond

Polycrystalline diamonds formed by either CVD of diamond films or high-pressure sintering of PCD have a rough surface. They must be polished often to attain a mirror finish before they can be used. Because diamond films contain nearly 100% diamond; and PCD, up to 95 V%, any diamond tool used to grind or polish them will have a very low life as most diamond tools contain only about 10 V% of diamond grits. Moreover, most volume of the diamond grits is lost due to the pullouts. Hence, it would be very inefficient and also costly to shape diamond by mechanical means.

A faster way to remove diamond is by dissolving it into a carbon solvent (e.g., iron) at high temperature. For example, during the mechanical polishing of a diamond film, it may be pressed against a rotating steel plate (Fig. 5.1). Such kind of polishers are employed in polishing gemstones, PCD, or CVD diamond. In this

Handbook of Industrial Diamonds: Diamond Films and Carbon Coatings
James Sung

ISBN 978-981-4310-00-0 (Hardcover), 978-1-003-18045-6 (eBook)
www.jennystanford.com

case, the polishing abrasive used is diamond powder that is added as suspended slurry. The size of diamond powder is dependent on the requirement of polishing rate and the surface finish. For rough polishing mesh size (e.g., 120/140) diamond powder may be used; for fine polishing, micron diamond is required.

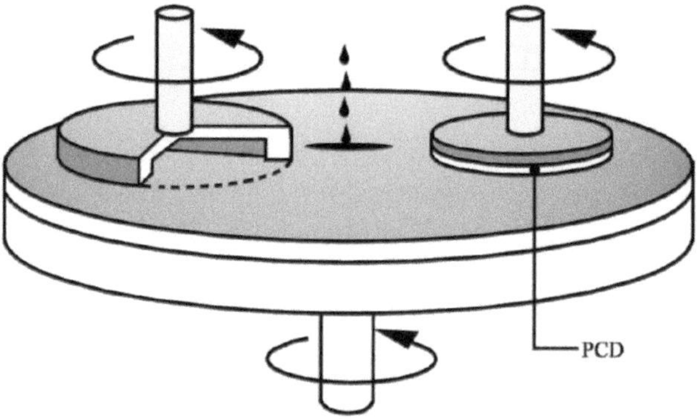

Figure 5.1 Diamond films of PCD compacts can be polished by pressing them against a rotating steel platen. The surface of the platen is coated with slurry that contains mesh or micron diamond.

5.2 The Diffusion Etching of Diamond Film

The pure mechanical polishing is extremely slow, so chemical means may be used to accelerate the process. A convenient method is to heat the rotating steel plate to high temperature (e.g., 950°C) in vacuum (e.g., 0.8 Pa). The high-temperature polishing by the rotating plate does not require slurry. Instead of abrading diamond away mechanically, the hot steel plate will back convert diamond to form amorphous carbon that can be rubbed off by the rotating platen. Alternatively, amorphous carbon is also dissolved in steel, which further accelerates the removal rate of diamond. In addition to steel, other carbon solvents may also be used to back convert and dissolve diamond at high temperatures. For example, the rotating plate can be made of nickel, cobalt or its alloy.

The back conversion polishing process of diamond by a carbon solvent of metal may also be performed statically without using a rotating plate. In this case, diamond film is pressed (e.g., 30 MPa) against the metal and heated (e.g., 900°C) in a vacuum furnace. The removal rate of this static process is slower (e.g., 2 microns/h) than using a rotating plate (e.g., 10 microns/h).

In addition to solid-state back conversion of diamond, liquid-phase carbon solvent can accelerate this attrition process. For example, manganese powder may be added to the interface between diamond and steel to speed up the back conversion and dissolution process of diamond; so are rare earth elements such as lanthanides (e.g., La, Ce). If the temperature is above the melting point of the eutectic composition, a thin layer of liquid carbon solvent is formed that can accelerate the transport of carbon atoms. An effective liquid carbon solvent is Ni-Ce alloy, which has a melting point of only 477°C at the eutectic composition (22 atom% of Ce). In this case, the polishing can be performed at a temperature as low as 500°C (Fig. 5.2).

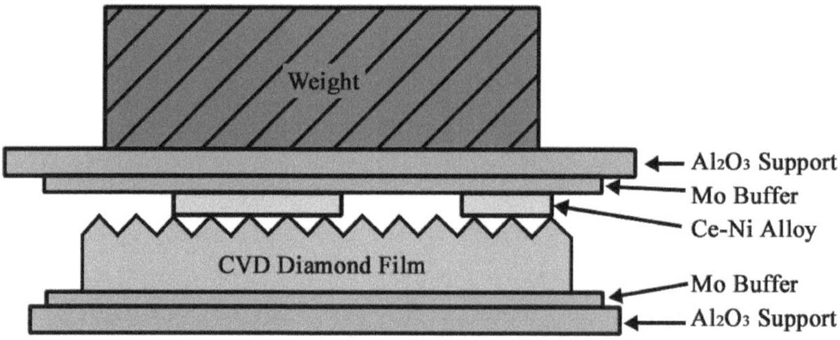

Figure 5.2 The hot diffusion etching of diamond film by Ni-Ce alloy performed at 500°C. In this case, the rough surface of diamond film may be flattened within 1 h.

For the diffusion etching process to continue, the carbon flux must not be blocked at the interface between diamond and the metal. Thus, carbide former metals (e.g., W, Mo) are not suitable for etching diamond because they will form carbide at the interface. In addition, cast iron is a poor etchant because its carbon content has already been saturated.

5.3 Converting Diamond Film to Methane Gas

The CVD process is to decompose methane in hydrogen atmosphere to allow carbon atoms to form diamond. The reverse process may take place by dissolving diamond with a carbon solvent in hydrogen atmosphere. In this case, the dissolved carbon atoms will not be saturated in the carbon solvent as in the case of cast iron. Instead, they will combine with hydrogen and eliminated as methane gas. In essence, diamond is gasified to become methane gas.

For example, under the hydrogen atmosphere, a hot steel wire can continually etch diamond so it can cut diamond or even penetrate diamond (Fig. 5.3). In this case, the steel wire is not consumed because it serves merely as a catalyst for diamond to gasify into methane.

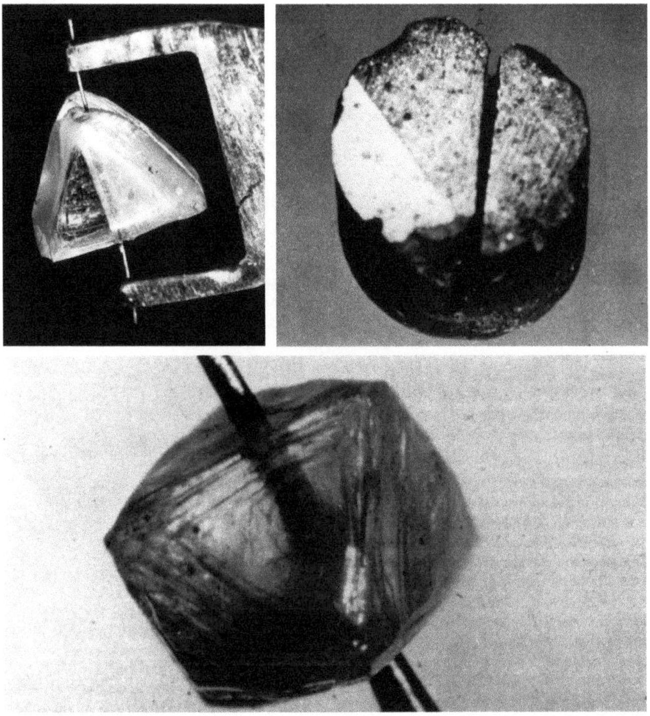

Figure 5.3 Diamond can be sliced with a hot steel or nickel wire in hydrogen atmosphere (Grigoriev & Kovalsky, *Indiaqua*, 1974, 39, 3).

Using a hot steel wire saw, diamond can be cut in various shapes (Fig. 5.4). Such a technique may be handy for making miniature diamond components for MEM (micro-electro-mechanical) devices.

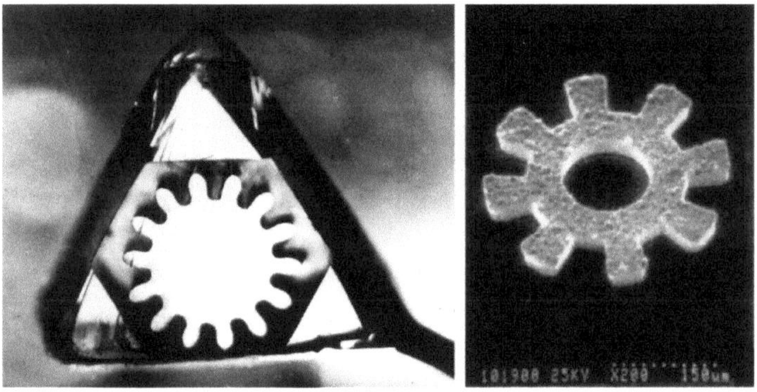

Figure 5.4 The miniature diamond gears in negative (left) and positive (right) forms.

Diamond films may also be etched with hot steel or other carbon metal solvent under hydrogen atmosphere (e.g., 10 KPa) to allow continual removal. The removal rate is slower (e.g., about half) than that in vacuum (Fig. 5.5). However, the increased polishing rate in vacuum may be due to the presence of residue oxygen (e.g., 0.8 Pa) that is absent in hydrogen atmosphere.

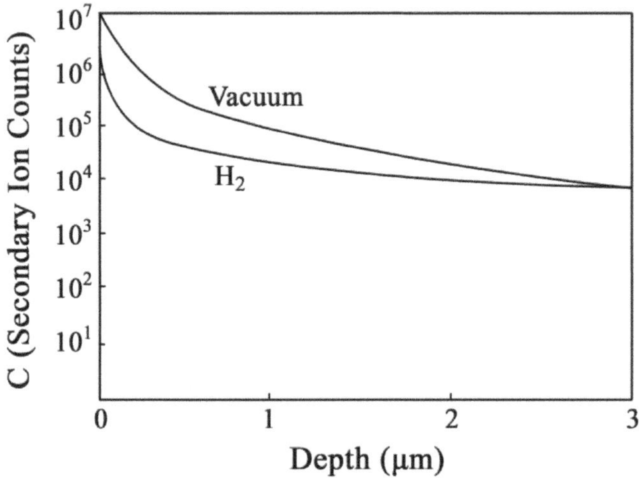

Figure 5.5 The depth profile of carbon in steel as measured by SIMS. The carbon atoms diffuse deeper in vacuum than under hydrogen atmosphere because the grain boundaries of iron are more open. This is why the former polishing rate is about twice of the latter condition.

In addition to be able to continue the polishing process of diamond film, the temperature required in etching under hydrogen atmosphere is lower (e.g., 700°C) than that for pure mechanical polishing (e.g., 900°C) or diffusion etching (e.g., 800°C). Because of the slower removal rate at lower temperature, diamond film polished under hydrogen atmosphere is smoother than that in vacuum.

The effects of diamond polishing under various conditions are summarized in Table 5.1.

Table 5.1 The effect of polishing conditions on polishing rate of diamond films

Polishing conditions	Polishing rate
Plate material	La > Ce > Fe > Ni >> Cast Iron > Mo=W=Ti =0
Atmosphere	02 > Air > Vacuum > H_2 >> Ar > He > N_2
Plate temperature (°C)	950 > 850 > 750 > 700 = 0
Applied pressure (KPa)	1000 > 100 > 10
Spinning speed (mm/s)	10 > 1

5.4 The Vibration Etching of Diamond Film

The bottleneck in etching diamond film by a carbon solvent metal is the deposit of amorphous carbon at the interface. This carbon deposit may be etched away with oxygen or hydrogen atoms. But alternatively, they can be abraded off by applying a vibration frequency (e.g., 200 Hz). With such a means to remove carbon deposition, the rotation of the iron plate may not be necessary. Moreover, the stress applied to the diamond film may also be greatly reduced.

5.5 Ion Bombardment Etching

Diamond films may also be etched by ion bombardment. For example, oxygen ions (e.g., 1.5 KeV) are effective etchants (Fig. 5.6). However, such echants may have strong local effect so the final surface cannot be smooth. Alternatively, clusters of argon ion may also be used to flatten the diamond film. In this case, the impact energy is spread out so the craters formed on the diamond surface, although larger in scale, can have more gentle slopes.

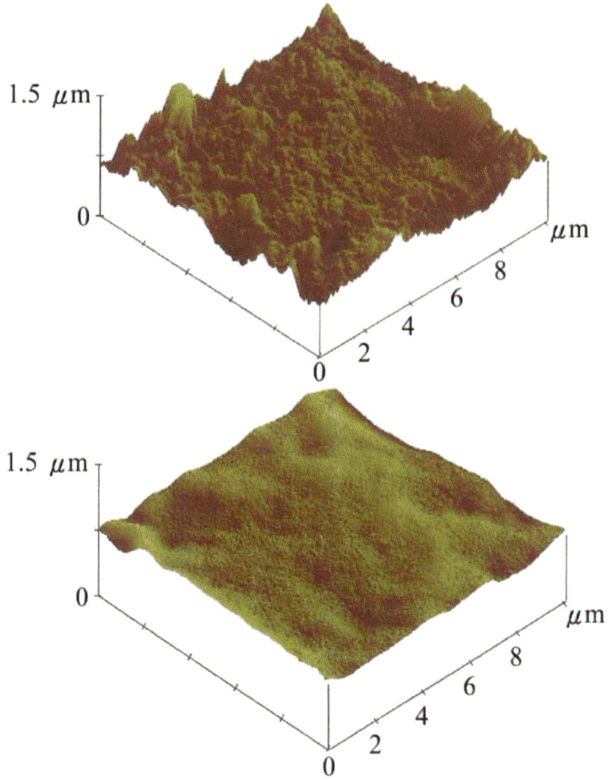

Figure 5.6 The surface roughness of a DC arc deposited CVD diamond film of about 1 mm thick (top diagram). The smoothed surface after etching of 100 nm with plasma of 95% O_2 and 5% CF_4.

The diffusion etching of diamond film will leave a back carbon coating on the protruded areas. They can be mechanically polished at room temperature using abrasives.

Chapter 6

Cutting Tools

Stone cutters were invented more than 2 million years ago. We have been improving them ever since, however.

6.1 Evolution of Cutting Tools

Cutting tools are the most primitive tool humans have ever used. Two million years ago, our ancestors, *Homo habilis*, began to use sharp stones to cut animal meat. When Egyptians invented copper, they used copper knives to building pyramids. Subsequently, Mesopotamians discovered iron, so it has since then become the standard cutting knife.

Cutting tools are the most basic tools for eliminating excess stock of a material. Material removal is necessary for processing most manufacturing parts. Typically, about 10% of the manufacturing cost is expensed for material removal.

In order to cut, the cutter must be harder than the cut material; hence, the history of cutting tools development has been the search of a harder material for making cutting tools. The modern history of cutting tools started in 1897, when high-speed steel (HSS) emerged as the standard machining tool. Although

Handbook of Industrial Diamonds: Diamond Films and Carbon Coatings
James Sung

ISBN 978-981-4310-00-0 (Hardcover), 978-1-003-18045-6 (eBook)
www.jennystanford.com

HSS had been a convenient cutter for softer materials, it was inefficient for cutting certain industrial materials such as gray cast iron.

In 1923, cemented tungsten carbide (cWC) was invented. In 1927, Krupp introduced Widia cutting tools, the first cWC cutting tools. In 1928, GE established Carboloy department (now under Sandvik) to mass produce cWC. cWC is much harder than HSS, so it can triple the cutting speed. Moreover, cWC can easily cut gray cast iron, for which HSS was found to be powerless.

cWC is a cermet that combined the hardness of ceramics (WC) with the toughness of metal (Co). The marriage of these two seemingly uncompromising attributes laid down the foundation for all future cutting tools, including the most advanced cutting tools used today, such as diamond-cutting tools.

In 1931, TiC, a ceramic harder than cWC, was added into standard cWC. The hardened cWC could cut ductile steel at a higher speed than cWC could without being worn excessively by the long threads produced. In 1968, TiC coating was applied to cover cWC's top surface where hard material is best suited. In 1973, new cermets were invented. Many contained carbides or nitrides of Ti, Ta, or Nb, and they may be cemented by either Co or Ni. Cermets are typically harder than cWC. Moreover, they are more thermally stable. Hence, cermets may cut at a higher speed than cWC.

Although ceramics (e.g., Al_2O_3, Si_3N_4, SiC, ZrO_2) and cermets are harder than cWC, as a substrate to support a hard or superhard coating, cWC is still the industrial norm. Hence, the advancement of cutting tools has become the development of harder coating materials for cWC. Thus, cWC tools are now coated with either a hard ceramic layer (e.g., Ti(CN), CrN, (Ti,Al)N, (Ti,Al,Cr)N, (Ti,Al,Si)N) or a superhard layer (e.g., PVD, PCD, or CVD diamond).

For the ceramic coating, multi-layers may be applied to further boost the hardness. One advantage of multi-layers is that the stress accumulated in the coating due to the thermal expansion mismatch may be relieved (e.g., by alternating hard and ductile layers). As a result, thicker coatings may be applied for longer cutting life. Moreover, by manipulating the coating intervals, a superlattice of nano-size may be formed. The formation of such a nanostructure may lock up the dislocation movement, so an apparent hardness may be boosted beyond the hardness of the crystal lattice itself (Fig. 6.1).

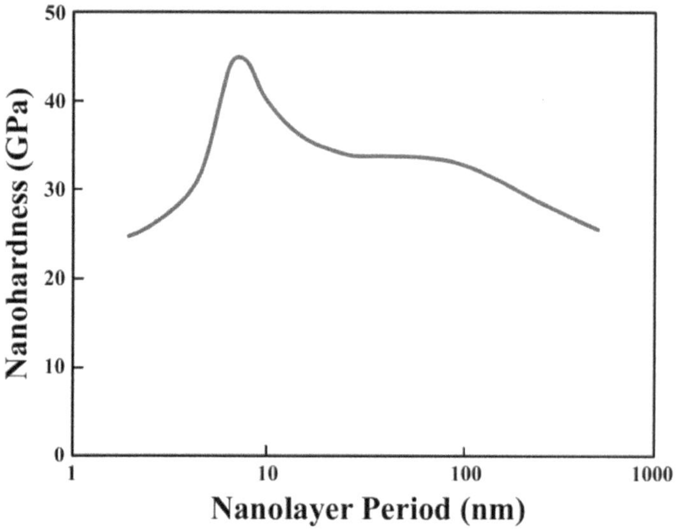

Figure 6.1 The alternative coatings of numerous TiN-CrN and AlN layers around 7 nm intervals can form a superlattice. This superlattice may double the apparent hardness of the coating and hence may also double the cutting life.

Although superlattice may boost the hardness beyond its intrinsic value, such hardness is not isotropic. In fact, if the stress is exerted parallel to these layers, it could shear them off along the planes of weakness. Hence, nowadays, a new form of superlattice is formulated in which the coating contains a cemented structure of hard nano-sized grains. For example, nanocrystalline (Ti,Al)N may be embedded in amorphous Si_3N_4 (Fig. 6.2). In this way, the dislocation movements are locked up in three-dimensions, so the boost of apparent hardness reveals no weak planes.

The nanocomposite coating may retain its hardness up to about 800°C. Such a thermal stability is important to withstand aggressive cutting action. Hence, nanocomposite-coated cWC has been demonstrated to double the life of conventional coated cutting tools.

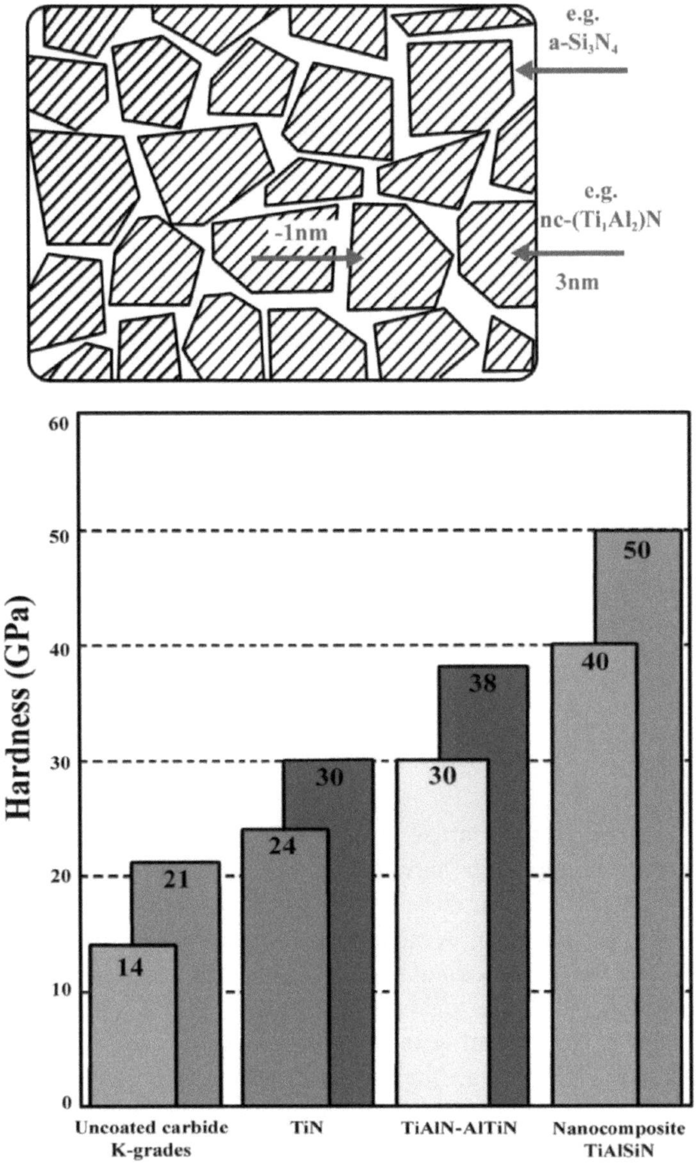

Figure 6.2 The nanocomposite (top) resembles the cementing structure of cWC but with a scale about 1000 times smaller. An example are nano-sized particles of (Ti,Al)N, which may be cemented in an amorphous matrix of Si_3N_4. Such downsizing of cementing structure are effective in boosting the apparent hardness (bottom).

6.2 Cutting Mechanism

When the tip of a cutting tool touches the surface of a work material, it will begin to exert a pressure. As the pressure increases, the work material will first be strained elastically, which means no permanent deformation will take place as the strain will recover as soon as the cutter tip leaves the spot. With the further increase of pressure, the work material will begin to yield with plastic deformation. This time, the deformed region will bounce back only slightly, leaving a permanent region of deformation. But still there is no loss of any material, i.e., cutting has not started yet. When the yielding of the material becomes excessive, rupture occurs and so cutting begins. Even at this stage, the ruptured material may not be lost as it may still be attached partially to the original body. However, with further sink of the cutting tip, eventually the elimination of work material becomes continuous. At this stage, the electrical power is used to break the atomic bonds in the material, and the cutting tool is used to transfer the energy. Figure 6.3 shows the three distinct stages of cutting.

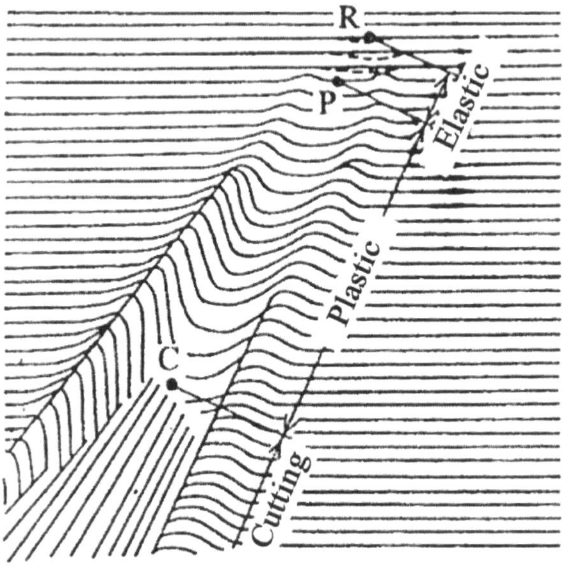

Figure 6.3 The cutting groove on the work material shows two distinct stages of surface deformation before cutting takes place.

The degree of precutting deformation will reduce with the following conditions:

1. Hardness difference between cutter and work materials
2. Sharpness of the cutter
3. Smallness of cutting angle
4. Speed of engagement and movement

The three stages of cutting are highly sensitive to the sharpness of the cutting tip. The smaller the radius of the cutting tip, the lower the stress required to cut (Fig. 6.4). This is why dulling cutting tool requires much higher energy to cut because much of such energy is spent in deforming the work material plastically. The hardness of the cutter is also important, a less hard cutting (e.g., HSS) will wear out faster. Moreover, it can be deformed under pressure; hence, more energy must be expended to penetrate the work material.

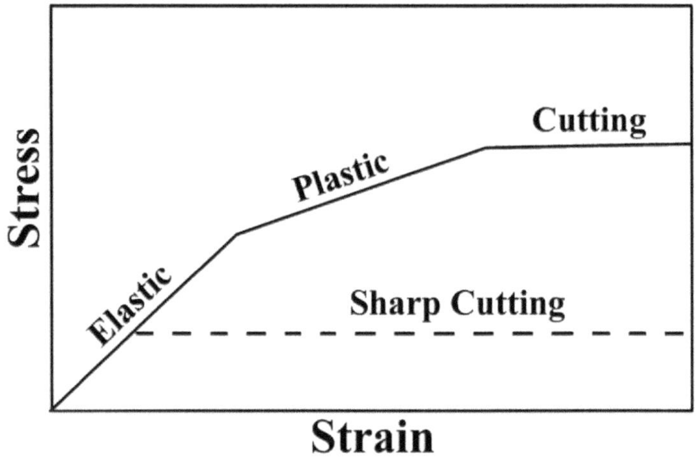

Figure 6.4 The stress–strain curve of a work material in responding to a cutting pressure. Note that a sharp cutting tool can penetrate easily because it may skip the entire range of plastic deformation.

The cutting action will produce debris. The mechanism of the debris formation is dependent on the cutting speed (surface speed and penetration depth), as well as the properties of the work material. High-speed cutting tend to produce brittle chips, and ductile work material will yield to long shreds (Fig. 6.5).

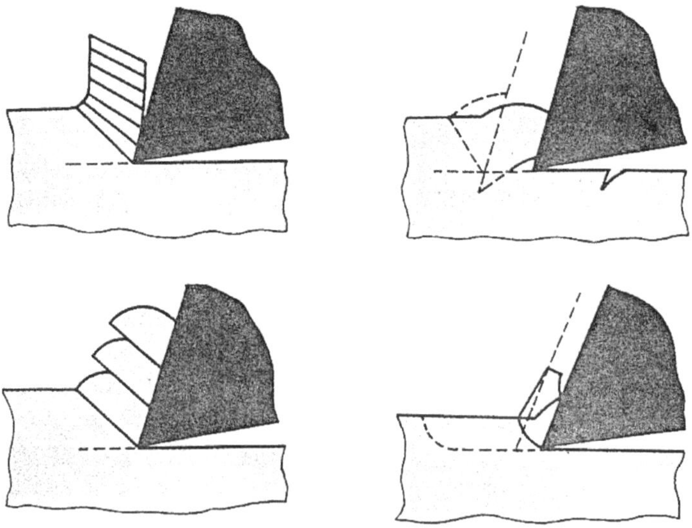

Figure 6.5 Different mechanisms of chip formation that are dependent on the parameters of cutting action and the properties of work materials.

During the cutting action, the deformation and the cutting will generate heat, and the heat output will increase with the cutting speed as well as the hardness of the work material. Most of the heat will be carried away by the breaking chips, but about 1/5 of the heat may be retained, and it is split between the cutting tool and the work material (Fig. 6.6).

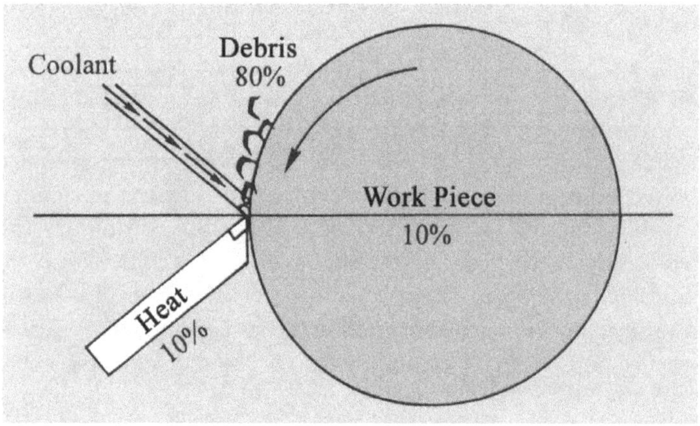

Figure 6.6 Heat dissipation during cutting.

Although the cutting tip generates most heat, the debris will carry such heat upward. Moreover, the friction of the debris against the cutting surface will generate additional heat. As a result, the hottest spot is not at the cutting tip, but slightly above this tip (Fig. 6.7).

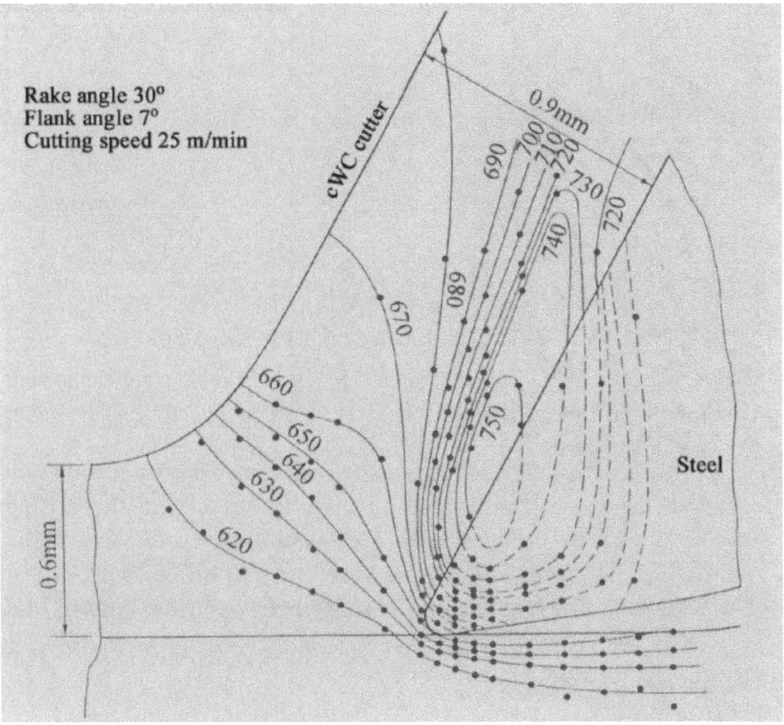

Figure 6.7 An example of temperature distribution around a cutter. Note that the highest temperature is located at a point above the cutting tip where the cutter will be worn the fastest.

With the hottest spot the most vulnerable to wear, consequently, a cutting tool will show a depression above the cutting tip. This is particularly noticeable in cutting ductile metal that may form long cutting debris. Because of this fast wearing problem, nowadays cutters are designed with a depressed area at this wearing point, known as chip breaker. The chip breaker not only can reduce the friction against the cutting debris but also may turn and break long shreds. In fact, without chip breaker, cutting of very ductile metals becomes difficult.

The wear mechanism of the work material is mainly mechanical, but it is the high temperature that consumed the cutter (Fig. 6.8).

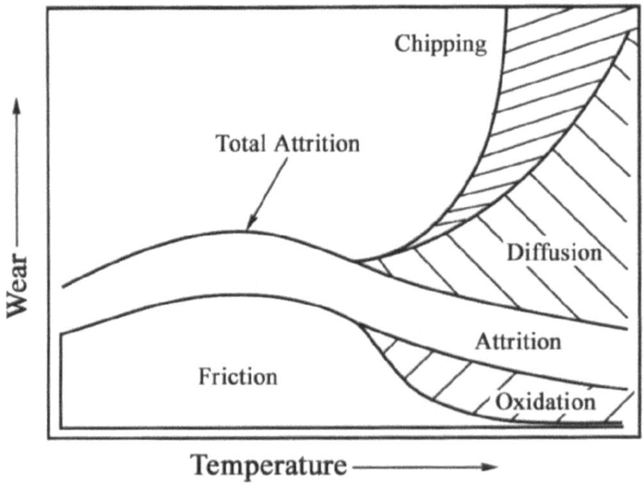

Figure 6.8 The mechanisms of wear and their temperature dependence. Note that at high temperature, micro chipping induced by thermal cracking is the main mode of wear.

Because most wear is caused by thermal degradation, the temperature control is important to minimize the wear rate. Of course by cutting slower will reduce the heat generation, but this will incur serious productivity loss. Hence, temperature is most controlled by using coolant. Indeed, by using a coolant, the tool life can be greatly lengthened (Fig. 6.9).

Although coolant can extend the tool life, most coolants contain emulsion or other hazardous ingredients, and the disposal of such liquid becomes an expensive task due to the ever-stringent environmental regulations. In fact, it may cost several times more to treat the coolant than paying for the tool. Hence, nowadays, more dry cutting operations are practiced. In order to cut dry with long life, the cutting tool must be able to maintain most of its hardness at high temperature. cWC tools may lose 2/3 of their already-low hardness when the tip reaches 800°C. In contrast, superhard tools may retain half of their high hardness. So, diamond and PcBN cutting tools are uniquely capable of cutting dry effectively. This is why they are in more demand than ever.

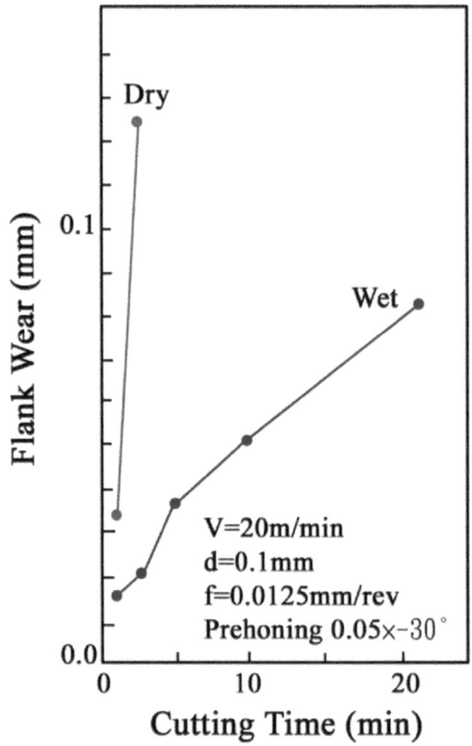

Figure 6.9 An example of cutting alumina by a diamond tool. Note that using coolant can effectively lengthen the tool life.

6.3 Evolution of Cutter Materials

With the improvement of yield and quality, the ability to maintain the tolerances becomes critical. This means that the cutter must not wear too fast. Consequently, the cutter's hardness must be upgraded. This trend of increasing hardness is reflected in the history of cutting tools advancement. Thus, HSS tools were replaced by cWC tools during World War II, and cWC tools have since been covered by ceramics and now by diamond or PcBN. With such an advancement, the cutting tools have now reached the realm of superhard materials (Fig. 6.10).

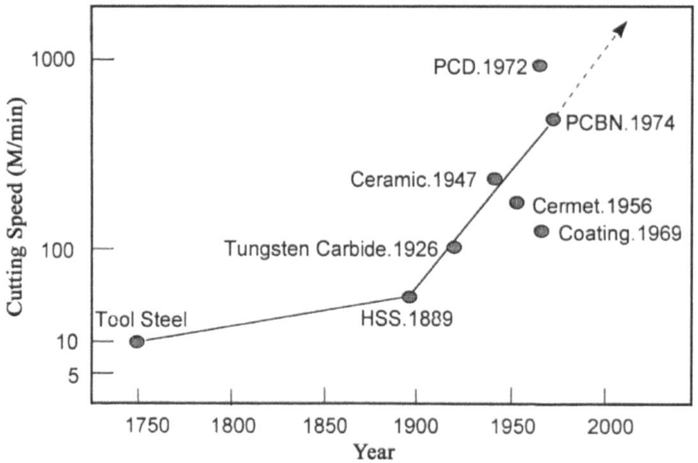

Figure 6.10 The history of cutting tools development is following the trend of increasing hardness as well as cutting speed of the cutting material.

With increasing hardness, the cutting materials become more brittle; hence, a new trend is to use composite material to support the hard material. One common practice is to coat a superhard material on a tough substrate such as cWC (Fig. 6.11).

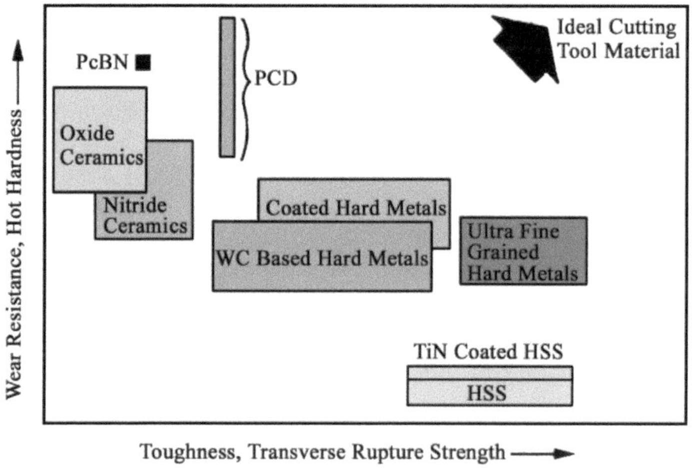

Figure 6.11 The trade-off hardness and toughness of cutters materials. The compromise is to use composite material with hard cutting tips that is backed up by tough support.

The above limitation of hardness versus toughness is reflected in another trade-off of cutting speed versus cutting depth. Harder cutters can stand high cutting temperatures, so they may cut fast but shallow. On the other hand, less hard cutters can have a high impact force, so they may cut slow but deep (Fig. 6.12).

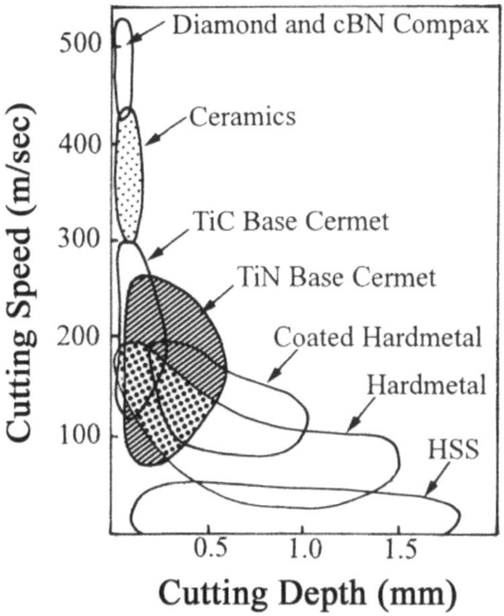

Figure 6.12 The trade-off of cutting speed and cutting depth. A harder cutter can skim the top fast, but less hard ones may plough deep slowly.

For a given material, the cutter's life is inversely dependent on its cutting speed. In order to break this rule, a harder cutter must be used to push the limit. Thus, with the increasing hardness of the cutter, both the cutter's life and the cutting speed have been increased (Fig. 6.13).

With the increase of hardness, the cutter can stand a higher cutting temperature. Hence, it can cut faster. Moreover, the high temperature generated may soften the work material, so the cutter's life may also lengthen. In addition, faster cutting often results in a smoother surface (Fig. 6.14). Hence, there are multiple benefits of increasing the cutting speed.

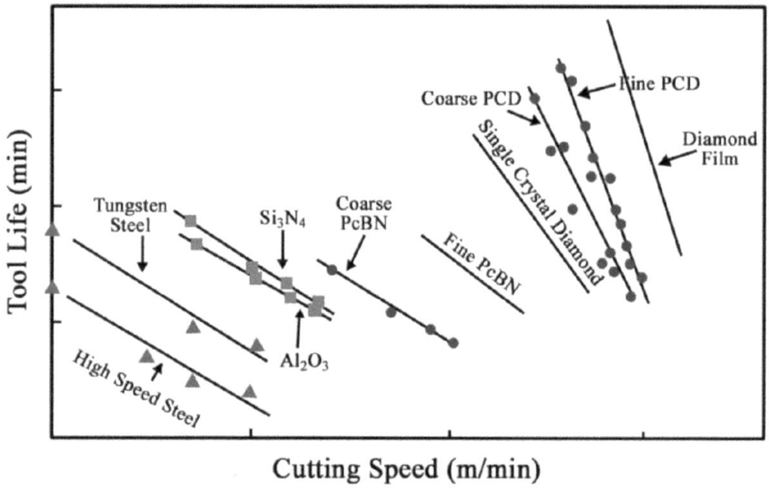

Figure 6.13 The inverse relationship between cutter's longevity and utting productivity. However, this relationship is moving toward longer life and faster cutting by using harder and harder cutting tools. In this diagram, the tool life is measured by the time to reach a flank wear of 0.3 mm.

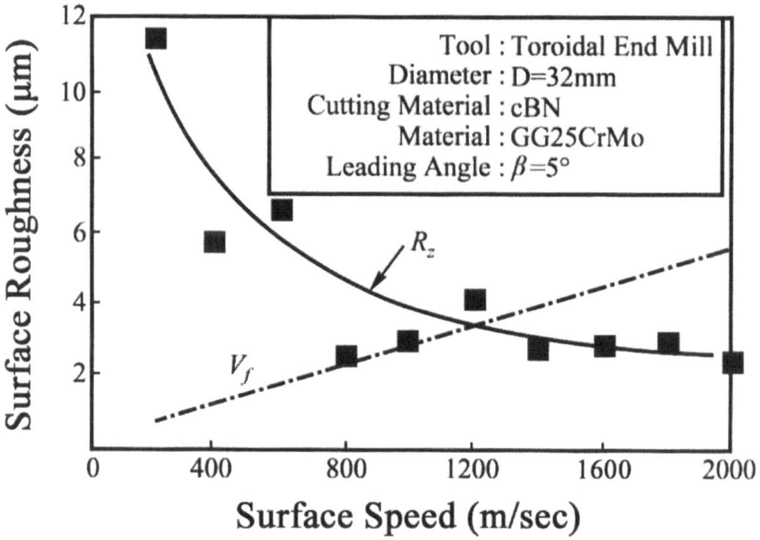

Figure 6.14 The cutting depth will decrease with the increase of cutting speed. As a result, the surface smoothness improves.

In summary, as the cutter becomes harder, it can allow a higher surface speed. The increase of surface speed of cutting therefore reflects the trend of the cutter's development (Fig. 6.15).

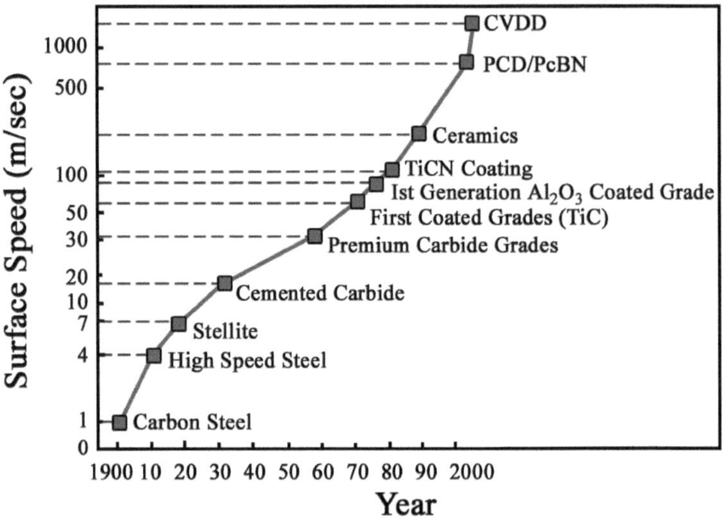

Figure 6.15 The evolution of cutting tools as reflected in the increase of surface speed of cutting by using materials of ever increasing hardness.

Cutting tools include inserts, mills, and drills. The world market of cutting tools is about $10 billion annually. About 60% of the market is served by cWC tools and about 1/3 by HSS tools. Various diamond tools account for about 5%. The rest are ceramic or cermet tools.

6.4 The Cutter's Geometry

The cutting tools may be engaged with the work materials in a variety of ways (Fig. 6.16).

In addition to the geometrical relationship between cutting tool and work material, the cutting parameters are also critical in determining the cutting performance (Fig. 6.17).

The cutting insert will be worn out in several modes (Fig. 6.18).

Sometimes, abnormal worn patterns may appear (Fig. 6.19); they are indicative of the defective material or wrong use of the tool.

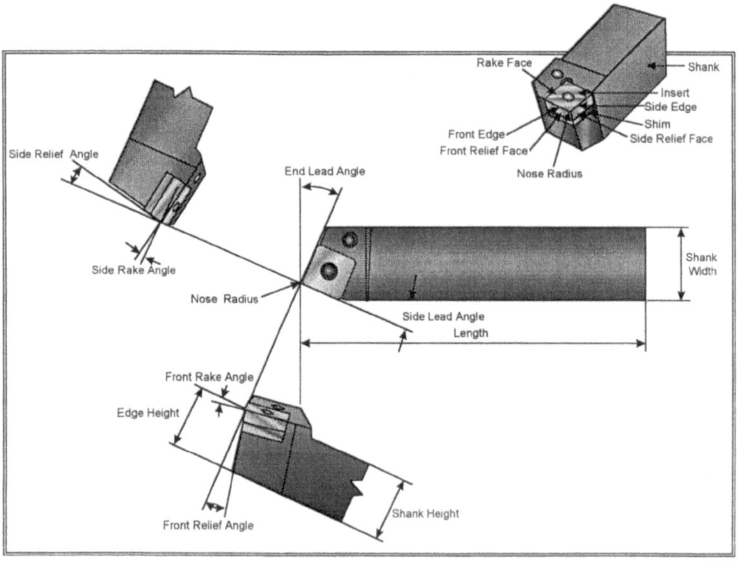

Figure 6.16 The cutting tool geometry include certain key angles.

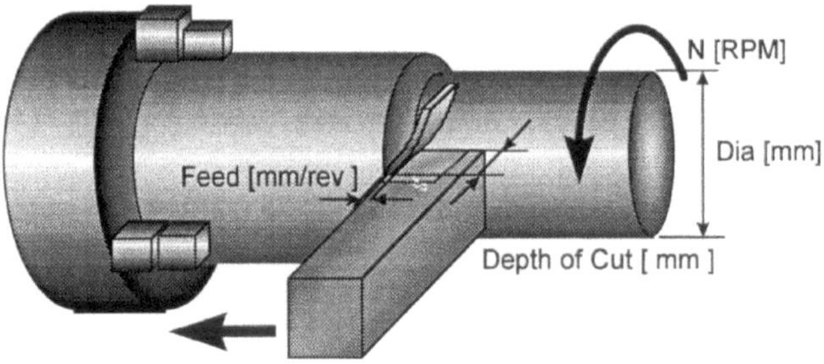

Figure 6.17 The typical cutting parameters for machining work materials.

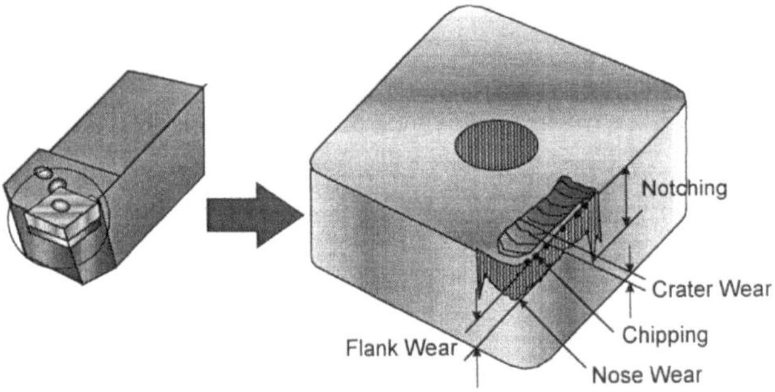

Figure 6.18 The wear location of a typical cutting insert.

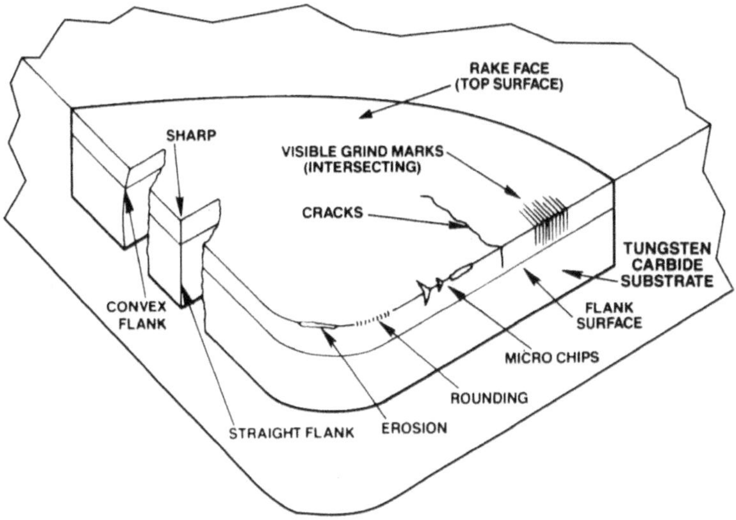

Figure 6.19 Cutting tool wear pattern may shed the light on the future direction of material or process optimization.

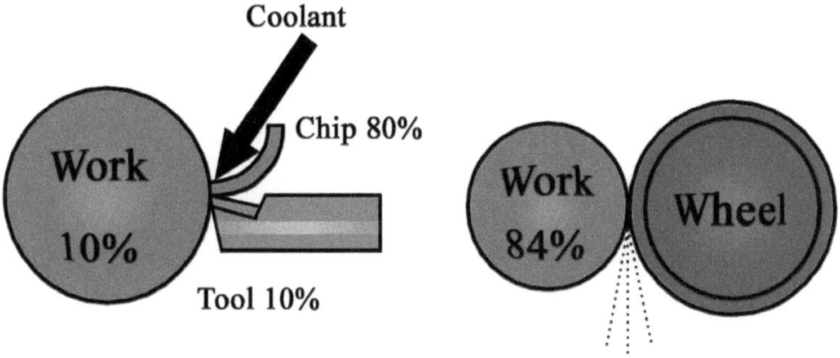

Figure 6.20 Heat distribution for cutting (left diagram) and for grinding (right diagram).

6.5 cWC Cutting Tools

Although cWC cutting tools' heyday is over, and they are replaced by harder coating materials, cWC is still being used as a substrate for supporting the harder coating. Moreover, cWC's material itself has also been improved, primarily by reducing the grain sizes. As the grain sizes decreases, cWC becomes stronger and more resistant to wear (Fig. 6.21).

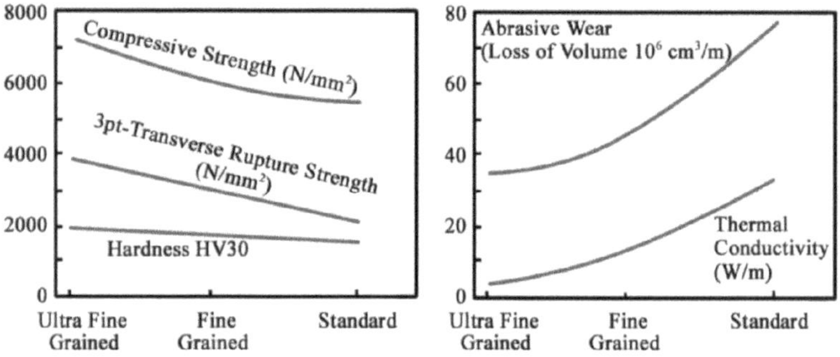

Figure 6.21 The trend of using smaller grain size will increase the strength of cWC. Moreover, it becomes a better thermal conductor. The above example is for cWC that contains 6% Co. Standard grains = 1.4–1.8 μm; fine grains = 0.8–1.0 μm, ultrafine grains < 0.5 μm.

When combined with the changes in grain size with the cobalt content, the property of cWC may be improved significantly (Fig. 6.22).

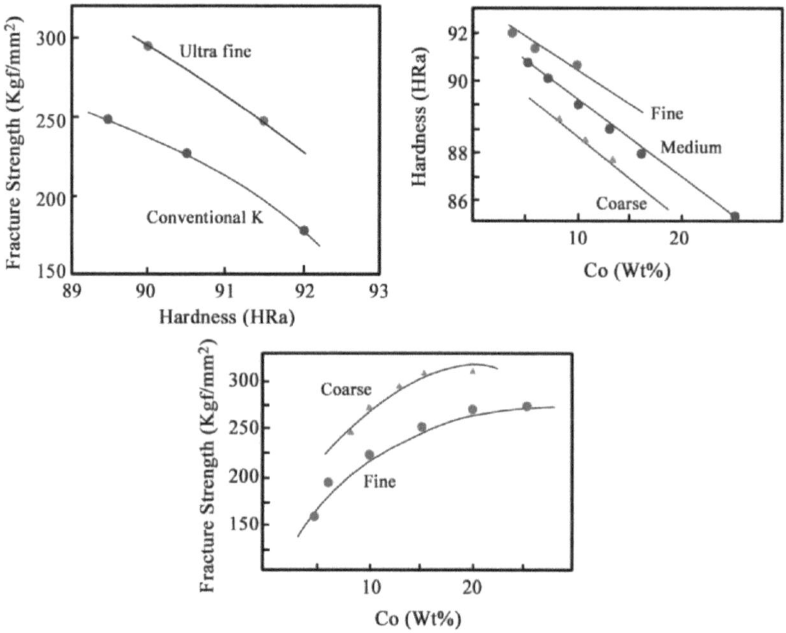

Figure 6.22 The dependence of hardness and strength of cWC on the cobalt content with various grain sizes of WC.

In addition to the grain size and cobalt content, the compositional changes may also alter the properties of cWC. In particular, adding TiC, TaC, or NbC will improve wear resistance. Moreover, the thermal stability may also be increased. If WC becomes less than other carbide, cWC gets converted to cermets (Fig. 6.23).

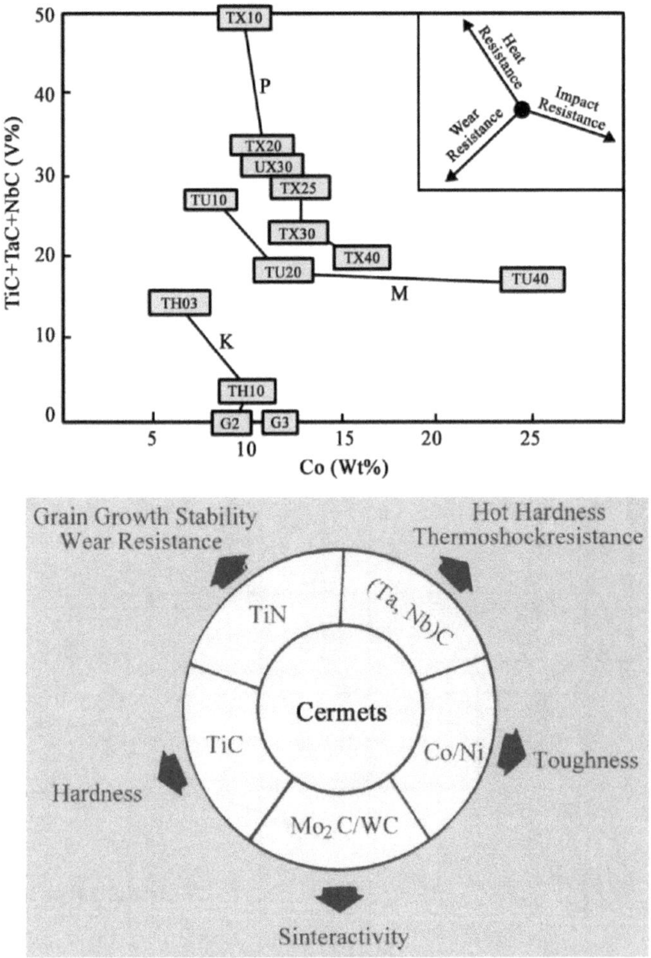

Figure 6.23 Example of properties manipulation of cWC by adding non-WC carbides (top diagram). The insert shows the trend of property change. If non-WC becomes dominant, a cermet is formed. The possible ways to adjust the properties of cermets are shown (bottom diagram).

6.6 Coating of cWC

WC is not the hardest carbide; there are other carbide or nitrides that can be harder (Fig. 6.24), and these harder ceramic materials may be used as the hard facing for cutting tools.

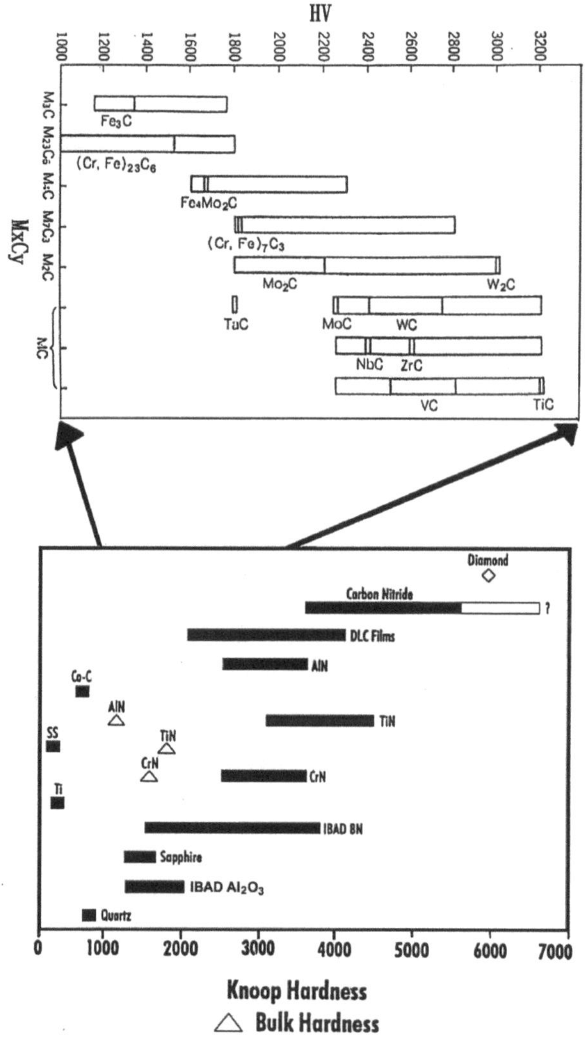

Figure 6.24 The ranges of hard materials of carbides and nitrides as well as oxides. The horizontal scale shows Vickers hardness.

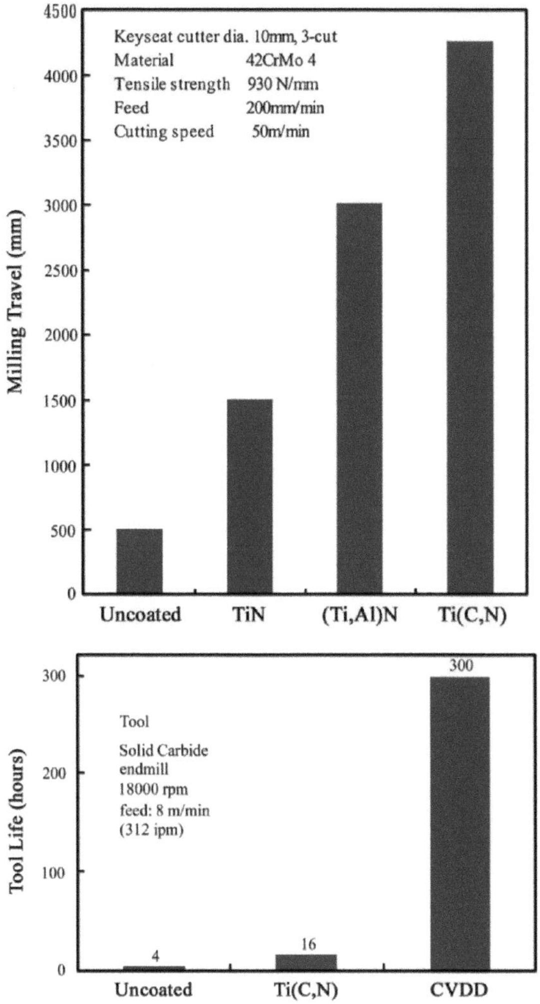

Figure 6.25 Tool lives can be greatly lengthened by coating high-speed steel mills with ceramic coatings (top diagram). However, the diamond film (e.g., 6 mm) coated cWC can last 75 times longer still (bottom diagram).

Most cWC cutting tools today are covered with a harder coating. Initially, the coating was mainly based on TiC or TiN, but harder and more specialized coatings have since been developed. Thus, Ti(C,N), (Ti,Al)N, CrN, (Ti,Al,Si)N, and (Ti,Al,Cr)N coatings have become popular (Fig. 6.25). These ceramic coatings may be applied by either low temperature PVD processes (e.g., cathodic arc) or high temperature CVD processes using reactive gases. The PVD processes deposit thinner films that are non-stoichiometric in composition. The CVD processes can lay down a thicker film with more equilibrium growth, but they are expensive processes and the substrate may be degraded by the high processing temperature.

Numerous compositions of carbide and/or nitride may be coated on substrates made of either HSS or cWC with significant gains of tool life. Moreover, multiple layers are incorporated to further improve the performance. However, no matter what combination of ceramic coatings is used, they are still trivial when compared to diamond coating that can easily outperform all non-diamond coatings (Fig. 6.25).

As an example of comparing CVD diamond film coating performance with conventional cWC tools, the common A390 alloy (Al + 18% Si) is used for testing. The result indicated that CVD diamond film coating can outperform conventional ceramic coatings by a wide margin in various conditions of cutting (Fig. 6.26).

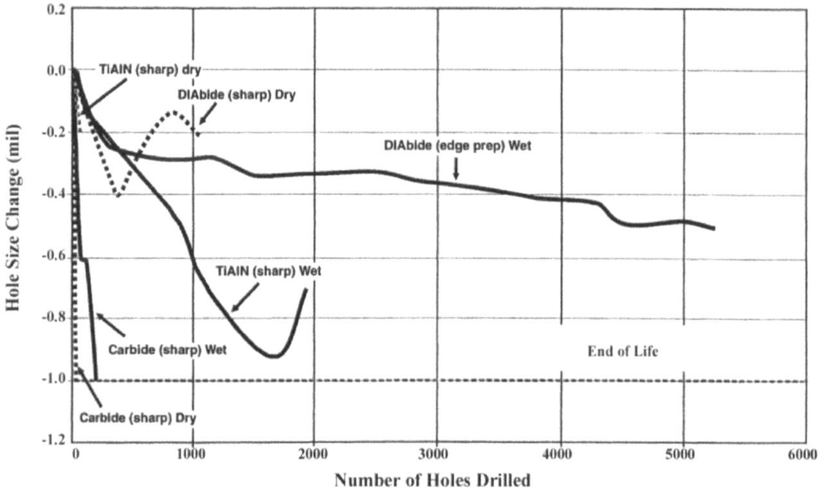

Figure 6.26 Performance comparison of dry step drilling at various speeds and feeds including continuous wet drilling. In this diagram, DIAbide is a CVD diamond-coated product provided by sp^3, Inc. The end of life of the above tests was set at 1 mil (0.001 inch) change of hole size or loss of cutting edge due to the micro chipping of underlying cWC (sp^3 brochure).

6.7 3D Diamond Coatings

Three types of diamond coatings may be referred to as "3D" coatings: physical vapor deposited (PVD) diamond-like carbon (DLC), chemical vapor deposited (CVD) diamond film, and polycrystalline diamond (PCD) composite.

As diamond is so much different from other materials in hardness and other properties, it would be difficult for diamond to stick on a substrate (e.g., cWC) without flaking off. In order to coat a substrate, diamond must become less diamond so its properties can be closer to the substrate. Moreover, it coating thickness must be very thin (e.g., 1 μm), otherwise, the stress at the interface can be enormous. Thus, DLC is such a thin coating that may be coated onto a substrate that may be harder than steel (e.g., HSS or cWC).

In the case of CVD diamond film, it will not stick to substrate less hard than cWC. In addition, its coating must be thin (e.g., < 40 μm). If thicker (e.g., 500 μm) diamond coatings are desired, it may have to be brazed onto a hard substrate (e.g., cWC, SiC, Si_3N_4). Alternatively, a thick diamond coating may be formed in situ with cWC by sintering diamond grains to form PCD. Thus, with increasing thickness, diamond coating may trend from PVD DLC, and CVD diamond, to PCD. This is also the trend of the processing temperature (Fig. 6.27).

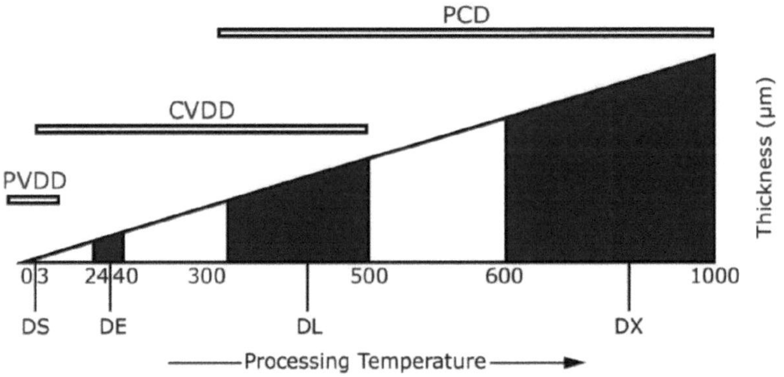

Figure 6.27 The thickness of diamond coating and its formation temperature. The darkened areas show commercial products available. Examples of process temperatures are <200°C for PVD DLC coating, 900°C for CVD diamond film growth, and 1300°C for PCD manufacture.

Diamond film is the hardest coating available. Besides being hard, diamond film may also be the most slippery with a frictional coefficient rivaling the very soft Teflon. Moreover, diamond film also conduct heat the fastest. Thus, the combination of hard, slippery, and not hot makes diamond film the ideal coating for all cutting tools (Fig. 6.28).

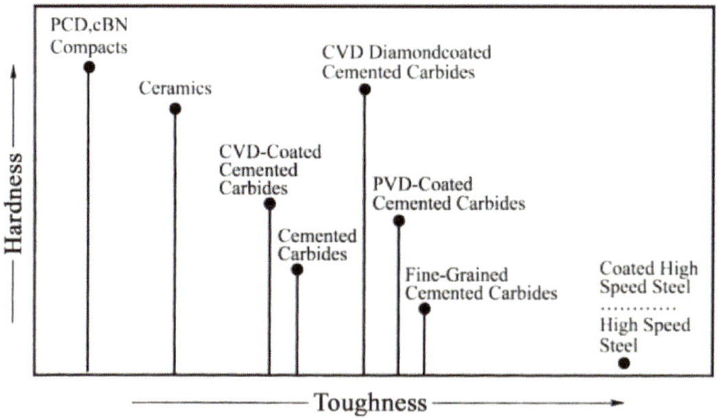

Figure 6.28 The relative hardness and toughness' of various coated cutting tool materials.

Figure 6.29 CVD diamond film-coated cWC cutting tools manufactured by Kinik Company.

CVD diamond coating can be applied to most cWC tools, in particular flat inserts and twisted drills (Fig. 6.29). These advanced cutting tools are now used to cut many hard-to-cut materials, such as silicon-impregnated aluminum (e.g., A390), metal matrix composite (MMC) (e.g., SiC dispersed aluminum), ceramics (e.g., alumina), or fiber-reinforced composite materials (e.g., graphite or glass reinforced laminated boards).

6.8 Diamond Adhesion

In order for two materials to join together with a strong bond in between, the joint must avoid to form a sharp boundary that can weaken the link of the two materials. In addition, the two materials' properties should differ as little as possible. Otherwise, their mismatch will create a stress on the boundary and this stress may increase with environmental changes such as with the increasing temperature. Additionally, there must be a strong bond, preferably chemical bond, formed at the boundary between the joining members.

Diamond is the extreme material; its properties are very different from the common substrates (e.g., HSS, cWC) for diamond to adhere. Consequently, the boundary between diamond and the substrate is sharp, so it is often the weakest link, subject to separation under stress. One of the common causes of the stress is the thermal expansion mismatch. Diamond, being the hardest material, has the lowest thermal expansion coefficient of all crystalline materials over a long temperature range. Graphite may stretch out less along the graphene plane, but its thermal expansion perpendicular to these planes is an order of magnitude higher. Certain materials may shrink (e.g., graphite, codierite) in certain direction in a narrow temperature range, but nothing is like diamond, which shows very little volume change throughout the temperature range (Fig. 6.30).

The thermal expansion mismatch is the main cause of joint failures between diamond and other materials. If the joining temperature is high, as in the case of CVD or PCD processing, the large thermal contraction of the substrate material can squeeze temperature-insensitive diamond so hard that crack may developed in the diamond layer. Alternatively, if the heating is fast, as in the case of brazing the tool or cutting the work piece, the fast extending substrate may delaminate from the little stretch diamond layer.

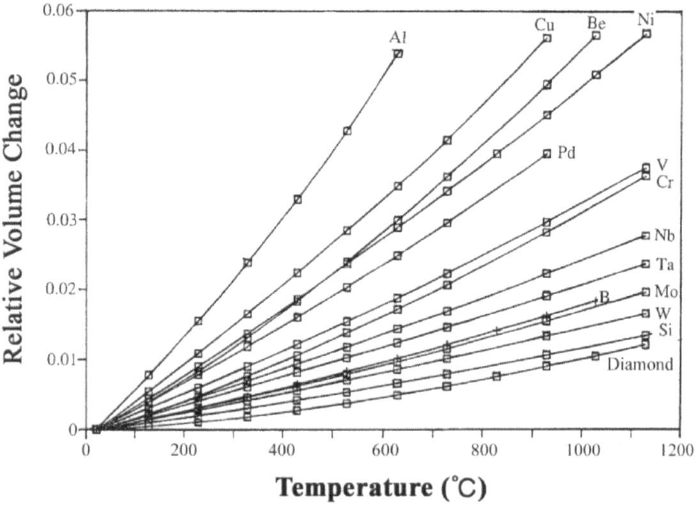

Figure 6.30 Volumetric expansions of different elements at high temperatures. Co line is not shown as it is similar to Ni line.

In order to adhere diamond on a substrate, either the substrate must be made harder so that its properties can be closer to diamond's, or diamond must be made less diamond-like so that its properties can be related to the substrate. In the case of coating real diamond, the substrate must be at least as hard as cWC. But for the more commonly available steel substrate, diamond-like carbon coating would be the best hope. Even so, DLC may not be too thick; otherwise the stress will build up too much as to flaking off the coating.

PVD DLC may adhere well on steep tools because of its low hardness and thin thickness. In addition, the low processing temperature (e.g., < 200°C) has minimized the thermal expansion mismatch between the two different materials across the joint. However, DLC itself is not as thermally stable as diamond. Diamond may hold on its structure up to about 1200°C. But with the presence of a catalytic metal, such as the steel substrate, it will back convert to amorphous carbon beginning at 700°C. But DLC is much less stable than diamond, even without the catalytic promotion, it will convert to amorphous carbon at a temperature <500°C.

Although DLC may adhere well to a substrate, it can peel off easily if the substrate is not prepared properly before coating. For example, the substrate must be thoroughly cleaned to remove all rust (e.g., by dissolving in acid) and grease (e.g., by cleaning in freon). It may have to be further etched by ion bombardment (e.g., by argon ions at −1000 V) to eliminate possible absorption of oxygen atoms on surface. The thoroughly cleaned substrate may further be planted with carbon ions (e.g., by an arc process) to make it more susceptible to DLC coating. Alternatively, an interlayer such as a carbide former (e.g., Ti, Si) may be deposited prior to DLC coating (Fig. 6.31).

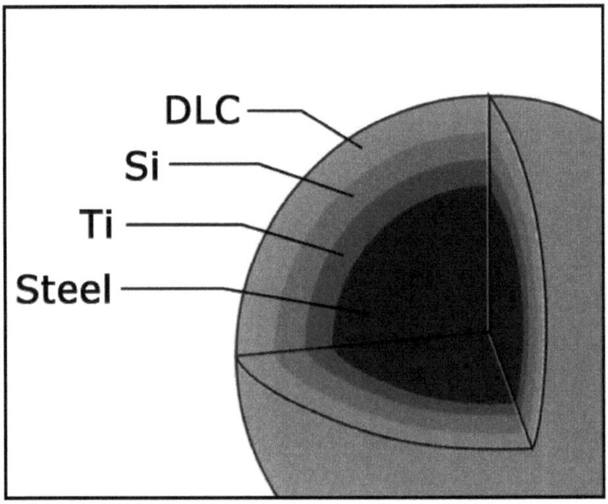

Figure 6.31 A DLC coating may not bond a steel substrate well, but it can hold on Si firmly. Si can stick to Ti, which in turn forms a diffusion joint with steel. Such a strong chain of links can prevent DLC from flaking off.

CVD diamond is much harder and thicker than DLC; moreover, its processing temperature is also high (700 to 900°C); hence, CVD diamond may not deposit on steel substrate. Instead, it is mostly coated on hard cWC, although some refractory metals (e.g., W, Mo), carbide (e.g., SiC), or nitride (Si_3N_4) ceramics may also be suitable.

In order for CVD diamond to adhere well on a cWC substrate, the surface cobalt that may contact the overlaying diamond have to be eliminated beforehand, otherwise cobalt may catalyze the back conversion of diamond at high temperature, either during the deposition process or at the cutting action.

An easy way to avoid cobalt is to use low cobalt cWC, although such cWC is too brittle for application that requires toughness (e.g., as a mold that may require stretching or as an impact-resistant surface), it is suitable as the support for cutting tools. In fact, low cobalt cWC themselves are used as cutting tools. However, in order to minimize the cobalt contact with the deposited diamond, the cobalt content may be reduced to less than wt%.

Moreover, the grain size of WC may not be too small; otherwise, they cannot expose high to receive diamond deposition. The combination of low cobalt and coarse WC is contrary to the trend of cWC development, which prefers using high cobalt and fine WC.

In addition to using low cobalt cWC, the surface cobalt may be further removed by acid etching. Alternatively, cobalt may be vaporized in a vacuum furnace. The heating may also coarsen WC grains. As a consequence, the enlarged WC grains may cover the entire surface. The CVD diamond deposited on such a rugged surface will have the benefit of both the chemical bonding and mechanical bonding. Hence, it may adhere well (Fig. 6.32).

However, because the interface is not smooth after the heat treatment, the coated diamond film is also very rough (Fig. 6.33). Such a CVD diamond tool may be used only for rough cutting applications.

Another way to strengthen the bond to diamond film is to seed cWC surface with microscopic or nano-sized diamond seed (e.g., by polishing cWC in an ultrasonic bath of acetone that contains suspended diamond particles). These minute diamond seeds may be embedded in the crevices of cWC grains and also attached to cobalt matrix. The subsequent CVD deposition of diamond can use such embedded diamond as anchors so the bonding of the diamond film to cWC is stronger.

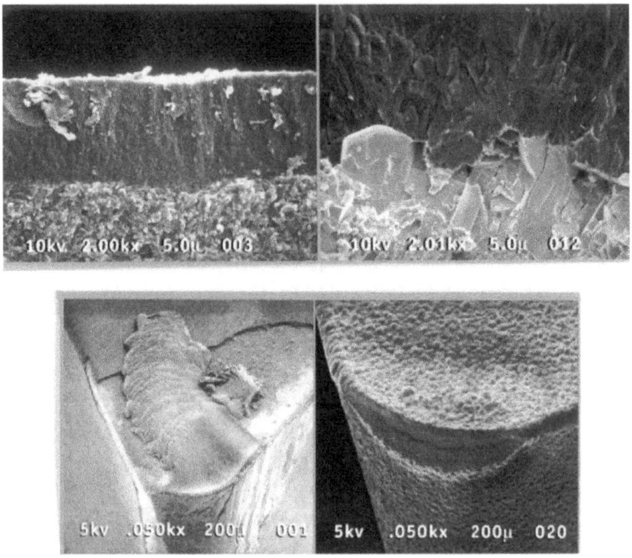

Figure 6.32 A smooth interface (top, left) may not hold CVD diamond firm. Hence, flaking of the diamond film from the cutter top is inevitable (bottom, left). On the other hand, the roughened interface (top, right) may bond the diamond film well, so the flaking of top layer is avoided (bottom, right) (Norton Diamond Film brochure).

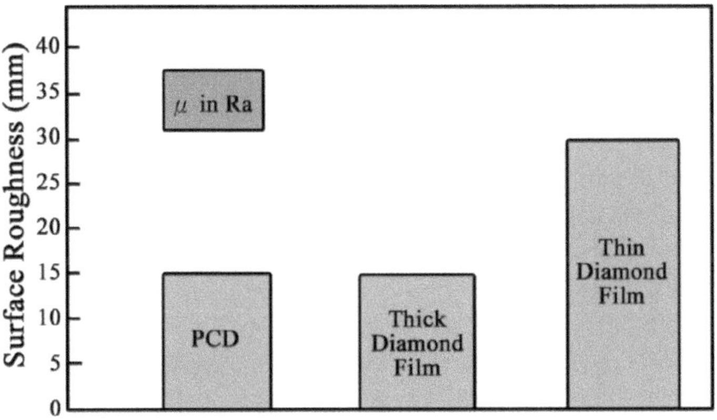

Figure 6.33 The surface roughness is dependent on the sharpness of a diamond cutter. If the CVD diamond film is not smooth, the finish of the cut surface will not be smooth (Norton Diamond Film brochure).

An interlayer may also improve the bonding of diamond to cWC. For example, cWC may be over coated by a thin layer of a carbide former (e.g., W, Si, Ti) that may be laid down conveniently by PVD sputtering. Subsequently, CVD diamond can be deposited onto this interlayer. The high-temperature processing of CVD diamond may also promote the diffusion bonding of the interlayer onto the cWC substrate so the weak link is removed.

In the case of a PCD composite, the interface is a boundary between diamond layer and cWC substrate. The two strata are actually bonded by the same cobalt that infiltrated from cWC body into PCD layer. Because of this continuation of cobalt, PCD composite is normally held together well. But for some stressful applications, such as the PCD composites used for percussion drilling, diamond table may still be delaminated. In this case, a transition layer may be needed to reduce the contrast of the compositions. For example, diamond and WC grains may be mixed to form a graded zone so stress may not be concentrated on a sharp boundary.

6.9 Diamond Film Coating

Although CVD diamond film also contains polycrystalline diamond grains, these grains are not bonded together by a metal, such as that existed in a PCD composite. As a consequence, CVD diamond film is harder than PCD diamond composite.

Moreover, CVD diamond film can stand a cutting tip temperature up to 1200°C. In contrast, PCD composite may be back converted to amorphous carbon starting 700°C because of the presence of cobalt catalyst. Such a back conversion will wear the cutting tip rapidly. Hence, when cutting gently to maintain a low tip temperature, the PCD cutter with its much thicker diamond table may outperform CVD diamond tools with a thin coating, but the tide is reversed in aggressive cutting or in dry cutting when the tip temperature may exceed the stability threshold of PCD (Fig. 6.34).

Because of the compositional difference, the pure diamond film can last much longer (e.g., 3 times) than cobalt sintered PCD. However, PCD has a thick diamond layer (>500 microns) so it can still match the performance of a thin (<25 microns) CVD diamond film. But for very abrasive objects, such as metal matrix composite (MMC), CVD diamond films can outperform

PCD composite with high margin (Fig. 6.35), particularly when the cutting is aggressive enough to heat the cutter's tip to above 700°C.

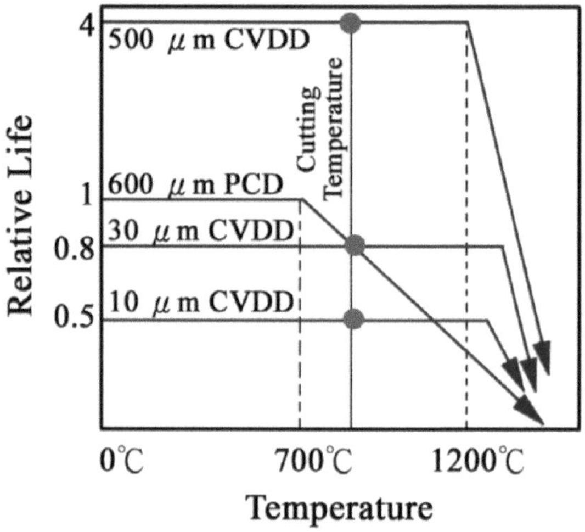

Figure 6.34 The relative cutters' lives of PCD cutters and CVD diamond tools. Note that the thick PCD layer may last longer than thin CVD diamond coatings at low cutting temperatures; the reverse is true when cutter tips become hot. At a normal cutting speed, a 30 micron-thick CVD diamond layer may match with a 600 micron-thick PCD layer.

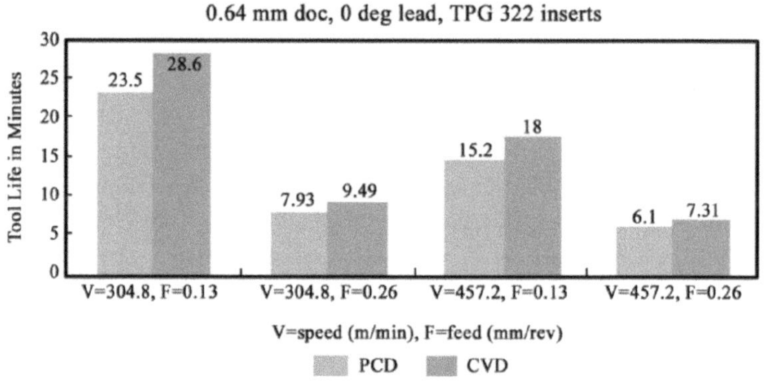

Figure 6.35 The cutting performances of coated CVD diamond film cutter and sintered PCD cutter in machining MMC Duralcam (SiC 20 V%). The tool life is measured by the cutting time to reach a wear flat of 0.2 mm. In the above comparison, the diamond film thickness is only 25 microns, but the PCD diamond table is thicker than 500 microns. Even so, the thin CVD diamond film still outperformed super thick PCD.

The superior performance of coated CVD diamond film is not limited to insert, but also to round tools, such as twisted drills. Although the round tools can only be coated to about 10 microns of CVD diamond, its effect is already impressive as it could double the life of a drill that contained solid PCD (Fig. 6.36).

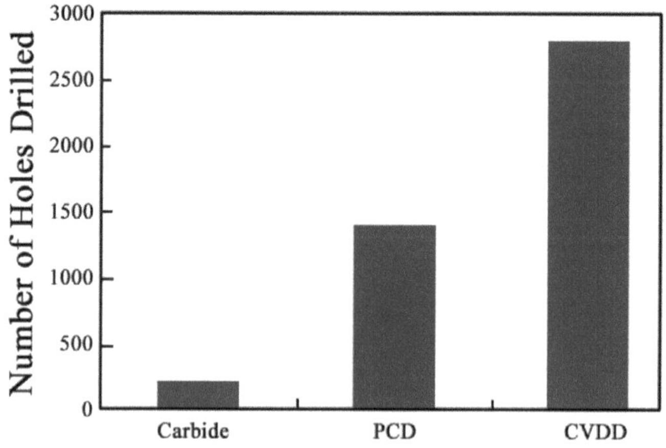

Figure 6.36 A coated CVD diamond drill can double the life of a PCD drill and it can replace 50 cWC drills when they are used to drill carbon fiber reinforced composite materials.

Because CVD diamond film does not contain metal, it can cut without smearing. Consequently, the cut surface is very smooth. If the cutting edge is sharp, it may produce a mirror surface without further polishing. Hence, CVD diamond film can be used as a sharp turning tool, similar to a single crystal diamond cutter. One of the applications of a CVD diamond cutter is to turn aluminum silicon alloys (e.g., A390) to make its surface shine. Another application is to shave an acrylic object to reveal its clarity. This unique capability makes sharp CVD diamond film cutter an ideal tool for making artistic displays.

In contrast, a PCD tool always drags on the cutting surface because its cobalt will diffuse into the work piece under high pressure and high temperature of cutting. As a result, the cutting surface is highly deformed. Thus, if a PCD cutter is used to shave an acrylic object, the surface will be foggy. Even with subsequent polishing with micron abrasive powder, it is still not as clear as that cut by a sharp CVD diamond cutter.

In addition to being more wear resistant and more thermally stable, because CVD diamond coating is relatively thin, its impact strength can be effectively boosted by the under layer of cWC. This is important because hardness and toughness are two non-compromising attributes. All cutting tools must make a pick on whether hardness is more important or toughness is more important, but coated CVD diamond tools can have both (Fig. 6.37).

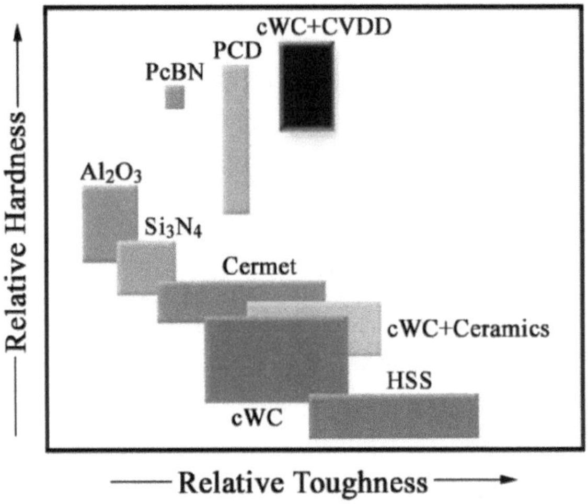

Figure 6.37 The reverse trend between hardness and toughness. Note that cWC backing has effectively raised the toughness of hard coatings for PcBN, PCD, CVDD and ceramics, but because CVDD is the thinnest, so the increase of its toughness is the most.

Although CVD diamond film can be coated on a hard substrate such as a cWC, but it cannot build up the diamond layer thicker than about 50 microns (most coating thickness for inserts is about 25 microns; and for drills, about 10 microns). This is because the cWC shrinks linearly about five times faster than cobalt in cWC. Hence, large compressive stress will build up in the diamond layer. This stress will increase with the increasing thickness so delamination of the diamond film is inevitable.

In order to bond a thick (e.g., 300 microns) diamond film to a cWC substrate, the author developed world's first brazed thick film diamond cutter (Fig. 6.38) in 1987 when he was responsible to establish a commercial diamond film technology for Norton Company. This brazed diamond tool proved to be the most durable cutting tools ever developed as it could last much longer (e.g., three times) than even PCD tools with a much thicker (e.g., 700 microns) diamond table (Fig. 6.38). Moreover, because diamond film is separated from cWC by a brazing layer (e.g., AgCuTi) that will not catalyze its back conversion, the thick diamond film can stand a cutting temperature of up to 1200°C. In contrast, a PCD tools may fail due to the catalytic back conversion by Co it contains if the cutting tip temperature reaches only 700°C. With all the above-mentioned advantages, thick CVD diamond film is the most advanced cutting tool ever invented.

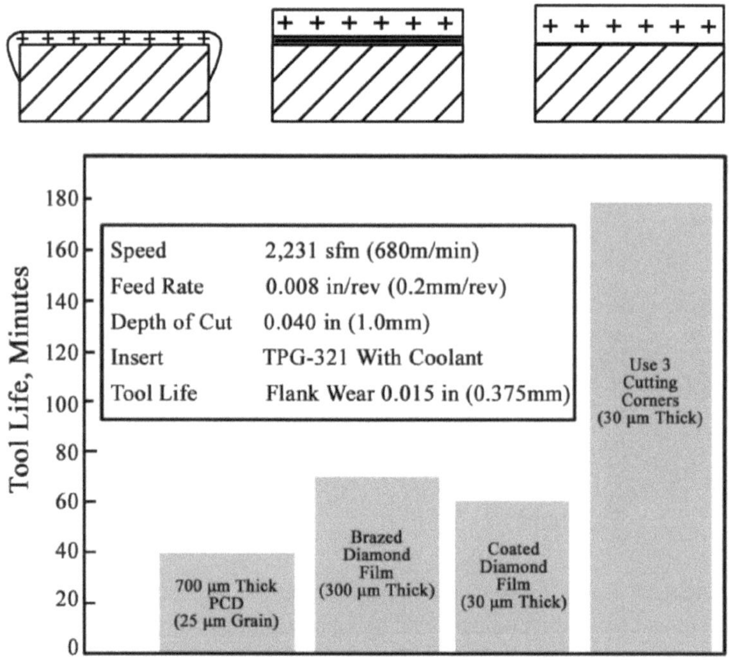

Figure 6.38 The design of coated thin CVD diamond film (e.g., 10 μm), brazed thick CVD diamond (e.g., 300 μm), and sintered super thick PCD compact (e.g., 600 μm) as cutting tools (upper diagram). Comparing these three in cutting the abrasive A390 Al-Si alloy, brazed thick diamond film is the most durable, but coated thin diamond film often come with multiple cutting tips that may combine to last the longest.

For a cutting insert, coated diamond film may not last as long with one cutting tip, but its three or four cutting tips can add up to beat the performance of a brazed thick diamond film cutter that often has single cutting tip. Consequently, the economics of cutting operation may favor to use a coated cutting insert (Fig. 6.39).

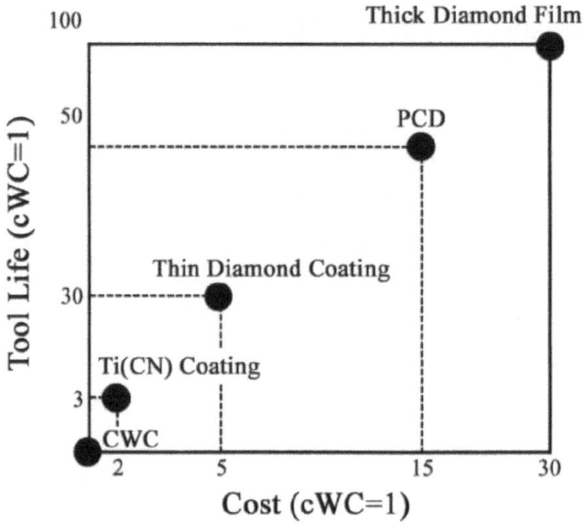

Figure 6.39 The comparisons of tool life and unit cost for various cutting tools. This chart compares only one cutting tip. If all cutting tips are added on, thin diamond coating may stand out as the winner in life and cost.

6.10 Applications of Coated Diamond Film Cutters

Because diamond can be back converted at high temperatures when cutting carbon solvent metals (e.g., Fe, Ni, Co or its alloys), or it may react with carbide formers (e.g., Ti, Zr, V), so coated diamond film cutter is best to work on non-ferrous metals or non-carbide formers, in particular those that are highly abrasive (e.g., MMC), so there is no suitable tools to cut them. However, if the work material is soft (e.g., Al or its alloy) or the cutting temperature is low (e.g., stainless steel at low cutting speed), diamond may still be used to machine such materials.

Typical work materials suitable to be machined by coated CVD diamond film cutters include green perform of ceramics or cWC, graphite or carbon blocks, graphite or glass fiber reinforced composites, polymer cemented composites, epoxy resins, acrylics, melamine, wood composites, non-ferrous metals (e.g., Al, Cu, Pb, Sn), metal matrix composites (MMC), Kevlar/metal composites.

Most of the above material examples are highly abrasive, so cWC or its ceramic coated tools are not capable of machining. Moreover, PCD tools can only form inserts so they are not practical to drill holes. In this case, coated CVD diamond cutters are most suitable to work on such materials (Fig. 6.40).

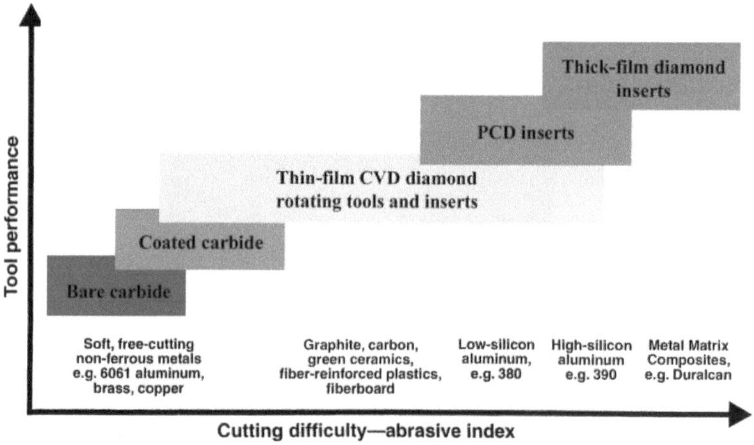

Figure 6.40 Cutting tools performance spectrum as portrayed by sp^3, Inc.

The CVD deposition of diamond film starts with nucleation on cWC with very small grains. These nuclei with grow big with time. Consequently, the growth side is much rougher than the start side (Fig. 6.41).

A thin coating of CVD diamond may increase the radius of already-sharp cWC; hence, it is not capable of producing a smooth finish (Fig. 6.42). This dulling effect compounded with the enlarged grain size makes the cutting edge very rough. Consequently, the surface finish of the work material will not be smooth.

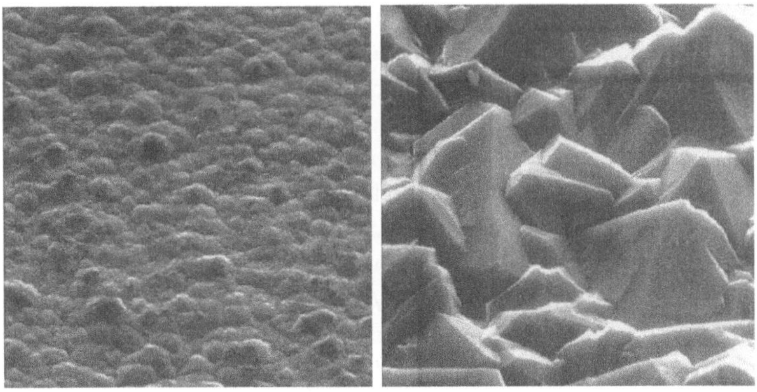

Figure 6.41 The contrast of nucleation enhanced growth (left diagram) and nucleation suppressed growth (right diagram) of CVD diamond films (sp[3] brochure). The magnifications of these diagrams are about 1000 times.

Figure 6.42 The cross section of a CVD diamond film-coated cutting tip reveals its larger radius (top). The rougher surface appearance (bottom, left) and its enlarged view showing the coarse diamond grains (bottom, right). (Balzers brochure).

A thin CVD diamond coating is difficult to sharpen because its adhesion to the underlying cWC may not be strong enough to support the grinding action. However, a thick CVD diamond film firmly brazed onto a cWC substrate can be ground and polished to make it very sharp. In fact, thick CVD diamond film cutters may be as sharp as single crystal diamond turning tools, but their polycrystalline nature will make them last longer. In contrast, PCD cutters are intrinsically not capable to be super sharp due to their micron-sized grains. Hence, their surface finish is intermediate between that of a single diamond tool and a thin CVD diamond-coated cutter (Fig. 6.43).

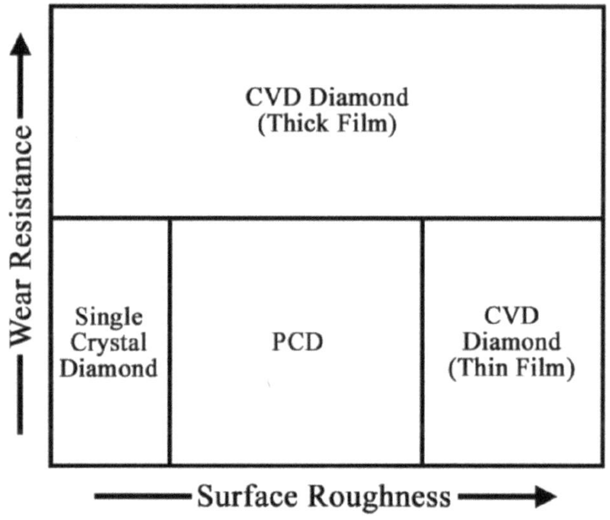

Figure 6.43 The tool life and surface finish of various diamond cutters. Note that the thick diamond film may outperform all other diamond tools in longevity of tool life and quality of finished product provided that the cutting tip is polished sharp.

Although CVD diamond films can make the best cutting tools, unfortunately, they are expensive to make. Because diamond is unstable at low pressures, the growth of CVD diamond film in vacuum is intrinsically slow (1–15 microns per hour). There are several options to grow CVD diamond films (e.g., by using hot filaments, microwave plasma, or arc jet reactors), but all of them have the same limitation of the amount of diamond to be deposited per unit energy consumed, roughly 0.02 gram or 0.006

cubic centimeter per kWh of electricity. Hence, the faster the growth rate, the smaller the area of deposition.

In addition to being energy limited, CVD diamond film processes are materials wasting. The diamond conversion rate from carbonaceous gas is only a few percent, and the faster the diamond growth, the poorer of material utilization.

Because of the poor efficiency of diamond film growth, its cost of deposition has been high, now about $200 per cubic centimeter or more than $10 per carat. In comparison, the sintering of PCD is about half of the above cost; and the growth of diamond grits is 1/10 of the above cost.

The most successful cutting tools were developed by former Norton Diamond Film (Fig. 6.44). They used DC arc jet developed by Technion for growing thick films, and microwave plasma developed by Wavemat for depositing thin films. The author was responsible for the acquisition of these technologies for Norton in late 1980s. Although the products were successfully implemented and well accepted in the industry, unfortunately, the operation was too costly due to the heavy investment in other applications (e.g., heat spreaders, radomes, etc), Saint-Gobain closed the business altogether in 2000.

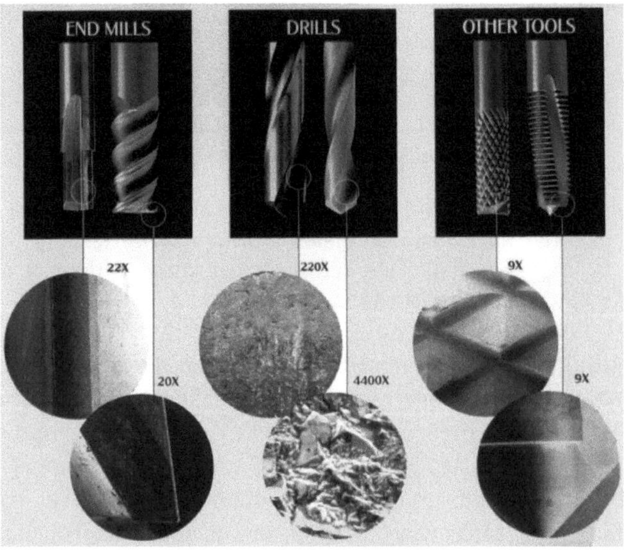

Figure 6.44 Former Norton Diamond Film's CVD diamond-coated round tools.

Although CVD diamond is expensive to grow, fortunately it is highly durable, so only a thin film is needed. In fact, most cutting tools cannot tolerate a wear flat of more than 25 microns, so any thicker deposition is wasting the expensive material. Moreover, thicker diamond table will accumulate the compressive stress so the deposited film is more vulnerable due to flaking or delamination. Hence, thin CVD diamond coating on a cWC substrate would be the best way to exploit the superior properties of diamond film.

Thin diamond films are best deposited by hot filaments method that, although depositing slowly, can cover a large area. Alternatively, the hot filaments may be replaced by a long arc of high current. These reactors have been used for commercial production of CVD diamond-coated cutting tools.

Numerous CVD diamond-coated cutting tools are available (Fig. 6.45). Typically, such tools can outperform cWC tools by 100 times in service life, twice in cutting speed, and use about half of the energy.

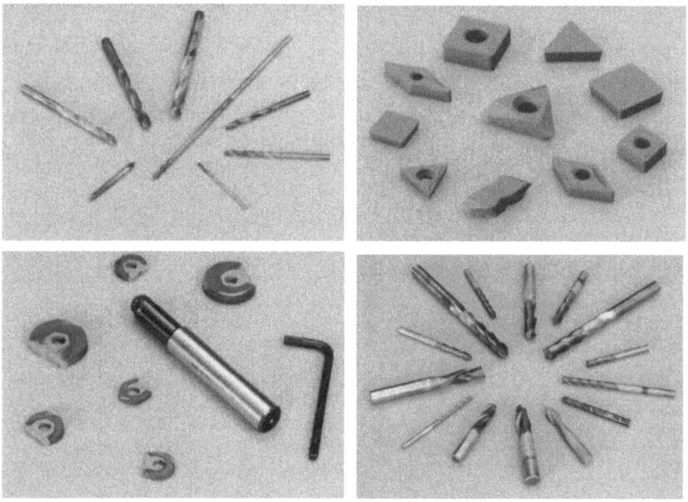

Figure 6.45 CVD diamond-coated cWC tools include drills (top, left), inserts (top, right), mills (bottom, left), and round tools (bottom, right).

In general, CVD diamond film cutters are much more wear resistant than DLC-coated tools, which are in turn more durable than ceramic coating on cWC substrate (Fig. 6.46).

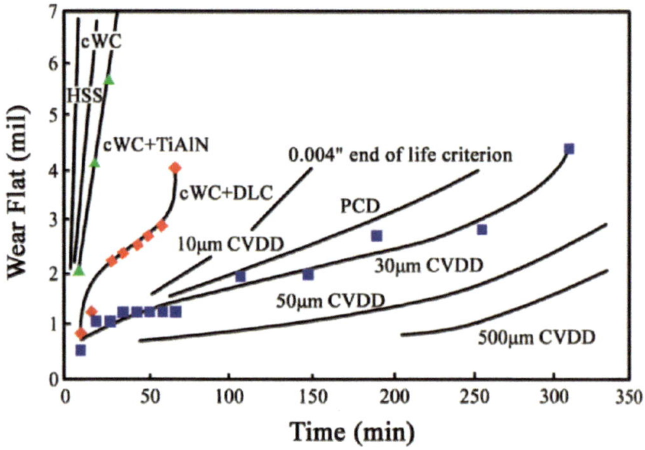

Figure 6.46 The wear resistance of cutting tools with various coatings indicates that CVD diamond films are most durable.

One good example of applying CVD diamond-coated cutters is in machining graphite molds for electrical discharge cathodes. Such graphite molds are quite abrasive. However, at a high cutting speed, graphite will not soften like most other materials, instead, it become brittle and easier to cut. CVD diamond film is uniquely capable of sustaining the high temperature generated at high cutting speed, so it is particularly suitable for cutting graphite (Fig. 6.47).

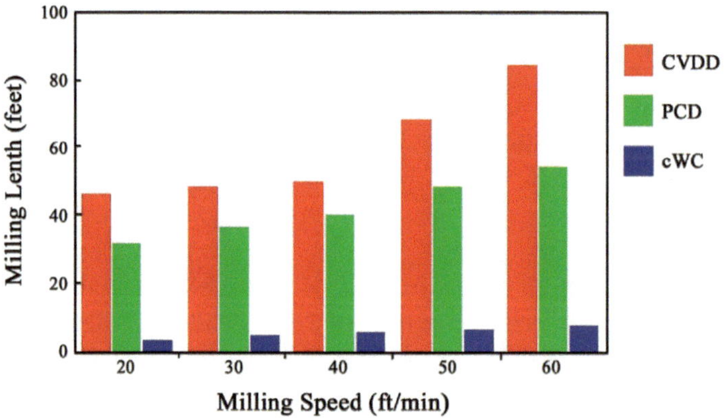

Figure 6.47 The milling of graphite shows that CVD diamond-coated end mills has the longest service life and this lead increases with the increasing cutting speed.

CVD diamond-coated cutting tools can not only last long but also cut fast. Consequently, the productivity would increase enormously. Thus, the prevailing use of CVD diamond-coated tools can reduce much of material waste and save considerable machine time. Hence, CVD diamond tools are green products that may reduce the environmental burden.

Another benefit of using CVD diamond tools is that they can stand high cutting temperature and therefore may not use coolant, which may contain emulsion or other harmful ingredients. Due to the increased environmental concern on the disposal of potentially hazardous coolant, dry cutting becomes a preferred method for material removal. CVD diamond tools are uniquely capable of cutting without using coolant, so their prevailing use will help preserve a clean environment. Moreover, the tool cost accounts for only about 3% of the total expenditure incurred in machining (Fig. 6.48). This cost is actually less than half of that may be needed to dispose of the coolant. Thus, drying cutting is not only environment friendly but also cost effective.

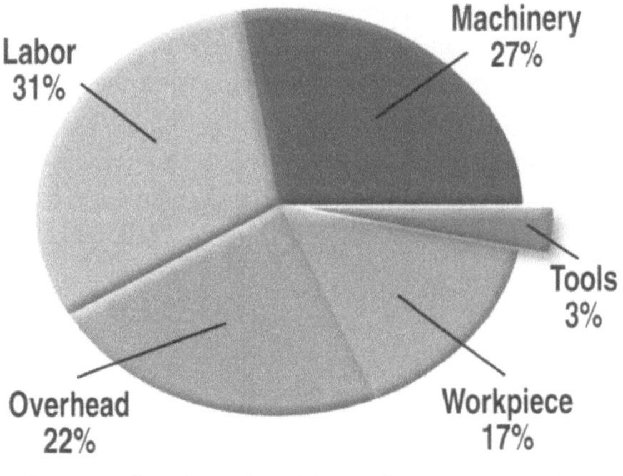

Figure 6.48 An example of cost breakdown for using CVD diamond-coated cutting tool.

6.11 Reverse Casting CVD Diamond

CVD diamond is a high-quality cutting tool, but its cost of deposition may be prohibitive for making thick diamond film cutters. Fortunately, the cutting tools require only about 30 microns thick as when it is worn out, the wear flat will be too big to use it anyway. However, the dilemma is that thin CVD films are difficult to machine so their surface is intrinsically rough. Hence, the durability of thin CVD diamond coating is hampered by its inability to achieve a mirror finish.

In order to solve this problem, the author has developed a reverse deposition technique by casting diamond films (patent pending) in a mold. In this way, the nucleation side of the diamond film is preserved as the cutting surface. The nucleation side is intrinsically smooth, so very little, in any, polishing work may be needed to finish the tool. Thus, this method can produce the cutting tool with minimal additional work (e.g., grinding). As diamond grinding and polishing often cost more than growing diamond, this diamond casting technology can not only save time but also reduce cost.

The diamond casting process starts with a mold that has the inverse shape of the diamond tool desired. The mold may be made by stamping thin layer of refractory metal such as Ti, Zr, Mo, W, and Ta. Subsequently, diamond is deposited in the mold by a CVD process to reach a desired thickness. The mold is then leached by using an acid (e.g., aqua regia) to leave a freestanding diamond film with one side mirroring the dissolved mold. This side is used as the cutting surface of the tool by brazing the growing side of the diamond film to a cWC substrate. The finished cutting tool has now a smooth surface that does not require further machining (Fig. 6.49).

All conventional diamond film tools contain only flat surfaces because a profiled surface is difficult, if not impossible, to form by any mechanical means. Thus, the diamond casting process has an enormous advantage by casting diamond film in the mold that may assume any shape, so chip breakers or other features (e.g., marking) may also be cast in situ. Hence, the cutting tool so made has the advantages of thick diamond film's mirror finish, thin diamond film's multiple cutting tips, and cWC inserts' chip breakers.

The diamond casting technique has a disadvantage, however. First, the nuclei formed have gaps in between. These gaps will be retained as the cutting surface of the tool (Fig. 6.50).

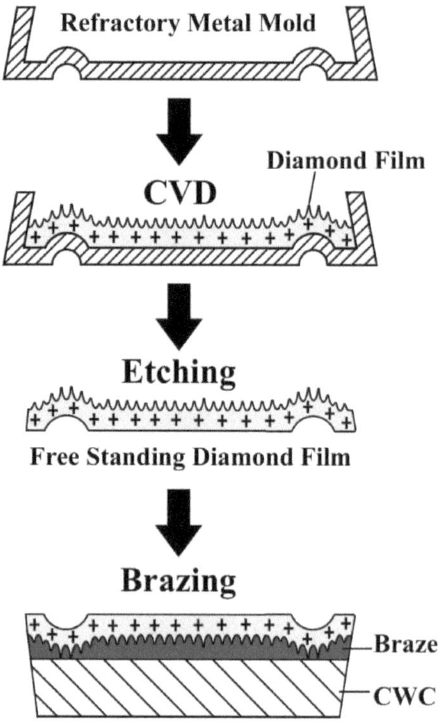

Figure 6.49 The casting of diamond film to form smooth cutting surface with chip breakers built in. This reverse deposition technique can produce diamond film cutting tools with low cost. The smooth surface and sharp corner can allow mirror finish of the work material.

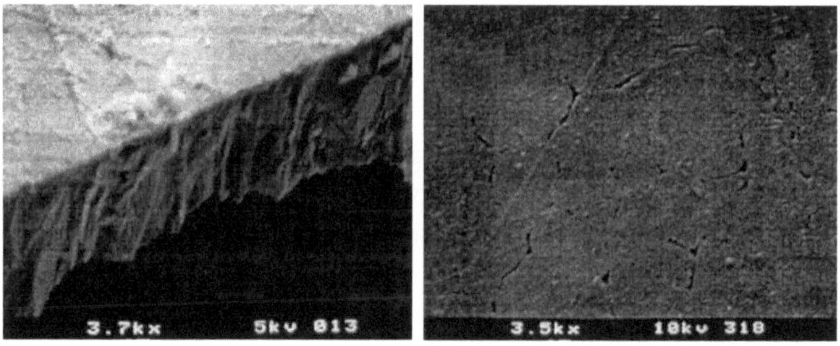

Figure 6.50 The nucleation side of the diamond film may reveal grooves that came from the original polishing scratches of the substrate (e.g., Si) (left diagram). The nucleation side of the diamond film also contains micro fissures between original nuclei (right diagram).

Moreover, the nucleation side of the diamond film is facing a non-diamond surface. The initial diamond so formed is not pure diamond; there may be amorphous carbon present, for example. Hence, the cutting surface may not be as hard as normal diamond cutter.

However, these drawbacks may be eliminated by enhancement of diamond nucleation. In this way, diamond nuclei can pack tightly with no gaps in between. There are several options to enhance the nucleation rate. For example, by applying a negative bias (e.g., 200 V) to the substrate, nucleation density may be increased to one million times. Further increase of nucleation density is possible by using nanom diamond (e.g., 5 nm synthesized by dynamite explosion) as seeds (e.g., polish substrate in nanom diamond suspended acetone with ultrasonic vibration).

In addition to pack nuclei tight to start with, higher hydrogen content (e.g., 99.5%) may be used to etch away (by forming methane) any and all non-diamond carbon formed. Of course, the nucleation side may still be polished gently to remove any residue of non-diamond material that may still remain.

The diamond casting technology can be used to make other diamond parts that may be difficult to form by mechanical means. For example, almost all semiconductor chips are now required to be polished by chemical mechanical planarization (CMP). The slurry used for CMP is often highly acidic (e.g., pH = 1 for polishing copper circuitry). During the polishing, a diamond dresser is needed to condition the polishing pad. However, if diamond dresser is made conventionally with a metal matrix, the acid will soon dissolve the metal and loosen the diamond. In this way, disaster will ensue as diamond falling out the dresser will be embedded in the soft pad. It may scratch the expensive wafer on the first touch.

In order to dress the pad using diamond grits, the grit can be covered by a thin layer of CVD deposited diamond film. In this way, there is no metal matrix so the grits can be securely attached (Fig. 6.51).

Although the above design of CVD-bonded diamond grit may survive the attack of an acid slurry, the grit distribution is random, so its dressing performance is erratic. Moreover, the thin CVD coating is not sufficient to hold the diamond grit firm, so wafer scratching rates remain very high.

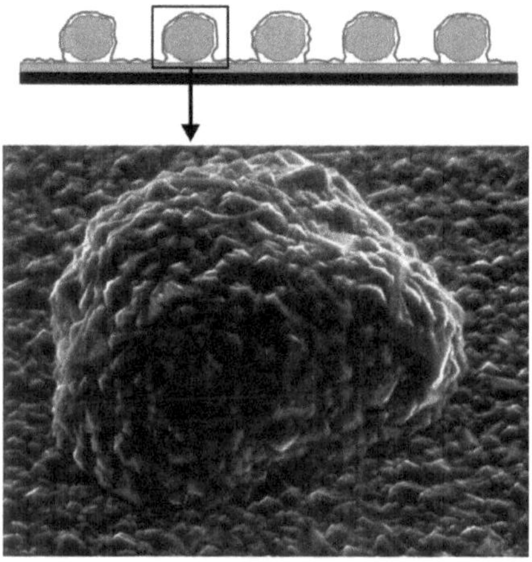

Figure 6.51 A diamond dresser that contains 50 micron diamond particles. These particles are coated by a CVD diamond film of a few microns thick on a silicon layer that is glued to a magnetic stainless steel substrate. The particle density is 15 particles/mm^2 (Diamonex brochure).

Using the reversed casting technology, robust diamond pyramids can be fabricated directly without using loose diamond grits (Fig. 6.52). The diamond pad conditioner so formed has the perfect pattern of diamond tips that may not be able to fabricate by other means.

The diamond pad dresser so formed has all cutting tips leveled at the same height. This is important as the diamond tip height will determine the depth of grooves to be cut. The perfect pattern combined with the exact height will allow the diamond pad dresser to condition the pad very uniformly, which will in turn to polish a silicon wafer very uniformly. Moreover, most diamond dressers contain diamond grits embedded in a metal matrix that may be dissolved away in an acid slurry used to polish the wafer. As a consequence, diamond grits may fall out and scratch the expensive wafer. In contrast, the diamond dresser formed by casting the entire diamond film contains no metal part on the surface so it can withstand any corrosion attacks without risking the dislodgement of diamond grits.

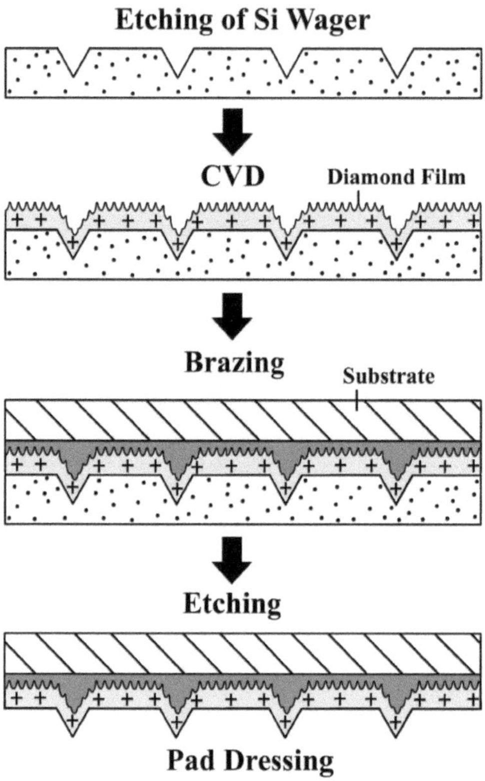

Figure 6.52 Diamond dresser with perfect pattern of sharp tips may be fabricated by casting diamond film in a silicon mold. The diamond film so formed may to too thin, so it needs to be supported by a suitable substrate by brazing. Subsequently, the silicon mold can be etched away to leave the diamond dresser that mirror the silicon mold.

The diamond casting technology can make more than just cutting tools and pad dressers, it can also be used to cast a hollow cylinder such as that may be used to make a wire-drawing die. Initially, a metal mandrel (e.g., W) is machined that has the exterior shape mirroring the interior contour of a wire-drawing die. The mandrel is subsequently deposited with CVD diamond film that covers the entire exterior profile. After then, the mandrel is leached away by an acid (e.g., aqua regia) to leave a hollow diamond cylinder that has the internal shape after a wire-drawing die. The hollow diamond cylinder is later placed at the center of a stainless steel collar and the space in between is filled up with carbide (e.g., SiC or WC) grains. Finally, a low-melting-point (e.g., 1000°C) alloy is infiltrated into the crevices of the carbide powder to bond carbide grains with the stainless steel collar. The infiltration may be performed in a vacuum furnace so diamond or metal parts will be no oxidized. After infiltration, the hollow diamond cylinder is inwardly compressed by the solidified infiltrant, so the entire mass becomes a consolidated wire-drawing die (Fig. 6.53).

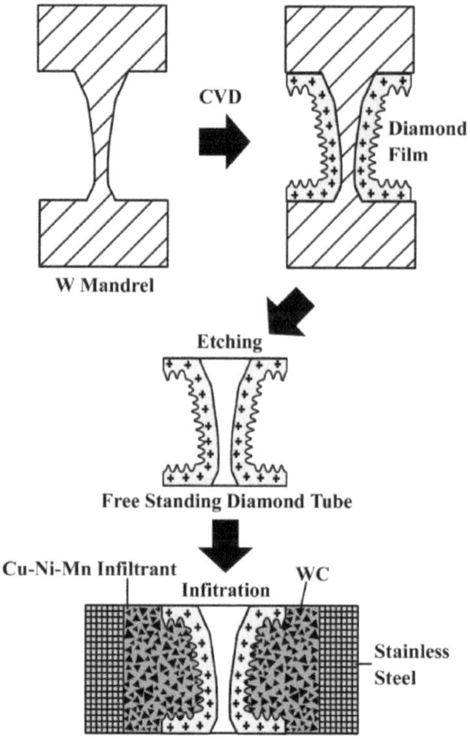

Figure 6.53 The wire-drawing die can be fabricated by casting diamond film on a mandrel and subsequently dissolved the mandrel to form a hollow diamond cylinder with the exact interior profile and surface finish of an intended wire-drawing die. This die can then be fixed at the center of a stainless collar by infiltrating an alloy into carbide powder that packed around the diamond cylinder.

The diamond wire-drawing die so formed by casting diamond films has the following features that are not available in the diamond wire-drawing dies commercially available:

1. No hole needs to be drilled in the first place.
2. No machining or polishing is required to smooth the internal profile.
3. Hole sizes can be much larger than single crystal wire-drawing die.
4. Unlike PCD wire-drawing die that contains cobalt on the drawing surface, the cast diamond wire-drawing die's surface is made of pure diamond.
5. The cast diamond is much more durable than PCD, so the hole size can be maintained for much longer time.
6. Because there is no dragging effect caused by the interaction of cobalt and the drawing metal under high pressure and temperature, the wire drawing will be faster.
7. The wire drawing also produces less heat so lubricant may not be needed for drawing copper wire.
8. The surface of the drawn wire will be smoother, also due to the absent of cobalt that may react with the passing wire.
9. The wire-drawing die is pre-compressed by the shrinking infiltrant due to solidification and cold contraction.
10. The cost of manufacturing is lower because there is no need to process the center hole.

Chapter 7

Diamond Heat Spreaders

Moore's Law is created by humans, but the Second Law of Thermodynamics is imposed by God; hence, diamond heat spreaders must be used to continue the progress of computers.

7.1 Thermal Conductivity of Diamond

Heat can be dissipated by two steps, first by spreading heat out, and then by sinking heat away. A heat spreader resembles a water hose; it can transmit heat like dispensing water. On the other hand, a heat sink is similar to a reservoir; it can store heat as holding water. A heat spreader does not retain heat, so it requires the combination of high thermal diffusivity and low heat capacity. In contrast, a heat sink may store heat, so it needs high thermal diffusivity and high heat capacity. Diamond is the best heat spreader; its thermal diffusivity is 10 times of copper, but its heat capacity is half. In comparison, copper is an excellent heat sink; it can absorb more heat than diamond and then release the heat slowly (e.g., by blowing with a cooling fan).

Handbook of Industrial Diamonds: Diamond Films and Carbon Coatings
James Sung

ISBN 978-981-4310-00-0 (Hardcover), 978-1-003-18045-6 (eBook)
www.jennystanford.com

Normally, a metal (e.g., Cu) can conduct heat much faster than an insulator (e.g., wood) because heat is carried by the movement of free electrons that are not available in insulators. Diamond is the best electrical insulator, so it has no free electrons to carry heat away. However, diamond's crystal lattice is so hard that it can transmit sound waves (phonons) with the frequency that has a higher energy than the kinetic energy of electron in copper. As a result, diamond's lattice vibration can provide a much faster heat path than copper's electron movement. At room temperature, a gem diamond (Type IIa) has a thermal conductivity of 2500 W/mK, which is more than five times of pure copper silver's 430 W/mK, the highest thermal conductor of all metals.

The high thermal conductivity can be used to test gem diamond non-destructively. For example, by placing a gem diamond on the tip of sensitive tongue, one can easily feel its coldness due to its high thermal conductivity. In contrast, an imitated gem (e.g., cubic zirconia, i.e., ZrO_2, or "Russian diamond") is actually a thermal insulator, so the tongue can immediately feel its warmness. Alternatively, by blowing a breath to a cold diamond (e.g., newly taken out of a freezer), the condensed moisture on its surface will soon evaporate, but an imitated gem will retain the condensed moisture for a much longer time. Another test is to place both diamond and an imitated gem on the surface of an ice cube and then watch the sinking speed of the two into the ice. Diamond being an excellent heat spreader can melt the ice underneath faster so it will sink deeper than the imitated gem. The test can also be accelerated by using a hair dryer or a heating lamp to speed up the melting of ice underneath the stone. One other test is to press diamond in either dry ice or liquid nitrogen; a real diamond will be heard as a sonic boom due to the immediate vaporization of the supercooled liquid around diamond. In contrast, the imitated gem will generate a much smaller sound, if any (Fig. 7.1).

However, most diamonds contain several PPM nitrogen atoms (e.g., Type Ia). These atoms are heavier than diamond's carbon atoms, so they will vibrate slower in the diamond lattice. As a result, the phonons will not move coherently. Consequently, the thermal conductivity will be reduced considerably.

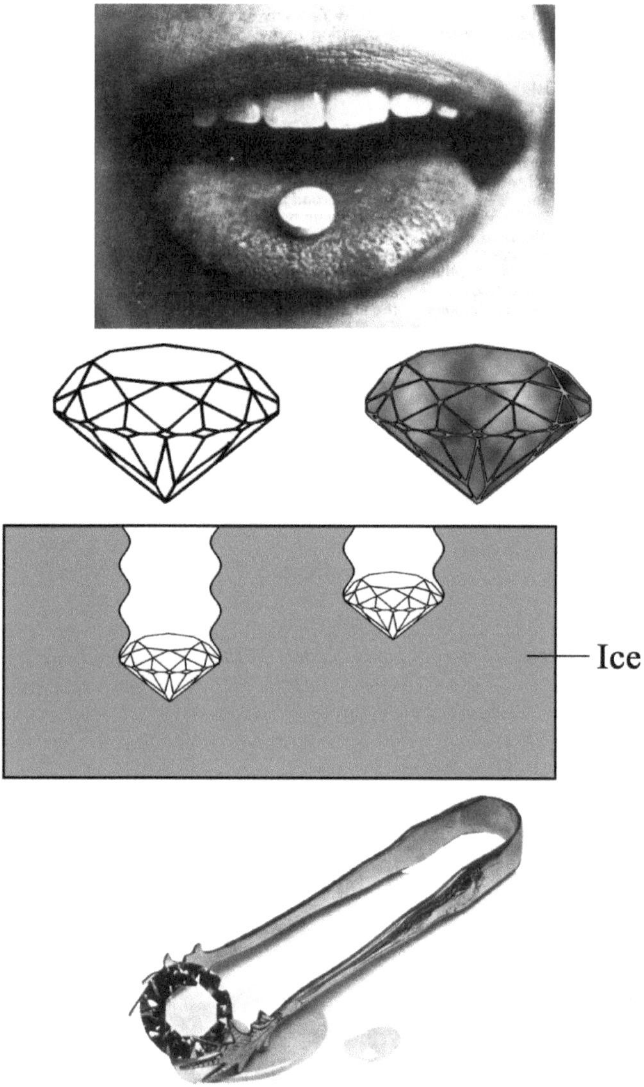

Figure 7.1 Thermal conductivity tests of diamond. A sensitive tongue can detect the coldness of a true diamond upon touching (top diagram). The quick evaporation of mist on the surface of a cold diamond reveals its identity (upper diagram). The fast sinking of a true diamond into ice reflects its exceptional ability to transfer heat to melt the ice underneath (lower diagram). The mini-sonic boom can be heard by pressing a genuine diamond into dry ice with a tong (bottom diagram).

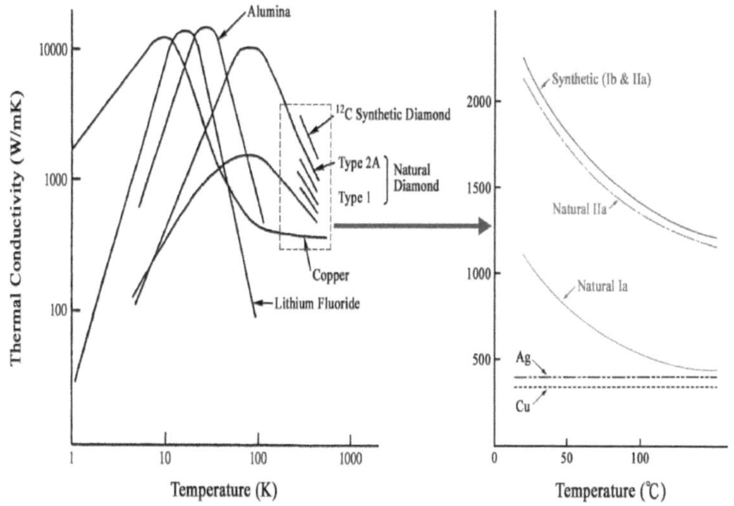

Figure 7.2 Thermal conductivities of various diamond types and other materials as a function of temperature. Note that diamond has the highest thermal conductivity at room temperature, and ^{12}C diamond can conduct heat much faster than even gem diamond (IIa). For each material, the decrease of thermal conductivity at low temperature is due to the presence of low energy phonons that can be effectively scattered by defects. The decrease of thermal conductivity at high temperature is because of the presence of high concentration of phonons that interfere among themselves.

In addition, most single diamond crystals contain defects or inclusions. In the case of polycrystalline diamonds, they incorporate a large amount of grain boundaries. As a result, the lattice vibration will not be fully synchronized, and hence, phonons are scattered around. It is because of these crystal imperfections that most diamonds (natural or synthetic) may not conduct heat faster than copper.

However, if diamond crystal is perfect, its thermal conductivity should be the highest of all materials. Moreover, this high thermal conductivity can be further increased. Carbon atoms in nature contain more than one isotope. Although they are predominantly ^{12}C isotope (the nuclei has six protons and six neutrons), but invariably they incorporate about 1.1% of heavier ^{13}C isotope (the nuclei has one extra neutron). The slightly heavier ^{13}C carbon atoms will vibrate a little bit slower than the bulk of ^{12}C atoms, as a result, the movement of phonons is not totally in phase. Consequently, there is a significant room for improving thermal conductivity beyond the best gem diamond.

Indeed, by using methane that contained only ^{12}C isotope, researchers at General Electric were able to make polycrystalline diamond film by CVD method. Such a diamond film contained negligible amount (0.055%) of ^{13}C isotope. Subsequently, the isotope pure diamond film was used as the carbon nutrient to grow single crystals of ^{12}C diamond under high pressure. It was found that the ^{12}C single crystal diamond could conduct heat more than twice than the best type IIa gem diamond that contained a minute amount of ^{13}C isotope (Fig. 7.3).

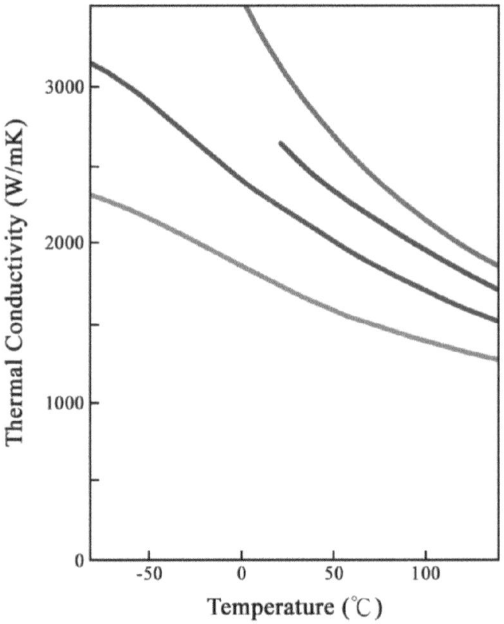

Figure 7.3 Thermal conductivity as a function of temperature for natural gem (IIa) diamond (bottom curve) with 1.1% ^{13}C isotope; perpendicular to the growth direction of a CVD diamond film with 0.055% ^{13}C isotope (lower curve); parallel to the growth direction of the same CVD diamond film (upper curve); and theoretical thermal conductivity limit of pure ^{12}C isotope (top curve). At room temperature, the measured thermal conductivity for the isotopic pure CVD diamond film was 26 W/mK along the growth direction; and 22 W/mK perpendicular the growth direction. The former data was the first experimental proof that isotopic pure diamond could have higher thermal conductivity than all gem diamonds.

CVD diamond film forms by first nucleating on a substrate (e.g., Mo). The nuclei will grow upward with increasing size. Hence, only part of the original nuclei will reach the top. The diamond film contains columnar diamond crystals with coarse grains on the top and finer grains at the bottom. The thermal path is most unimpeded by grain boundaries along the growth direction, and it is most interfered along the plane of nucleation. Consequently, the thermal conductivity is the highest in grains, and the lowest against grains along the nucleation side (Fig. 7.4).

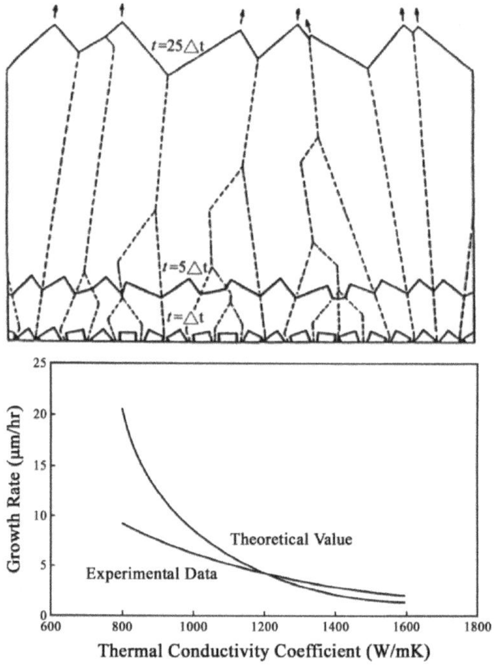

Figure 7.4 The cross section of a typical diamond film grown by CVD (top diagram). The thermal conductivity is the highest along the direction of diamond grain growth (arrow direction). It is the second highest across the coarse diamond grains (top layer). The slowest heat path is crossing all diamond nuclei (bottom surface). Δt measures the initial separation of diamond nuclei. This diagram indicates that diamond grains on the top of diamond film can be 25 times larger than the original size of diamond nuclei. The thermal conductivity of a diamond film is inversely related to its growth rate (bottom diagram); the slower the growth rate, the fewer defects the diamond contains, and hence, the higher the thermal conductivity.

7.2 Diamond Heat Spreaders

The early diamond heat spreaders were made of natural diamond (e.g., mackle) single crystal. Modern diamond heat spreaders can be made by either the high-pressure method (single crystal) (Fig. 7.5) or CVD (polycrystalline) (Fig. 7.6). However, high-pressure sintered PCDs are not suitable as heat spreaders. Although they may contain more than 90 V% diamond, their thermal conductivities are generally lower than copper due to the high contents of diamond defects and metal inclusions (e.g., Co).

Figure 7.5 Heat spreader made by high-pressure synthesis of diamond. The top diagram shows cleaved plates; the middle, mounted pieces; and the bottom, final products (Sumitomo Electric brochure).

Figure 7.6 CVD diamond film polished blanks (top), and the finished heat spreaders (bottom) (former Norton Diamond Film brochure).

Normally, diamond heat spreader must be metallized so that it can be brazed on a semiconductor chip or its substrate (Fig. 7.7).

A typical set of specifications of diamond heat spreader is listed in Table 7.1.

Table 7.1 Specifications of diamond heat spreader

Thermal conductivity (W/mK)	
Parallel direction	10–13
Perpendicular direction	14–17
Surface roughness (Ra, nm)	<200
Surface flatness (μm/mm)	0.5
Thickness (μm +/−25 μm)	200, 300, 500
Edge length (mm)	0.5–50
Metal coating	100Ti/200Pt/100Au
Braze layer	80Au/20Sn

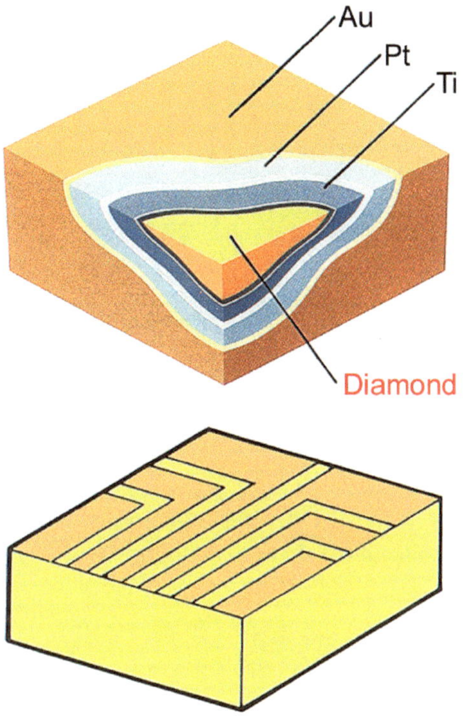

Figure 7.7 The metal coating of a diamond heat spreader (top) and its coated exterior (bottom) (Sumitomo Electric brochure).

7.3 Heat Spreader Geometry

Diamond heat spreader is indispensable for dissipating heat away from high power and high frequency semiconductor chips. For example, laser diode for boosting light in optical fibers. The way to dissipate heat is to spreader the heat over to a larger area where heat can be carried away by a slower means such as by air conduction. Thus, the heat from a semiconductor chip is dissipated by a heat spreader to a heat sink and eventually it is removed to ambient surrounding (Fig. 7.8).

Computer modeling of temperature distribution in a heat spreader indicates that there is an optimal thickness to area ratio for effective dissipating heat away from a heat source (Fig. 7.9). This ratio may be 1/4. The width of the heat spreader should be five times the size of the heat source, however.

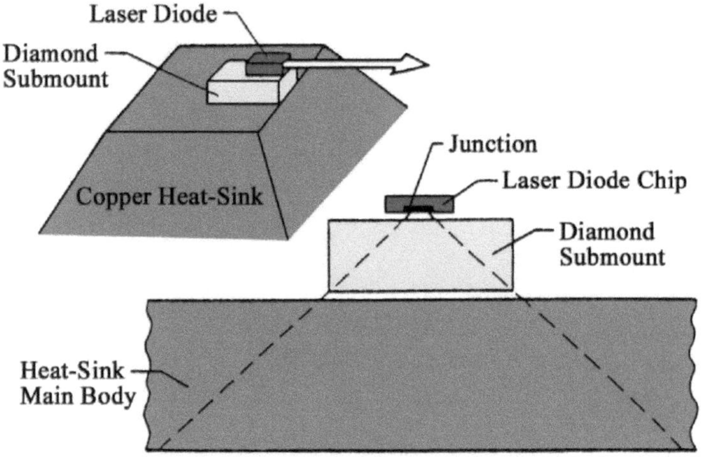

Figure 7.8 The assembly of laser diode with heat spreader/heat sink. Note that the heat is spread out fast in a smaller area and slow in a large area. This is the massive support principle of heat dissipation.

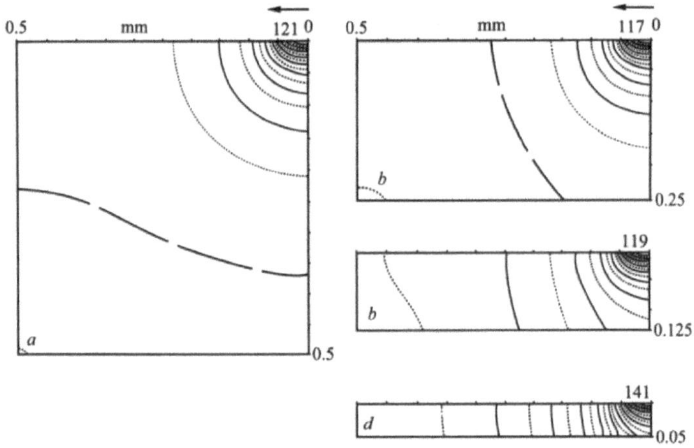

Figure 7.9 The temperature contours of diamond heat spreaders with different thickness. The heat source is located at the top right (shown by arrow) that represents the center of the heat spreader. Only the left half of the cross section is shown in the diagram. The edge length of the diamond heat spreader is 1 mm and it placed at the center of a copper heat sink of 10 mm wide. The isotherms shown are 50°C apart. The top right figure indicates the maximum temperature (centigrade) attained when the steady state is reached. Note that the above data suggest that diamond heat spreader may attain a lower temperature with a width to thickness ratio of 4.

Computer modeling further compares the diamond size and thickness with respect to the dimension of the heat source and the cooling effect it may achieve relative to diamond (Fig. 7.10). The results suggest that if both the width and thickness of the diamond heat spreader can be four times that of the heat source, then the best cooling result will be attained with temperature suppressed four times that of a copper heat spreader.

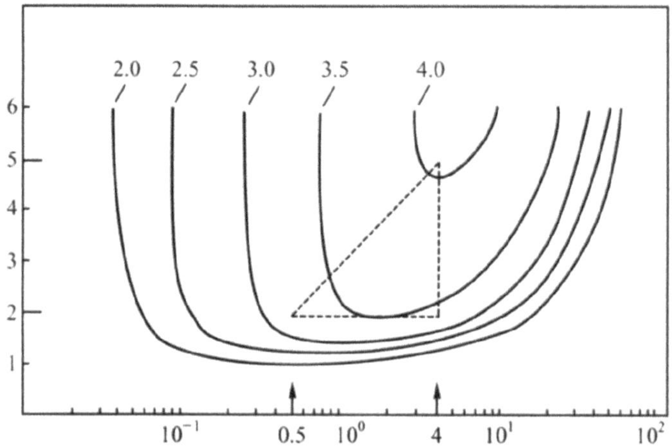

Figure 7.10 The cooling effect of diamond heat spreader versus copper ones. The cooling effect is indicated as the ratio of temperatures at the interface of the heat spreader and the heat source between that can be achieved by copper and diamond. The modeling indicates that by making thickness and width four times of the heat source, temperature cooled by diamond will be only 1/4 of that cooled by copper. In this diagram, the coordinates are labeled by the dimension of the heat spreader to the heat source. The abscissa is for thickness, whereas the ordinate is for width of the heat spreader.

As the price of diamond heat spreader will rise sharply with the enlargement of the size and the increase of the thickness, it may not be practical to use a diamond heat spreader with a thickness larger than 0.5 mm. Moreover, the thickness should not be larger than the dimension of the heat source, otherwise, it may be too costly. In order to compromise the performance and the cost, the width of the diamond heat spreader should be about three times of the thickness.

For some computer chips, the heat spreader used is a ceramic substrate (e.g., alumina, beryllia). The effect of using a diamond heat spreader to replace these ceramics substrate is demonstrated in Fig. 7.11.

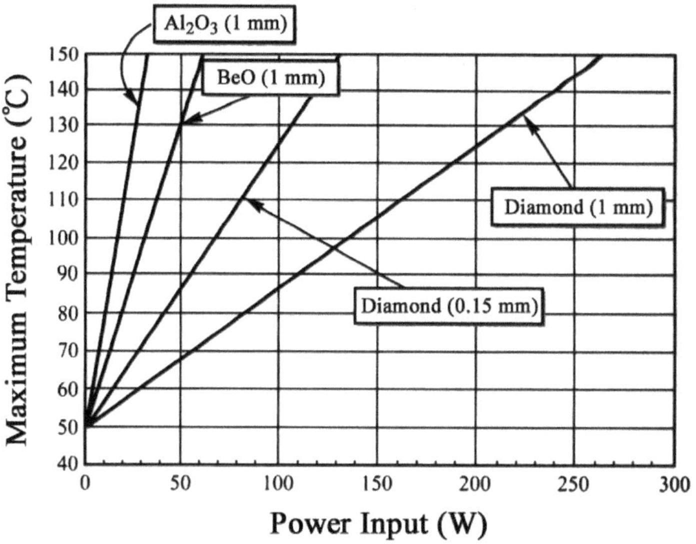

Figure 7.11 The cooling effect of diamond heat spreader versus ceramic substrate for computer chips.

For some light-emitting diodes (LED), the chip is very small (e.g., 250 μm). In this case, very thin diamond layer may be needed to dissipate heat. However, LED is made of III-V semiconductors. For example, the blue light is emitted from GaN/GaAs. The heat spreader used is typically AlN, which has a thermal conductivity of only 20% of copper. But the reason such a poor heat spreader is chosen is that it can match with the thermal expansion of the nitride semiconductor (e.g., 5.2 PPM). Hence, although diamond can conduct heat 20 times faster than AlN, it still cannot be used. In addition, diamond cannot be brazed onto the nitride chip because the latter is not thermally stable at the brazing temperature (e.g., 900°C) of diamond. What's more, the substrate may also be served as an electrode so the heat spreader should also be electrically conducting.

A novel diamond-copper heat spreader that can meet all the following requirements has been designed: high thermal conductivity, high electrical conductivity, tunable thermal expansion coefficient, metallized brazing contact (Fig. 7.12).

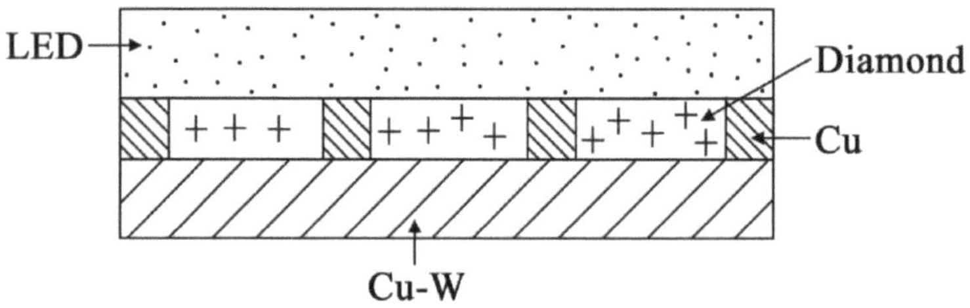

Figure 7.12 The diamond–copper composite heat spreader for LED chip. The heat spreader is also used as an electrode.

The above heat spreaders are mainly used for dissipate heat from a small semiconductor chip. However, computer central processing unit (CPU) is much larger, so it will require a larger diamond heat spreader.

7.4 The Miracle of Moore's Law

In 1964 Gordon Moore, the cofounder of Intel, proclaimed that the transistor number on a semiconductor integrity circuit (IC) (e.g., for computer's central processing unit or CPU) would double every year (latter revised to 18 months), since then, this trend has been holding on miraculously for almost four decades (Fig. 7.13). This has been the longest sustained technology progress ever achieved. In 1997, Moore declared "we are making one transistor for every ant" as Intel manufactured one hundred trillion transistors that year (Harvard biologist E. O. Wilson estimated that there are about 10^{17} ants on the Earth). By the turn of this century, humans had produced more transistors than all the neurons (there are about 10^{11} neurons for each brain) combined for the entire human race.

Although there were several technology breakthroughs (e.g., by adapting chemical mechanical planarization to smooth the surface and by using copper and low dielectric insulator to reduce the electron barriers) that have prevented Moore's Law from derailing in the past, but this time nothing can rescue it because the impenetrable Second Law of Thermodynamics is now lying squarely in the middle of its path. According to the Second Law, entropy (i.e., noise or garbage) in a closed system must go up and the more information is processed, the faster it will increase.

Moore's Law requires the miniaturization of the circuitry on silicon chips from a line width of 15 microns of the first generation (1960) to 0.13 micron of 2003. This unprecedented trend of downsizing has increased drastically the heat generated as the result of the Second Law (Fig. 7.14). It would seem that Moore's Law will soon hit the point one (0.1 micron or 100 nm) barrier.

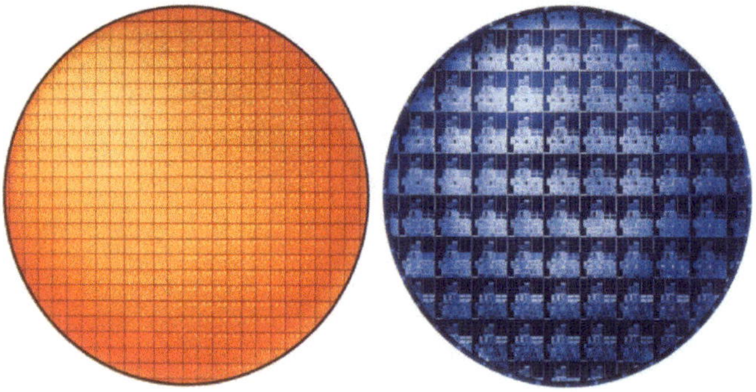

Figure 7.13 Intel's first-generation Pentium (1992) had 2300 transistors (left), and their Pentium IV (2002) contained 13 billion transistors (right).

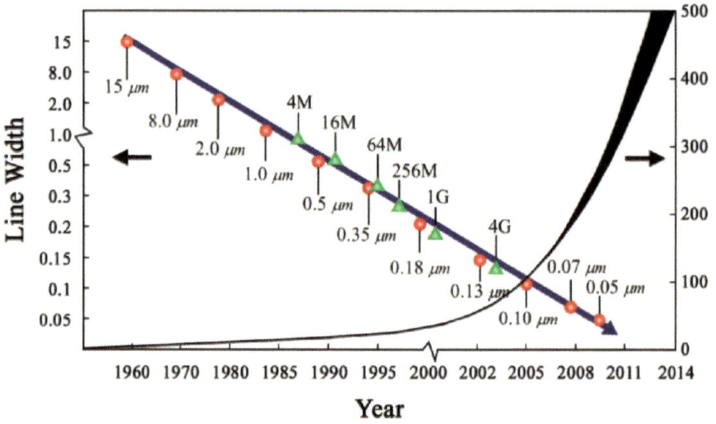

Figure 7.14 The relentless packing of circuitry with the consequence of exploding heat generated.

The power increase of CPUs chip is now accelerating (Fig. 7.15), whether Moore's Law can survive or not is dependent on how efficiently the waste heat can be dissipated from the surface of the chip.

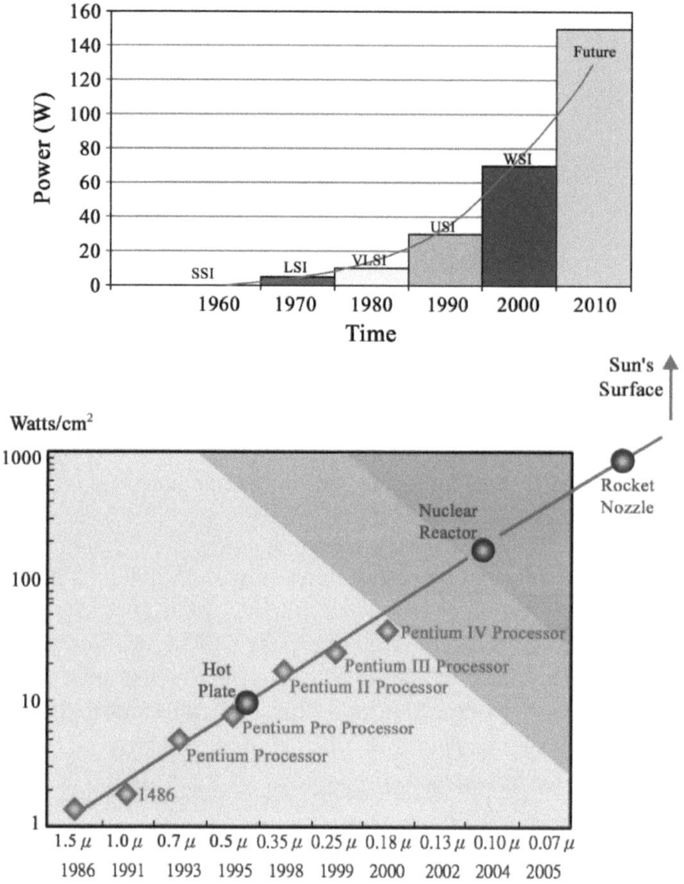

Figure 7.15 The increased pace of power surge in computer's CPU (top). The trend of increasing power density (bottom). Note that computer chips have long exceeded hot plate for cooking and may attain the energy density of nuclear reactor.

The consequence of densified circuitry and increased power can create hotspots in the chip during the operation, as soon as the temperature exceeds 100°C, it will begin to "burn out" (Fig. 7.16). Moreover, as the metal (e.g., Cu) circuitry will expand much faster than surrounding insulating layers (e.g., silicon oxide or nitride), microcracks may be induced in the latter even the temperature is as low as 70°C. These microcracks may propagate with repeated cycles of heating and cooling, and eventually, the chip may fail by fatigue.

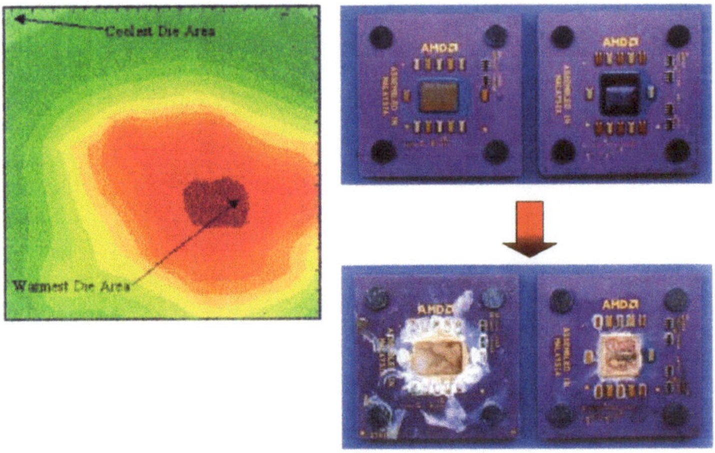

Figure 7.16 The hotspot generated in a computer chip (left). The "burning" of computer chips (right).

The heat dissipation of the chip is now handled by a three-step process. First, the heat is spread out by a heat spreader typically made of copper. Subsequently, the heat will be drained by a heat sink typically made of aluminum fins, finally the heat is blown off these fins by a fan. The heat sink and fan combination is the bulkiest device incorporated in a computer chassis (Fig. 7.17). If they can be downsized, the volume of a computer chassis will be greatly streamlined.

Figure 7.17 The appearance of a computer chassis. Note the conspicuous size of the aluminum heat sink and the cooling fan.

In order to for a heat spreader to dissipate heat from the CPU effectively, its thermal conductivity (the product of thermal diffusivity, heat capacity, and mass density) must be high. The current heat spreader is made primarily of copper that is the second highest thermal conductivity of all metals. Silver is only 8% more heat conductive but a lot of more expensive. Hence, it is very difficult to find a practical way to improve thermal conductivity beyond copper.

There are only a few choices to beat copper in thermal conductivity. Cubic boron nitride may conduct heat twice as fast, but its price is prohibitive. Another choice is to use graphite as the heat spreader. Graphite can conduct heat more than five times faster than copper (even faster than diamond) along its graphene planes, but it is actually a thermal insulator in the perpendicular direction. All industrial graphite contains microscopic crystallites, so its thermal conductivity is no better than copper. Hence, there is no improvement by replacing copper heat spreader with graphite ones.

Diamond is the best heat spreader as it can conduct heat at least four times faster than copper and the heat path does not have to be limited on plane. However, diamond is also very expensive, which would make it impractical as a commercial heat spreader unless it is very thin (e.g., <0.5 mm). But unlike heat spreaders for smaller chips (e.g., LED, laser diode) that can use thin disks, a CPU would require a heat spreader that is more than 2 mm thick. Currently, CVD diamond cost more than $10/cm^2 and their thickness is limited to less than 1 mm.

Facing with the above constraints, it appears that copper heat spreader will set the limit of CPU power at <100 W. However, there is a practical way out, i.e., by using a diamond composite material as the heat spreader. One way to do so is to use high-pressure sintered polycrystalline diamond (PCD) as the heat spreader. However, this option is not feasible because PCDs are also very expensive (about $5/cm^2). Worse still, although PCD may contain more than 90 V% of diamond, its thermal conductivity is actually at par with that of copper, which costs less than 1/10 of the PCD. The reason PCD cannot conduct heat fast is mainly that it contains fine diamond grains required by mechanical applications (e.g., cutting tools, wire drawing dies). The small diamond grains contain large amount of grain boundaries, slip planes, and metal inclusions, and all of them can slow down the transfer of heat by damping the phonon. In addition, the cementing materials (e.g., Co, Si) among diamond grains are also poor thermal conductors.

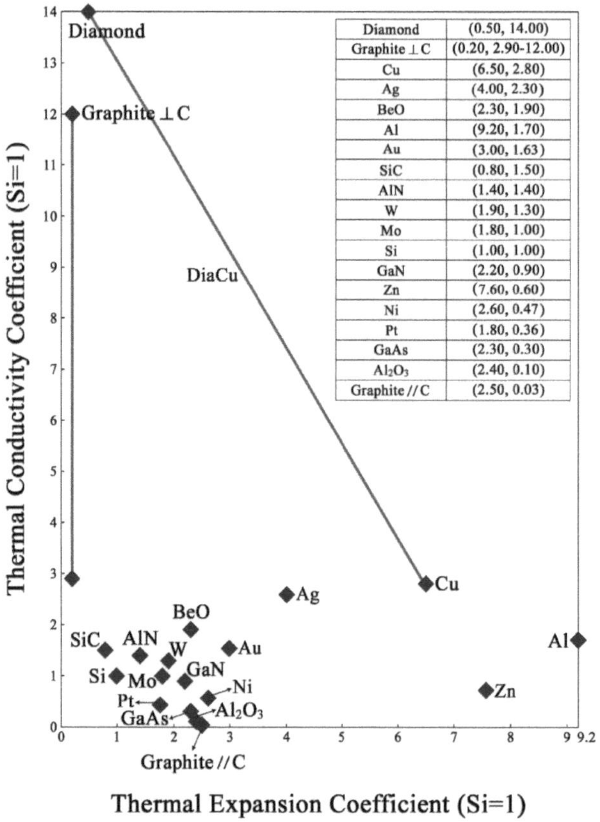

Figure 7.18 Thermal conductivity coefficient versus thermal expansion coefficient of various heat spreaders and substrate materials for semiconductor chips. Note DiaCu is capable of adjusting the two key parameters required for heat spreaders over a wide range of applicability. In the given plot, thermal conductivity coefficient is normalized by silicon's value of 139 W/mK; and thermal expansion coefficient, by 2.6 PPM/°C.

By coarsening diamond grain sizes and using copper as the cementing material, thermal conductivity can be doubled or tripled (Fig. 7.18). The novel DiaCu heat spreader is also electrically conducting, so it can be machined and shaped conveniently by wire EDM. In addition, the cost for making DiaCu is substantially lower than that for conventional PCD because it no longer requires the stringent control of mechanical properties as well as surface finish that may limit the yield and throughput for PCD production.

DiaCu (patent pending) is now manufactured under the trade name of HeaThru™ heat spreader. When it is incorporated in a conventional heat sink, it shows a noticeable decrease of temperature as well as its gradient (Fig. 7.19). The ability of DiaCu to eliminate steep temperature is particular important as the local heat surges from time to time during the computation of a CPU can ruin the chip.

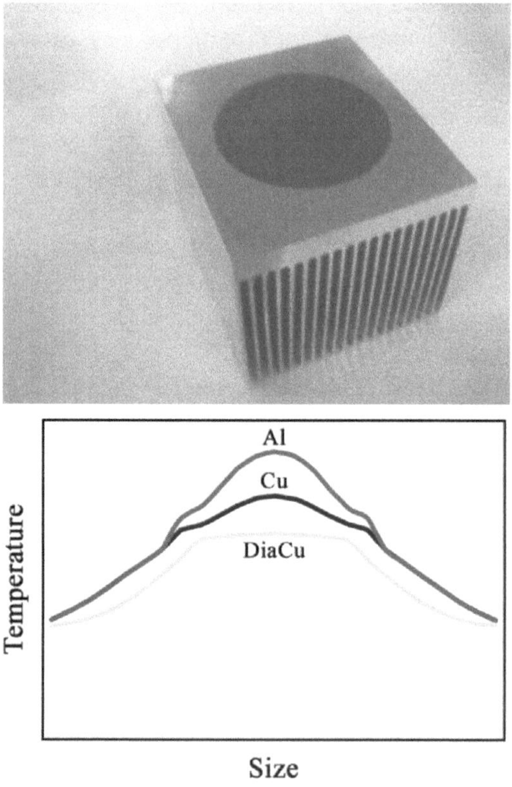

Figure 7.19 HeaThru™ DiaCu inserted aluminum heat sink (top diagram). The temperature profiles of aluminum heat sink, copper inserted aluminum heat sink, and DiaCu inserted aluminum heat sink.

It is interesting to note that by inserting DiaCu heat spreader in the aluminum heat sink, the bottleneck of heat flow has moved from the heat spreader to aluminum fins, so the design of aluminum heat sink becomes incapable of cooling future CPU sufficiently for normal operation. Hence, more effective heat sink mechanism must be designed.

Many future generation heat sinks are currently being developed. For example, porous graphite that is impregnated with paraffin may be used as a much more efficient heat sink. Not only graphite can conduct heat faster, but also the melting of paraffin can absorb a tremendous amount of latent heat that can be dissipated with time.

However, the most immanent solution of heat extraction is to use a heat pipe or a heat chamber. The heat chamber is a vacuum chamber that contains a wick path (e.g., by cementing copper powder that has high porosity) to allow water to circulate by capillary force. Thus, as heat is applied to one side (e.g., with the contact of a CPU), water in the wick path is vaporized that will absorb a large latent heat. The water vapor will dump the heat at the other side that is kept at a lower temperature. The condensed water will circulate back by following the wick path.

However, any heat sink will need a heat spreader; otherwise, the heat generated by transient hotspots cannot be released fast enough to avoid overheating. DiaCu with its larger thickness than CVD diamond film would be ideal to channel the heat immediately into the heat chamber (Fig. 7.20). Moreover, with its copper content, DiaCu can be readily brazed to the copper container of the heat chamber. The joining of DiaCu heat spreader and heat chamber not only will dissipate heat much faster than the current tandem of copper-aluminum but also may eliminate the bulky space occupied by system of the aluminum fin and the cooling fan.

Currently, a computer CPU is based on a surface platform so its extension of Moore's Law will be limited as a planar circuitry will become increasing awkward to integrate as the architecture continue to spread out. Hence, it is conceivable that Moore's Law may have to be further extended by using a three-dimensional architecture of circuitry. Such a volumetric design can take many forms. One of them is to rely on planar platforms of multi-chip module (MCM) and integrate them to form a three-dimensional structure. By stacking planar platforms together to form a CPU cube, diamond film heat spreaders can be sandwiched in alternation (Fig. 7.20). In this way, the much higher heat flux can be channeled outward to the edges and corners of the cube. At these intersections, heat can be carried out effectively by a circulated fluid. This may be the ultimate design for future computer chip that will continue to surf on Moore's Law.

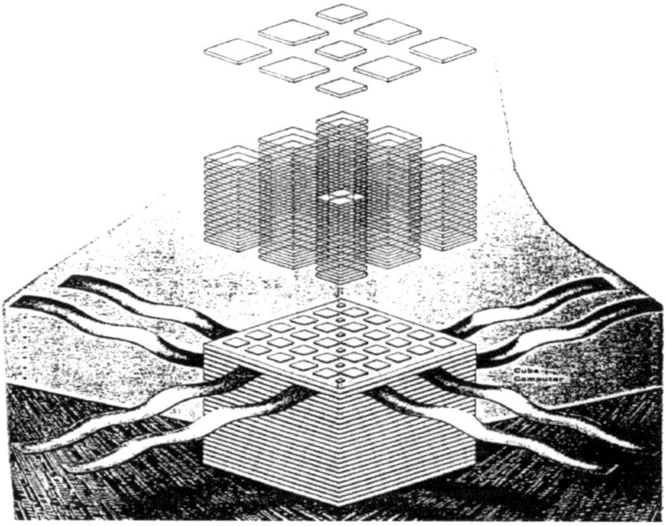

Figure 7.20 Planar MCM platforms are stacking together to form a three-dimensional cubical chip. Diamond films may be stacked in alternation to channel out heat that may be pumped away from the corners of the cube (former Norton Diamond Film brochure).

7.5 Diamond Stamping Heat Spreader

A heat spreader is normally dumping heat to a heat sink where the heat is further disposed. But diamond heat spreader can be used to deliver heat for performing work such as to bond lead frames to an IC's electrode pads. Tab automatic bonding (TAB) is the process for bonding electrical wires to an IC (inner lead frame bonding) or for bonding electrical wires to the printed circuit board (outer lead frame bonding). During the bonding process, lead frames (Cu-Au) are delivered to ICs by a continuous tape. When the lead frames are matched with the electrode pads that are coated with a solder alloy, a bonding tool with a diamond pressing face will stamp the lead frame against the electrode pads. The diamond face is kept at a constant temperature (e.g., 550°C) because a resistor heater that is set at a higher temperature (e.g., 700°C) is inserted at its center of the bonding tool. During the brief pressing period by the diamond face, heat is delivered to solder lead frames onto the matching electrode pads.

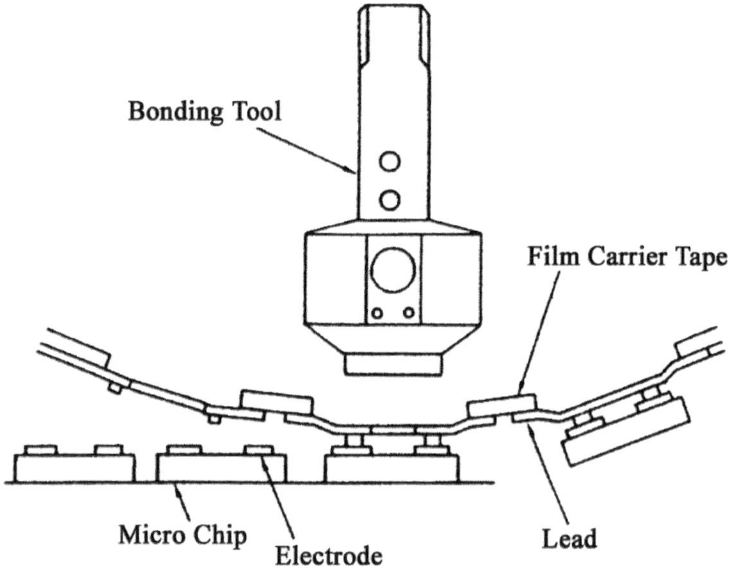

Figure 7.21 The layout of TAB with a bonding tool to solder electrical wires of a film carrier to electrical electrodes on chips.

There are two ways to solder electrical wires to electrode pads. The heat can be delivered by dumping from a continuous source, or the heat can be generated instantaneous by pulsing electricity through a heater (Fig. 7.22). Because the solder must melt and freeze again, the temperature profile during the soldering is controlled by stamping time. If heat is being continuously delivered, the freezing of the solder must take place after the stamping. Alternatively, if heat is generated by the pulse electricity, the freezing of the solder can occur during the pressing. Both types are employed in production.

Various geometries of bonding tools are available (Fig. 7.23).

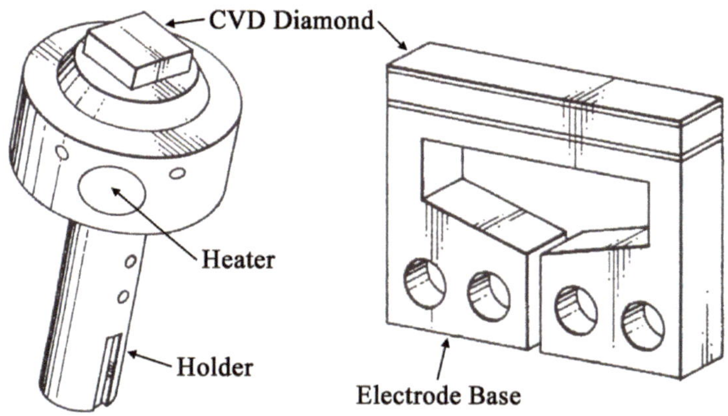

Figure 7.22 The bonding tool that is heated continuously by a resist heater inserted at the center hole (left). The bonding tool that is heated instantaneously by pulsing with electricity (right).

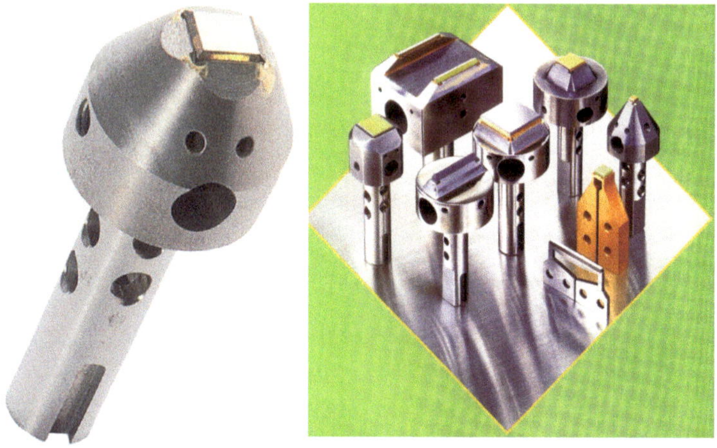

Figure 7.23 A conventional design of bonding tool (left) and newer designs of bonding tools (right) (Sumitomo Electric brochure).

The critical part of the bonding tool is the diamond face. Previously, other materials (e.g., BeO, cBN, PCD) were used, but CVD diamond serves the best purpose. First, it is the best heat spreader that can deliver heat fast and uniformly. Second, diamond is hard. When it presses the wire against the electrode, the high pressure can be applied evenly. Third, diamond's surface is highly inert, so it will not be corroded, nor will it be stuck with solder (e.g., tin alloy). In addition, CVD diamond can be coated to a suitable substrate (e.g., SiC) with different geometry. Finally, CVD diamond can spread thinly to cover a larger surface area without incurring much cost. With all the above advantages, CVD diamond-coated bonding tools are now the mainstream for practicing TAB. Their stamping frequency is high and their service life is long. Most importantly, the solder bonding so produced is highly reliable.

In pressed bonding, it is important that the diamond surface is flat when heated to the suitable temperature (e.g., 550°C). However, diamond's thermal expansion rate is much smaller than its substrate's; hence, at room temperature when the diamond surface is finished, it must be slightly concave. Upon heating, the larger thermal expansion of the substrate will flatten the diamond surface (Fig. 7.24).

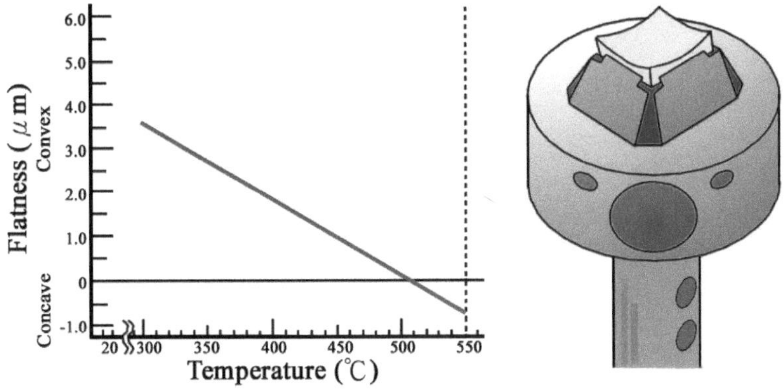

Figure 7.24 The design of the CVD diamond surface is to assume a convex curvature at room temperature. Upon heating, the extended substrate will flatten the diamond surface (left), so it will eventually become slightly concave (e.g., –1 μm) at the service temperature (550°C) (right) (Sumitomo Electric brochure).

Large chips (e.g., LSI) are also bonded to the printed circuit board, but instead of pressing wires to the chip, the chip is reversed and it is pressed against the wire. This reversing process is called flip chip assembly. The advantage of reversing the bonding process is that it combines both inner bonding and outer bonding of lead frames. Moreover, it eliminates the need of using lead frames altogether; hence, this is a much more efficient bonding process.

In practicing flip chip bonding, the chip must be vacuum chucked on the bonding tool, so the CVD diamond is pierced with a hole to allow air to be sucked in (Fig. 7.25).

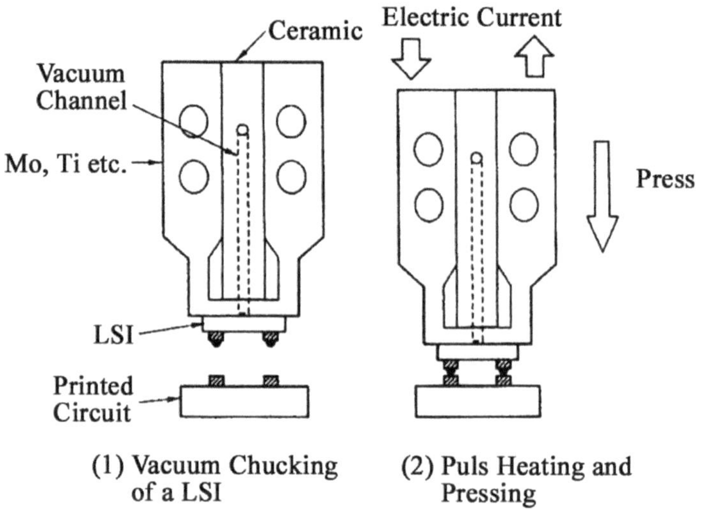

Figure 7.25 Flip chip bonding of LSI. The chip is vacuum chucked by the bonding tool and it is then pressed against electrodes located on printed circuit board.

During the flip chip bonding, the heating is provided by passing a pulse of electricity through a resistor heater. This resistor heater can be made by boron-doped diamond that is also used as the substrate to support the pressing CVD diamond. The heating assembly is tucked in with electrodes and heat sink and supported by a base holder (Fig. 7.26). The bonding tool can accommodate different geometries of the pressing surface. During the bonding process, flash heating is provided by sending pulses of electricity through resistor heater.

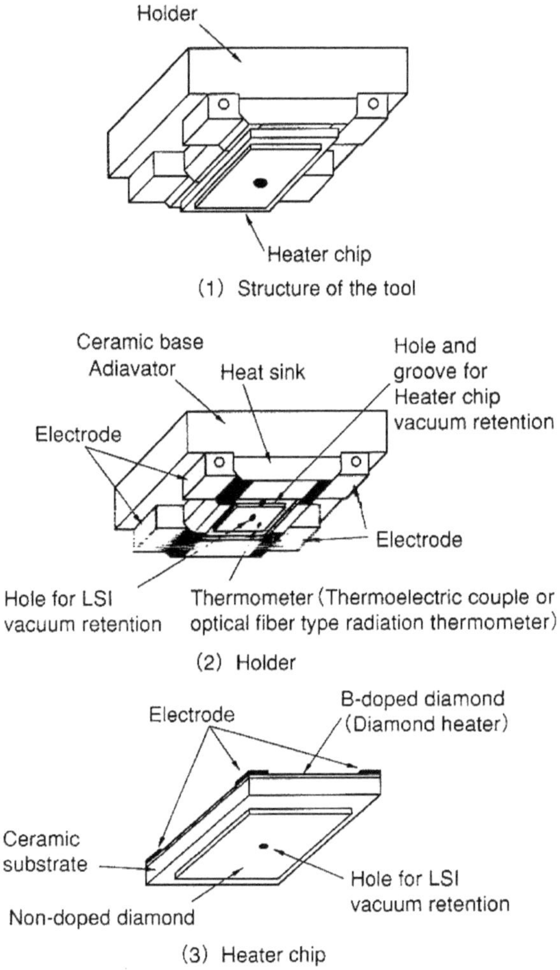

Figure 7.26 Bonding tool assembly for flip chip processing. The assembly is shown in the top diagram. It contains a base holder subassembly (middle) and the pressing diamond subassembly (bottom).

Chapter 8

Diamond Radiation Windows

Diamond's beauty is in its transparency, not just in visible light, but in invisible as well.

8.1 Optical Transmission

When light impinges on an object, part of it will be reflected back and the rest may enter the material. The entered light will be scattered, absorbed, or transmitted. Some of the energy may be absorbed by electrons to allow them to move to the excited state that may either instantly released as fluorescent light or latter released as phosphorous light. Other energy will be absorbed by atoms and manifested as heat. The heat is stored as kinetic energy of vibrations. The transmission of lattice vibration (phonon) is reflected as the conduction of heat. The vibrated atoms will also radiate low energy radiation due to the movement of electrons in these atoms. Such radiation is known as black body radiation that is temperature dependent (Fig. 8.1).

Handbook of Industrial Diamonds: Diamond Films and Carbon Coatings
James Sung

ISBN 978-981-4310-00-0 (Hardcover), 978-1-003-18045-6 (eBook)
www.jennystanford.com

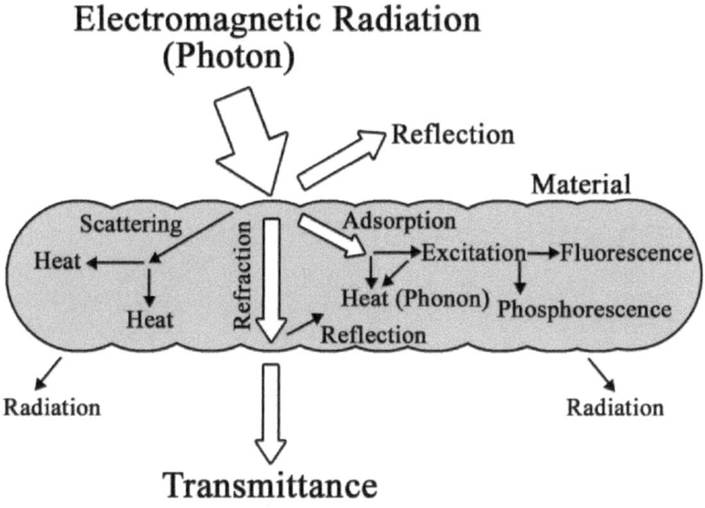

Figure 8.1 The distribution of light as it impinges upon a pile of atoms.

For an optical window material, the amount of transmitted light should be maximized. However, there is always part of light that is reflected before entering the material, and the higher the refraction index is, the more of the entered light must bend to enter the material, and consequently, the larger the amount of light is reflected at the surface. For diamond, because of its unusual large refractive index (η = 2.417), the amount of reflected light is high. As a result, 28.8% of the incident light is reflected before the light enters diamond. Hence, even with the best-quality diamond with almost no absorption, its transmittance can only be as high as 71.2%. However, if the surface of diamond is coated with anti-reflection layers, the amount of reflection can be reduced and hence more light may be transmitted through.

Although diamond's total transmittance may be smaller than most optical window materials (e.g., quenched quartz, LiF, CsI), diamond has the highest mechanical strength of all materials; hence, it can be much thinner than other window materials. As a consequence, the absolute amount of light transmitted is the highest of all materials. Moreover, diamond has the widest range of transparency. Except for the absorption from 2.5 to 6.7 μm due to multi-phonon scattering, diamond is transparent in almost all wavelengths that include nuclear resonant γ-ray, atomic bonding X-ray, atomic vibrating UV, visible light, molecular agitating IR, air penetrating microwave, to kilometer long radio wave, etc. (Fig. 8.2), so it can be used as a window material for almost all applications. For example, for microwave generation, diamond can replace sapphire as exit window, or it may substitute alumina as substrate.

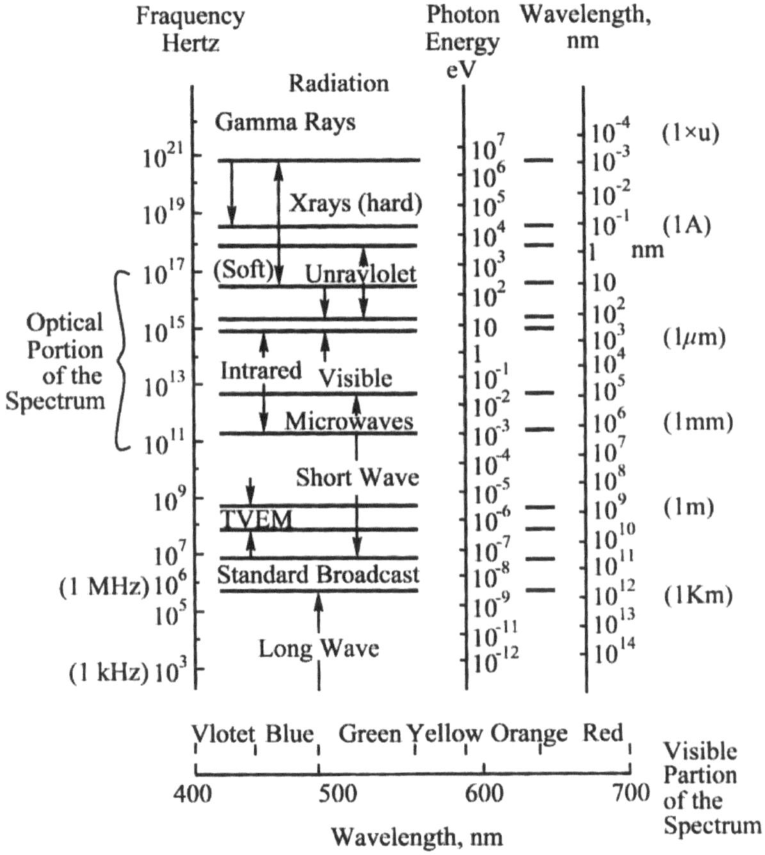

Figure 8.2 Electromagnetic radiation energies and wavelengths.

8.2 Optical Window Materials

Common optical window materials and their applicable ranges are summarized in Table 8.1.

Table 8.1 Optical window materials and their applicable range

Optical range	Thermal range			
	L	M	H	VH
UV	MgO, Al_2O_3, α-SiO_2, g-SiO_2	MgO, Al_2O_3, g-SiO_2	Al_2O_3	Al_2O_3
VIS	MgO, Al_2O_3, α-SiO_2, g-SiO_2	MgO, Al_2O_3, g-SiO_2	Al_2O_3	Al_2O_3
IR1	MgF_2, Al_2O_3, MgO, α-SiO_2, ZnS, ZnSe, g-SiO_2	MgF_2, Al_2O_3, MgO, g-SiO_2	Al_2O_3	Al_2O_3
IR2	MgF_2, MgO, CdTe, SiZn, SGe, GaAs, ZnSe	MgF_2, MgO		
IR3	GeZnS, GaAs, CdTe, ZnSe			
IR4	Ge, CdTe, ZnSe			
Rader	Pyroc, Al_2O_3, MgF_2, α-SiO_2, ZnSe, g-SiO_2, Si, CdTe, Ge	Pyroc, Al_2O_3, MgF_2, α-SiO_2, MgO, g-SiO_2	Al_2O_3	Al_2O_3

Code	Wavelength (μm)	Wave number (cm^{-1})
UV	0.2 → 0.4	25000 → 50000
VIS	0.4 → 0.7	14290 → 25000
IR1	0.7 → 3	3330 → 14290
IR2	3 → 5.5	1820 → 3330
Air Opaque	5.5 → 7.5	1330 → 1820
IR3	7.5 → 10	1000 → 1330
IR4	10 → 14.5	690 → 1000
Air Opaque	14.5 → 1000	10 → 690
Rader	1000 → ∞	0 → 10

Temperature (°C)	Code
20 → 500	L = Low
20 → 1000	M = Middle
20 → 1500	H = High
20 → 2000	VH = Very High

Diamond's high transparency in wide ranges can be combined with other unique properties (e.g., hardness, thermal conductivity) to make it the ideal optical window material.

Not only it can replace all current window materials, but also it can fill up the gap of current deficiencies due to the lack of suitable materials.

For example, because of high absorption, low mechanical strength, and slow thermal conduction, conventional window materials will burn out soon when used for generating high energy X-ray, kilowatt laser light (e.g., CO_2 laser with wavelength of 10.6 μm), or megawatt microwave (e.g., gyrotron tube with frequency of 70 to 170 GHz). Such intense radiation must be generated under very high vacuum so a window material must be used to seal the air extremely tight, but at the same time the window material must allow the exit of intense radiation. Diamond's unsurpassed strength can hold a vacuum of 10^{-9} torr for extended period of time, even with a thickness many times less than ordinary window materials (e.g., ZnS, ZnSe). In addition, diamond's strength can support such vacuum even at a temperature as high as 450°C. The low absorption coupled with high thermal conductivity will ensure that diamond window will not be overheated.

The superb diamond properties are compared with other window materials as listed in Table 8.2.

Table 8.2 The comparison of IR window materials

Property	Diamond	ZnS	ZnSe	GaAs	GaP	Si	Ge	Sapphire
Density	3.52	4.08	-	5.317	4.13	2.33	5.232	3.98
Flexure Strength	2944	97	-	55	103	127	93	153
Young's Modulus	1050	74.5	-	85.5	102.6	130.1	103.3	380
Thermal Expansion Coefficient	1.0	6.8	-	5.73	5.3	2.3	6.0	5.3
Thermal Conductivity	20	0.17	-	0.53	0.97	1.412	0.60	0.34
Thermal Shock Resistance	2500	2.8	-	9.6	-	-	10.5	-
Laser Destruction Threshold Energy Density	30	5	-	4	-	-	6	-

8.3 Diamond's Optical Classification

Diamond can be classified chemically based on its impurities contained. This classification is reflected in its optical properties, such as color. Because diamond's carbon atoms are very small (1.54 Å), they can be substituted only by very few other elements' atoms, notably boron and nitrogen. Hence, the optical classification of diamond is primarily based on the type of dopants present in diamond (Table 8.3).

Table 8.3 Diamond classification based on impurities

		Type I			Type II	
Category		**Ia**	**Ib**		**IIa**	**IIb**
Natural diamond		>98%	0.1%	Scarce	>2%	Scarce
Synthetic diamond		None	>99%	>1%		
Impurity	N	2000	100–1000	1–100	1	
	Form	Platelets	Substitutional	Magnetism Antimagnetic	Paramagnetic	
(PPM)	Element		Metal			B
	Concentration		10^3–10^5			100
	Magnetism		Ferromagnetic			
Color		Yellowish Transparent	Greenish Yellow	Yellow	None	Blue
Absorption (μm)		7.8	8.85			3.57
Resistivity (Ω-cm)		$10^4 \sim 10^{16}$	$10^4 \sim 10^{16}$	10^{16}	10^{16}	$10 \sim 10^4$
Thermal conductivity (W/mK)		9000	9000	20000	20000	
Refractive index					2.417	

The solubilities of boron and nitrogen atoms in diamond are actually very small in the temperature range of diamond formation (Fig. 8.3).

Diamond has the widest transparent window of all materials, from the absorption edge of band gap near the wavelength of 0.2 μm through visible, infrared, microwave all the way to radio wave (Fig. 8.4). In addition, diamond is transparent in much shorter wavelengths as well, such as X-ray and γ-ray.

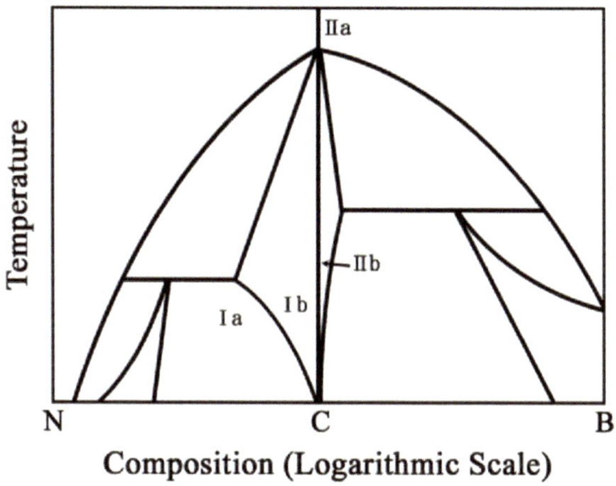

Figure 8.3 The phase diagram of diamond with either boron or nitrogen. The composition of either boron or nitrogen in the vicinity of diamond is intentionally expanded by using a logarithmic scale so the different types of diamond can be discerned.

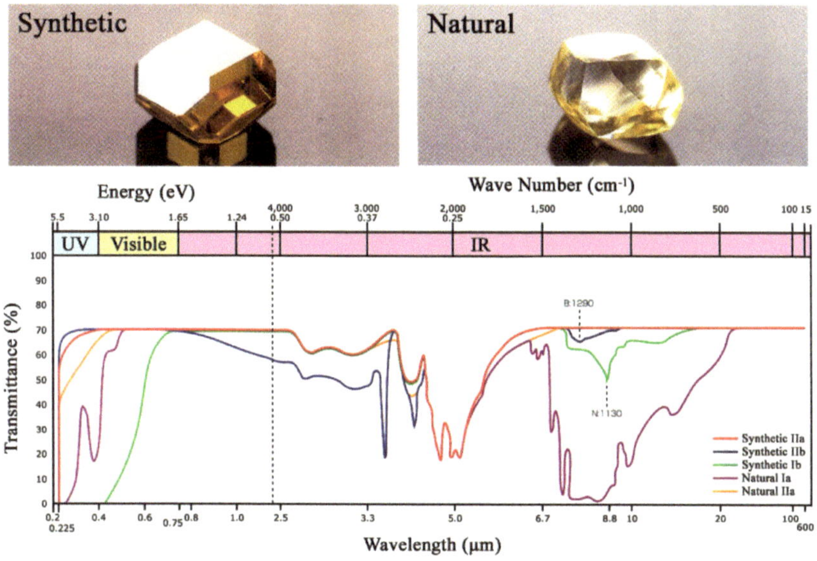

Figure 8.4 The absorption spectra of diamond across a wide range of wavelengths. Note that the transmittance of diamond is limited to 71.2% because of the total refraction caused by its high refractive index (2.417).

The amount of nitrogen content and its form (e.g., substitution or aggregate) will absorb different wavelengths. For example, Type IIa gem diamond absorbs only in the region of about 5 μm. But Type Ia natural diamond contains platelets of nitrogen; it has a strong absorption in the vicinity of 7–9 μm. Other than that, it is nearly transparent. All synthetic Type Ib diamond and a small amount of natural diamond contain substitutional nitrogen atoms absorb only slightly at 8.85 μm (Fig. 8.5). However, there is a strong absorption of blue light; hence, these diamonds will appear as greenish yellow.

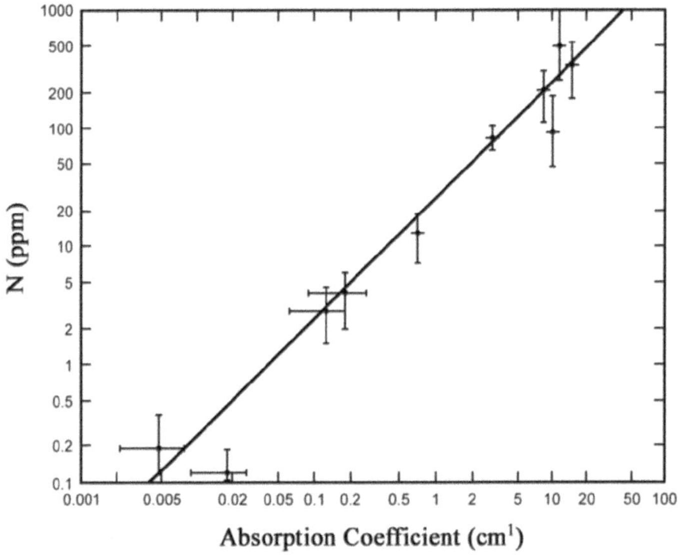

Figure 8.5 The correlation between the amount of substitutional nitrogen content and the absorption at wavelength of 8.85 μm.

8.4 The Transparent Diamond

Diamond has a very wide range of transparency (Fig. 8.6).

There are two types of synthetic diamond: High-pressure synthesized diamonds are typically single crystals that belong to Type Ia. The CVD diamond films are polycrystalline in nature.

The early CVD diamond films are opaque due to the extensive presence of grain boundaries and other defects. These diamond films are suitable for making cutting tools so they are referred to as tool grade. Subsequently, diamond films were grown at slower speeds. The results were larger grain sizes with less defects. Such diamond films are suitable to for making heat spreaders so they are called thermal grade. Eventually, diamond films were deposited very slowly so there is almost no defect. In this case, the film becomes transparent that are known as optical grade.

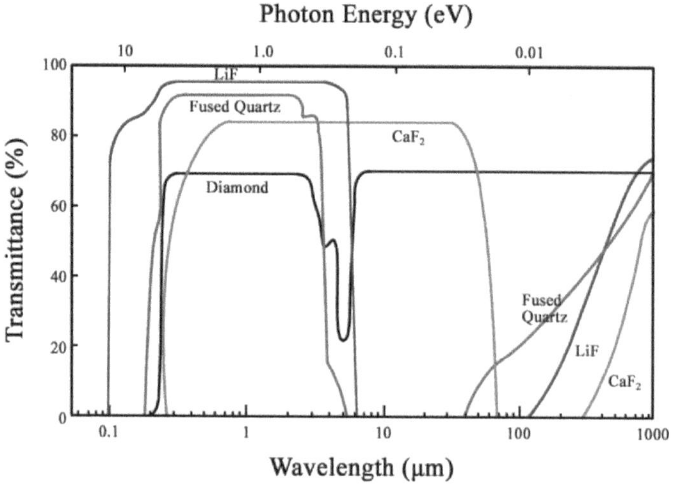

Figure 8.6 The ranges of transparency for various optical materials.

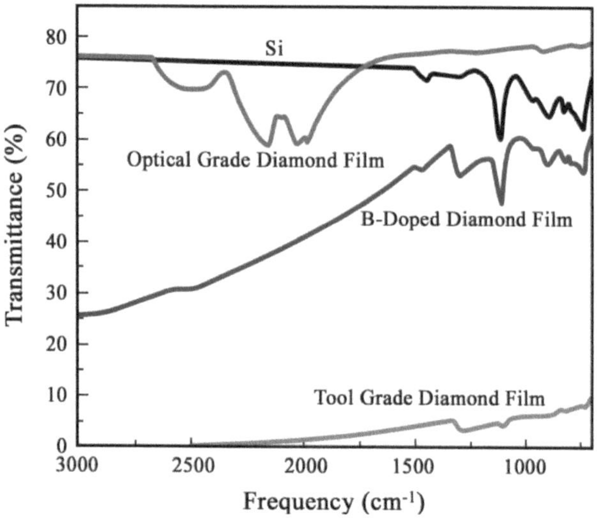

Figure 8.7 The transparency of CVD diamond films of various grades.

The transparencies of various CVD diamond films are shown in Fig. 8.7.

Although CVD diamond films contain numerous diamond grains, the optical-grade actually has the similar transparency as single crystal (Type IIa) gem diamond (Fig. 8.8). As the latter is relatively small and very expensive, the optical-grade CVD diamond film will be the major industrial window materials.

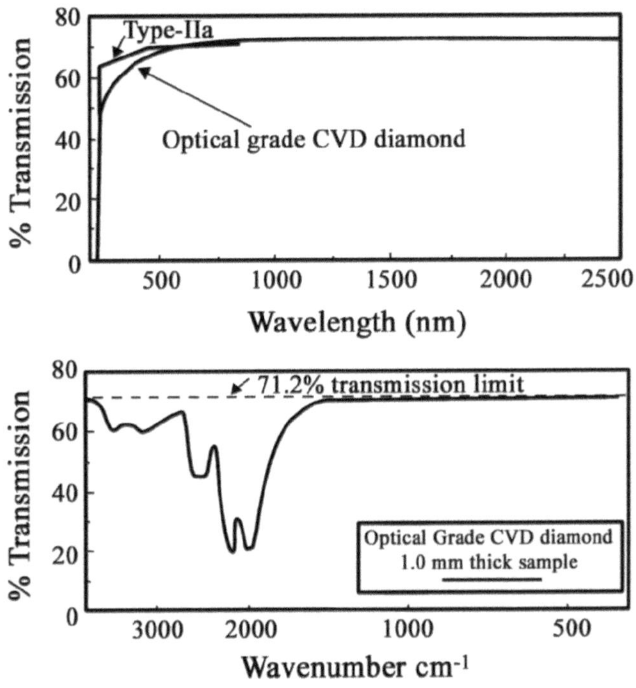

Figure 8.8 The comparison of optical transparencies between transparent CVD diamond film and natural gem diamond.

8.5 Diamond's Optical Windows

Although diamond's light transmission is limited to 71.2% due to its high refractive index, the highest mechanical strength can allow the use of a much thinner diamond; hence, overall diamond is still the most transparent material. Moreover, diamond's wide range of optical transparency has made it best window material applicable for almost all wavelengths (Fig. 8.9).

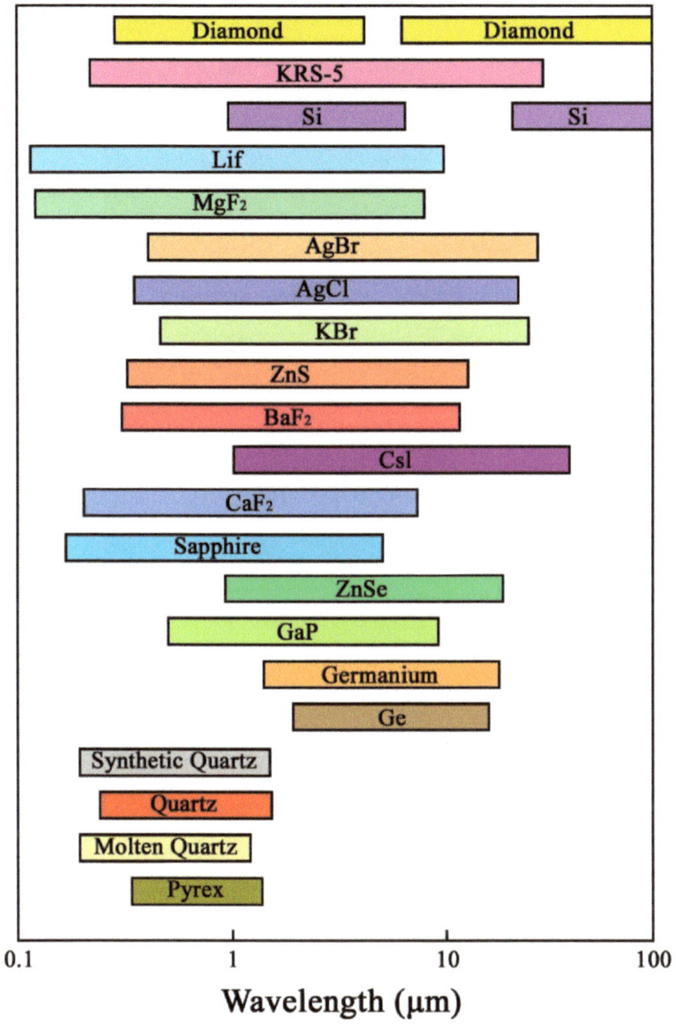

Figure 8.9 The optical transparent ranges for window materials.

Diamond windows may be made of natural diamond (Type IIa) or CVD diamond (optical grade). The former are typically very small (e.g., 3 mm) and they are expensive (e.g., $400) (Fig. 8.10).

Alternatively, less expensive optical windows may also be made from diamond films grown by CVD. For example, diamond lenses have been produced by this method (Fig. 8.11).

Figure 8.10 Natural diamond window (3 mm in diameter) for IR imaging.

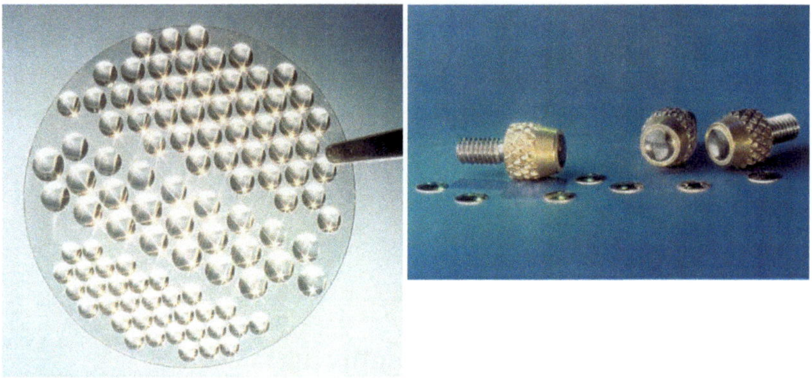

Figure 8.11 The array of CVD diamond lenses (left) and their possible use as length gauges (right) (Aixtron brochure).

Transparent CVD diamond films are typically grown by microwave plasma that can be kept with minimal impurities. CVD diamond films produced by arc jet may contain vaporized electrode metals (e.g., Cu), whereas that grown by hot filament often incorporates W atoms. These films are translucent at the best (Fig. 8.12).

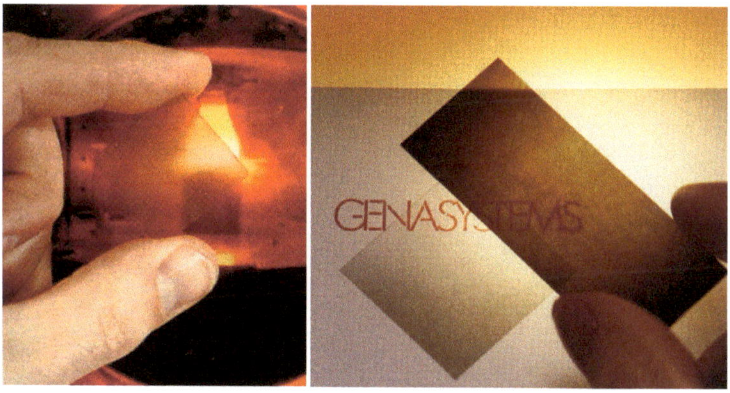

Figure 8.12 Former Norton Diamond Film's translucent diamond film (left) and former Genasystems of translucent diamond film (right).

Freestanding CVD diamond films of excellent quality are now grown using microwave plasma (Fig. 8.13).

Figure 8.13 CVD diamond films grown by De Beers (top left), by Sumitomo Electric (top right), and by Aixtron (bottom).

A highly impressive glass looking large CVD diamond was displayed by De Beers (Fig. 8.14). It was said that the diamond was so low in defect concentration that it was used to transmit 1 MW of microwave without overheating. All other materials would instantly vaporize under such an intensive bombardment of electromagnetic radiation.

Figure 8.14 A large transparent CVD diamond grown by De Beers. Similar diamond windows can be as large as 16 cm in diameter and more than 2 mm in thickness (e.g., Raytheon's). The amount of absorbed energy is less than 1%. The absorption coefficient is less than 0.1 cm^{-1}, or even as low as 0.027 cm^{-1}.

When light propagates in a medium, it may absorbed by atoms, in particular those loosely bonded. The light may also be absorbed by the presence of impurities, grain boundaries, dislocations, voids and other crystal irregularities. However, light may also be absorbed at the interface of entrance and exit, in particular when the surface has dense asperities. Hence, it is critical to polish the diamond surface to high finish (e.g., < 5 nm). By using coarse-grained CVD diamond and make the surface highly smooth, the total absorption of incident light can be kept below 1% for a thickness of < 1 mm.

8.6 Diamond X-Ray Windows

Diamond with almost no absorption in very high energies is ideal for transmitting X-rays. Because of its high mechanical strength, a window of about 1 cm in size would require a thickness of only 250 μm to seal a high vacuum (Fig. 8.15).

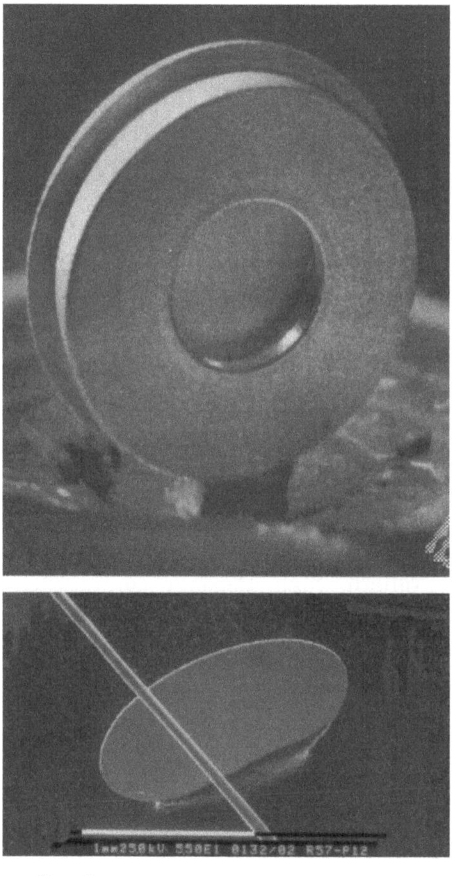

Figure 8.15 A smaller (6 mm) X-ray window with a thickness of 50 μm can support a pressure of 4 atmospheres (top diagram). Another even smaller (1.4 mm) window has a thickness of only 17 μm. A human hair (80 μm) is superimposed for comparison of size.

X-ray windows are indispensable to generate X-ray or for X-ray instrumentation (Fig. 8.16).

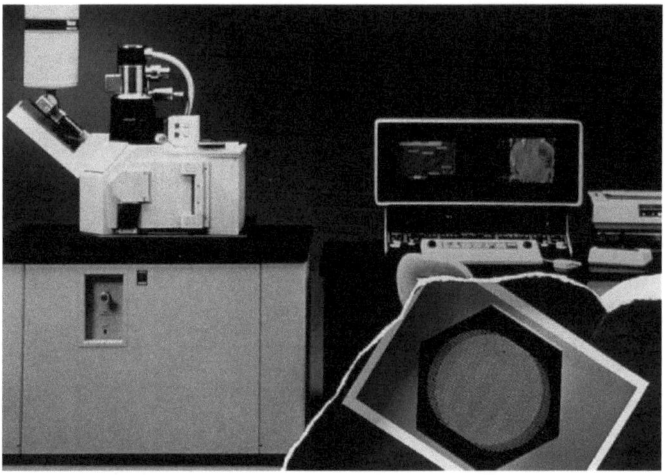

Figure 8.16 The EDAX window of a scanning electron microscope (SEM) contains a X-ray window (inset) that is less than 2 μm thick.

An even thinner diamond window may be used to mask photoresist exposed to X-ray lithography. The downsize trend of interconnected circuitry in ULSI (e.g., gigabytes of dynamic random access memory) will require line width to penetrate the point one (0.1 μm), such as 90 nm. Hence, the wavelength of current UV laser beams will have to move to the range of X-rays in order to achieve the required resolution. The strong mechanical strength, high thermal conductivity, and extreme radiation hardness have made diamond the undisputed first choice (exceeding boron nitride, silicon nitride, silicon carbide, and other ceramics) as the mask material.

The X-ray mask contains a diamond membrane that is mounted on a metal absorber (Fig. 8.17).

The X-ray mask is so thin (<2 μm) that it can be deposited without further polishing. For example, transparent diamond may be formed on copper substrate by a regular hot filament CVD process using methane and hydrogen (e.g., 99%). Alternatively, nanocrystalline diamond may be deposited by microwave CVD process using methane and nitrogen (e.g., 99%).

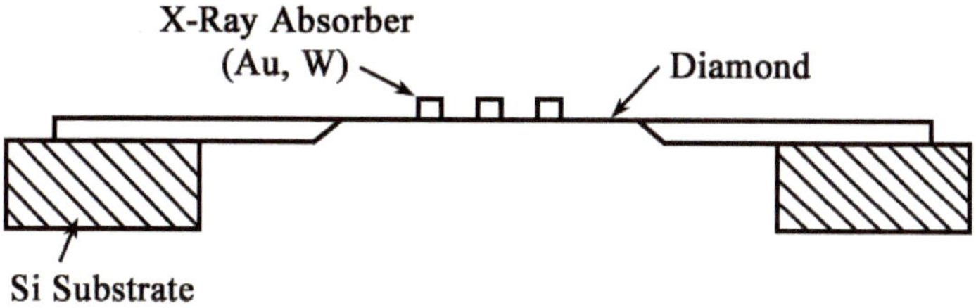

Figure 8.17 The X-ray mask in operation (top) and its cross section (bottom) in schematic.

The diamond X-ray windows described above are used in vacuum. At another extreme, at pressures higher than that at the Earth's center (375 GPa), diamond may be used as windows for measuring spectra of infrared, visible, X-ray window, or even γ-ray (e.g., Mössbaur effect). Because of its extreme compressive strength, diamond has been used as Bridgman anvils for ultra high-pressure research. For example, the color changes due to the phase transitions in AgI can be observed in real time under an optical microscope.

In addition, table salt has been used as an internal pressure gauge for a low pressure range, by measuring in situ diffraction patterns using X-ray (e.g., synchrotron radiation source), the pressure is reflected in the closing of its d-spacing. For higher pressures, the IR fluorescence spectra is used for pressure calibration. Numerous materials (minerals, ceramics, metals, polymers) have been studied at pressures of tens of GPa. There optical properties are measured by visible, IR or UV spectra, their crystal structures detected by X-ray diffraction or from Mössbaur spectra. In 1974, the author pioneered the research of using diamond anvils to study the kinetics of high-pressure phase transformations under both high pressure and high temperature. One extreme example for using diamond anvils is to squeeze hydrogen molecules (H_2) to form a metal. By pressurizing cryogenic solidified hydrogen molecules to extreme pressures, dissociated atomic hydrogen will assume closest packing with metallic characters.

Figure 8.18 The high degree of transparency in the entire range of electromagnetic radiation has made diamond anvils the most valuable research tools for studying ultra high-pressure phenomena.

8.7 Diamond Corrosion-Resistant Window

Diamond's another unparalleled property is its ability to resist all chemical attacks. Without the presence of oxygen, diamond cannot be etched by any other chemical below 1000°C. Even with oxygen around, diamond may resist the oxidation below 500°C.

The chemical inertness makes diamond the ideal window used in corrosive environments. For example, diamond window may be mounted on endoscopes to allow a doctor to examine a patient's stomach with high acid content. Alternatively, diamond windows can be used in corrosive environments (e.g., molten polymers, hot fluorine gas, hydrogen chloride fume, etc.) for temperature measurement, process monitoring or safety control.

An extreme example was to use a natural diamond window for Pioneer satellite to explore Venus (Fig. 8.19). The atmosphere of Venus is 200 times denser than that of Earth. The atmosphere contains a high concentration of sulfuric and nitric acids. Worse still, the temperature is as high as 500°C due to the runaway greenhouse effect. Only diamond could survive such corrosion for an extended period before the failure of the pioneer satellite itself.

Figure 8.19 Diamond window (18.2 mm in diameter) used during the exploration of Venus by Pioneer satellite.

8.8 Diamond Radomes

The development of supersonic apace vehicles (fighters, missiles, rockets, shuttles) requires an upgrade of radome materials that cover the nose cone for the flying object. In order for the vehicles to receive messages from the exterior, the radome must be highly transparent in the range of the radar wave. Moreover, the high speed of penetration in air would require the radome to be mechanically strong to withstand large g-force generated (e.g., at taking off) and shock waves created in air, wear resistant against the erosion of dust and rain drops that hit the radome surface at a speed faster than bullets (e.g., Mach 2), frictionally slippery to minimize the air drag and heat generation during high speed of penetration, thermally conductive to minimize heat accumulation, thermally stable to maintain its integrity at high temperature, thermal shock resistant (e.g., with very low thermal expansion coefficient) to tolerate sudden quenching of hotspots by ice cold rain or snow droplets, and radiation hard to retain its atomic lattice after bombardments of solar winds or cosmic rays without changing properties.

All existing radome materials (e.g., ZnSe) are relatively soft, so they lack many of the above desirable attributes, and hence, they will not be able to stand the high speed of penetration (e.g., beyond Mach 4) of future space vehicles. Diamond is the only material capable of meeting all these extreme challenges as it combines radar transparency with hardness, mechanical strength, surface slipperiness, thermal conductivity, thermal stability, thermal shock resistance, and radiation hardness. Hence, it is the ideal radome material that can allow space vehicles to move up penetration speed that is currently infeasible (e.g., Mach 10).

Modern CVD technology can grow freestanding diamond films to assume radome shapes (Fig. 8.20). Moreover, the surface finish can be smooth to sub-nanometer. Using a reversed casting technology the author invented (e.g., 2003 UK Patent GB2,366,804, Cast diamond tools and their formation by chemical vapor deposition), mirror finish diamond film with complicated shape may be formed without further requirement of polishing.

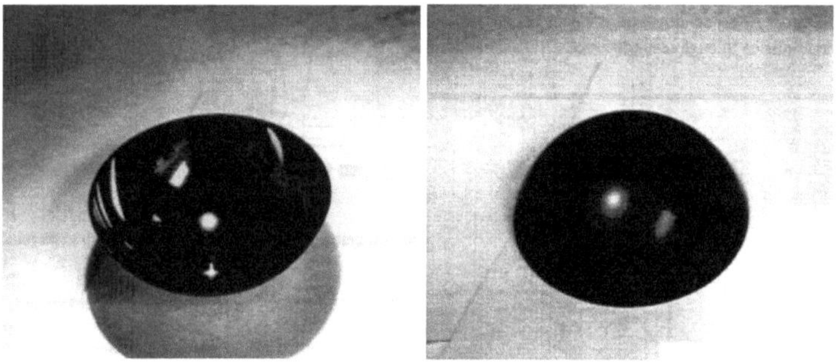

Figure 8.20 A radome of diamond film formed by CVD method.

8.9 High Power Laser Windows

High-power lasers (e.g., several kilowatts) are used to drill holes, cut metals, and weld alloys. Most of such lasers are CO_2 systems that emits 10.6 μm in wavelength. The commonly used ZnSe windows to exit the laser beam can only stand a power of a few tens kW. Even if it survives the absorbed heat, the high thermal expansion of the material and the thermal lensing effect due to the decrease of refractive index will defocus the laser beam. In addition to many desirable optical properties of diamond, its thermal expansion and thermal lensing are also the lowest of all commonly used window materials.

The absorption coefficient of a perfect natural single diamond (IIa) crystal can be as low as 0.04 cm^{-1} at the wavelength of 10.6 μm. The optical grade of polycrystalline CVD diamond film has attained an absorption coefficient of 0.07 cm^{-1}. However, such absorption is still 100 times higher than ZnSe. However, diamond is seven times stronger than ZnSe, the gap of absorption can be much narrower as a much thinner diamond may be used as window material than ZeSe. But the superiority of diamond window is demonstrated by its much better thermal properties.

Although diamond may still absorb more light than ZnSe, the heat is mainly coming from the anti-reflection coating of the window material, so the intrinsic absorption is not relevant. What is more important is the ability to dissipate heat, diamond's thermal conductivity is 120 times higher than that of ZeSe, so it will be much colder when used to exit high-power laser beams (Fig. 8.21).

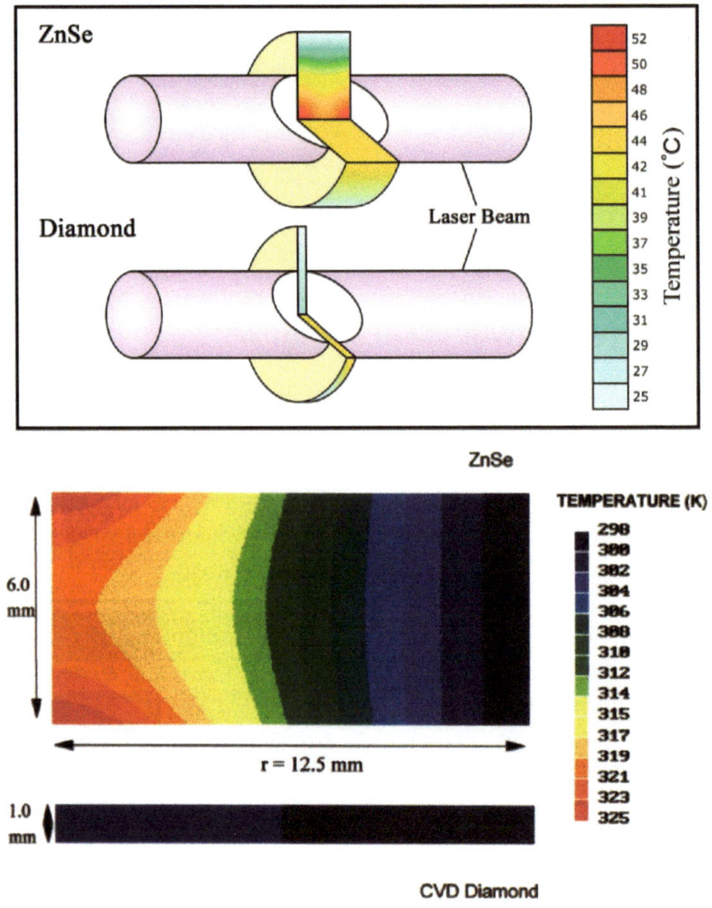

Figure 8.21 Diamond is a much cooler laser window due to its high thermal conductivity. The CO_2 laser with a power of 5 kW was used for the computer modeling (Element Six brochure).

The cold diamond has a thermal lensing effect that is more than a hundred times lower than ZnSe. Hence, laser beams can be much better focused by using diamond window.

Even if the thermal lensing is tolerated, the laser window will eventually crack due to the thermal stress created (Fig. 8.22). With the further increases of temperature, the crystal lattice of the window material will eventually melt or evaporate. Anti-reflection coated ZnSe cannot survive a laser power intensity of 0.1 kW/mm (energy divided by beam size), in contrast, diamond can tolerate at least two orders magnitude more of such power intensity. Thus, diamond window will be the only game in town for future generations of high power lasers.

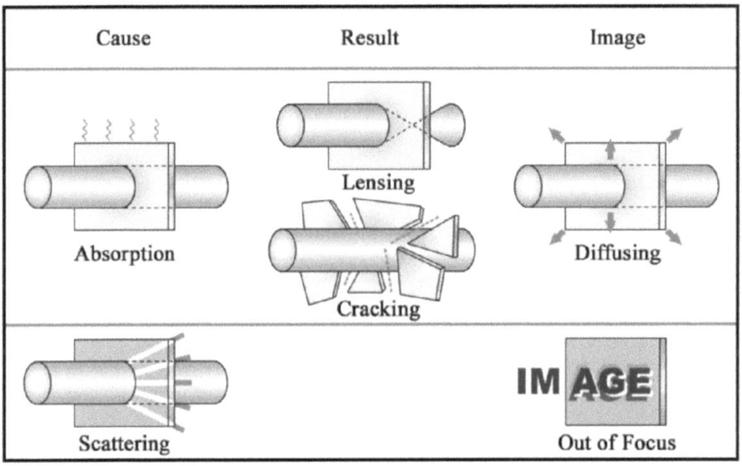

Figure 8.22 The absorption of laser beam by the window material may cause thermal cracking. Even if the window is still intact, its image will be blurred due to thermal lensing effect (top diagram). On the other hand, the scattering of light may cause double images (bottom diagram) (Element Six brochure).

8.10 DLC Optical Coating

Freestanding diamond windows, particularly those with a large size, are extremely expensive. Hence, certain optical applications may use a thin diamond coating to enhance the property of the window material. But diamond cannot be coated on optical materials due to the high processing temperature and poor adherence to the substrate. Hence, amorphous diamond-like carbon (DLC) is used to coat the optical material instead.

DLC contains both graphitic sp^2 bonds and diamond-like sp^3 bonds. It may be applied at very low temperatures (<150°C). In order to adhere well on the substrate, the coating is typically very thin (<1 μm). Because DLC contains graphitic bonds that may absorb light, the coating may be as thin as tens of nanometers.

DLC coatings are mainly used as a shield to protect the softer optical material (Fig. 8.23). In order to minimize the surface absorption, it is critical that the coating surface is very smooth. In many cases, this smoothness may be achieved by incorporating non-carbon elements (e.g., H, Si). The incorporation of non-carbon elements may also reduce the hardness. However, such reduction has a side benefit that is the adherence of the coating is enhance.

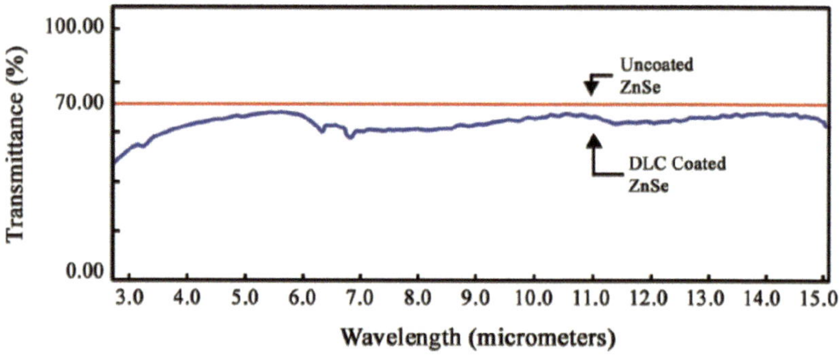

Figure 8.23 The DLC coating on ZnSe window may reduce the light transparency, but the softer substrate can be protected by the coating's hardness. This is a particular useful feature for night view IR glasses (e.g., for soldiers in duty).

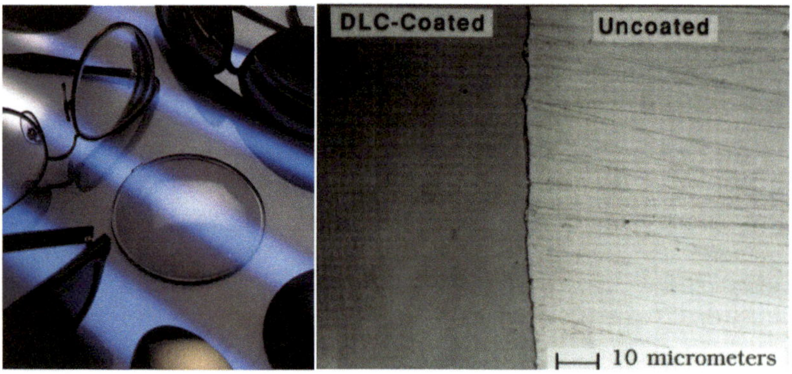

Figure 8.24 DLC coated eyeglasses (left diagram) and its ability to prevent these eyeglasses from scratching (right diagram).

DLC has also been used as a protective coating on sunglasses, eyeglasses, or optical lenses. Such a coating can indeed protect the soft substrate from scratching (Fig. 8.24). In addition to be a protective shield, DLC coatings are also used as an anti-reflection layer.

DLC may also be used to protect transparent glass on copying machine, scanner, or barcode machine. For example, DLC-coated glass cover on barcode machine can greatly extend its surface life (Fig. 8.25). However, most barcoders are now portable so the need to coat on glass panes is greatly reduced.

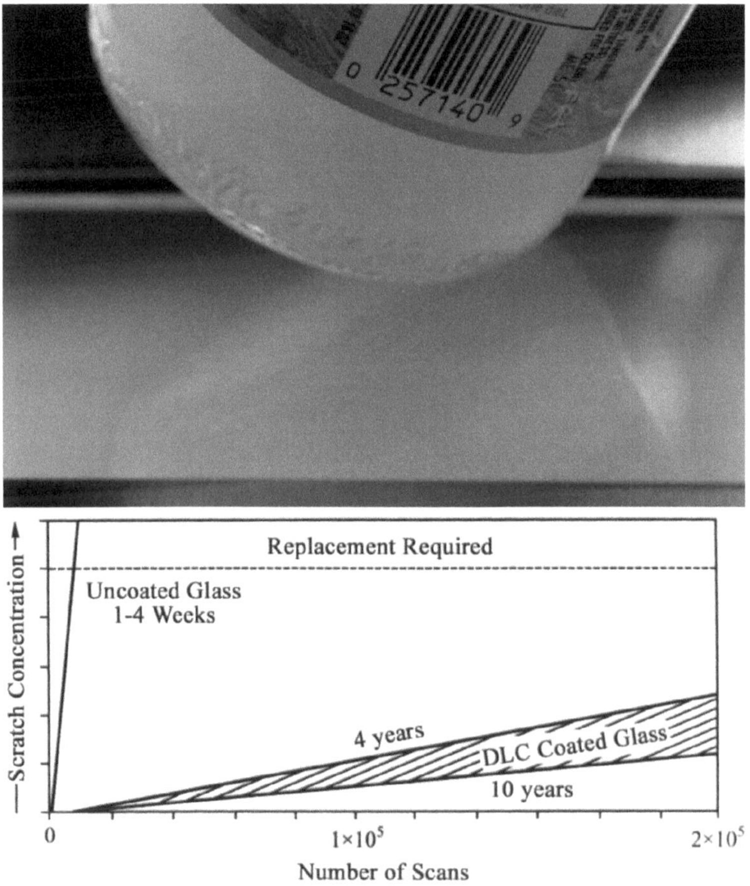

Figure 8.25 DLC protected barcoder's cover glass in action (top) and the dramatic extension of its surface life as noted (bottom).

Chapter 9

Diamond Sound Transmitters

Diamond can reproduce musical sounds with the highest fidelity. It may also transmit messages with the richest information.

9.1 Sound of Diamond

Acoustic waves are the manifestation of elastic deformation of lattice planes in succession. Such time-dependent movement of lattice planes may contain both information (e.g., signal, message, frequency) and entropy (e.g., noise, heat, phonon). The former may be exploited for sound transmission of music (e.g., by tweeter diaphragm) or wave transmission of message (e.g., by acoustic filter).

There are three modes of acoustic wave transmissions: primary wave (longitudinal wave, direct wave), secondary wave (transverse wave, indirect wave), and surface wave (Raleigh wave, circular wave). The primary wave is similar to the sound wave in air. It transmits by densifying and rarefying (push-pull) lattice planes in the direction of wave propagation. The secondary wave is like the light wave. It transmits by shifting laterally lattice planes

Handbook of Industrial Diamonds: Diamond Films and Carbon Coatings
James Sung

ISBN 978-981-4310-00-0 (Hardcover), 978-1-003-18045-6 (eBook)
www.jennystanford.com

perpendicular to the direction of wave propagation. The surface wave resembles water wave. It transmits between two media (e.g., water and air) with atoms moving in circular motion. In general, the primary wave travels about 1/3 faster than the secondary wave. The surface wave normally arrives the latest along the interface of two media.

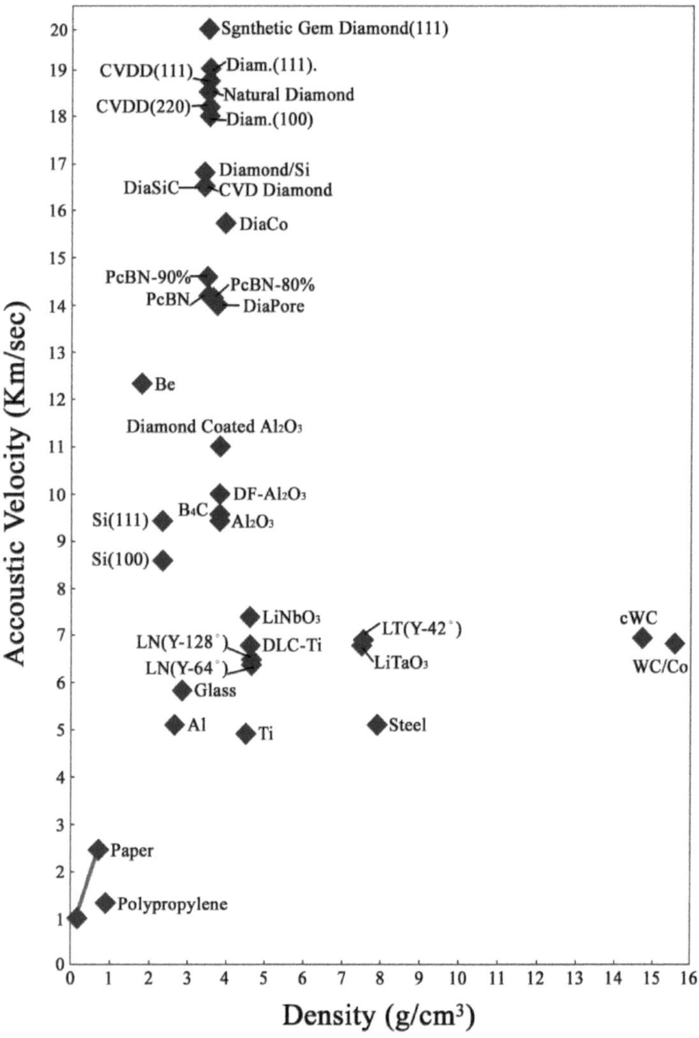

Figure 9.1 Sound speed (primary wave) versus density of various materials.

The speed of acoustic wave of a material is proportional to the square root of the material's modulus opposing to the atomic movement, but it is inversely proportional to the square root of the material's mass density. Diamond has the highest modulii (Young's, bulk, and shear) of all materials. Hence, it can transmit all three modes of acoustic wave the fastest (Fig. 9.1).

Diamond's primary wave is 18.5 km per second that is much faster than the sound wave in air (0.33 km/sec), steel (5.1 km/sec), alumina (10 km/sec). However, diamond's sound wave is not isotropic. It moves the fastest (19.0 km/sec) along (111) planes that contains the densest carbon atoms. In contrast, diamond's sound wave is much slower (18.0 km/sec) along (100) planes (Table 9.1).

Table 9.1 The sound speed of various types of diamonds and other materials

Sample	**V_L (km /s)**	**ρ (g/cm^3)**
Synthetic gem diamond (111)	20.0	3.5
Diamond (111)	19.0	3.5
CVDD (111)	18.8	3.5
Natural diamond	18.5	3.5
CVDD (220)	18.2	3.5
Diamond (100)	18.0	3.5
Diamond/Si	16.8	3.4
CVD diamond	16.5	3.4
DiaSiC	16.5	3.4
DiaCo	15.7	3.9
PcBN-90%	14.6	3.4
PcBN-80%	14.2	3.6
PcBN	14.2	3.5
DiaPore	14.0	3.7
Be	12.3	1.8
Diamond-coated Al_2O_3	11.0	3.8
DF-Al_2O_3	10.0	3.8
B_4C	9.6	3.8

(Continued)

Table 9.1 (*Continued*)

Sample	V_L (km /s)	ρ (g/cm^3)
Al_2O_3	9.4	3.8
Si (111)	9.4	2.3
Si (100)	8.6	2.3
LN (Y-128°)	7.5	4.6
$LiNbO_3$	7.4	4.6
LN (Y-64°)	7.4	4.6
cWC	6.9	14.7
WC/Co	6.8	15.6
$LiTaO_3$	6.8	7.5
LT (Y-42°)	6.8	7.5
DLC-Ti	6.8	4.6
Glass	5.8	2.8
Steel	5.1	7.9
Al	5.1	2.7
Ti	4.9	4.5
Polypropylene	1.3	0.9
Paper	1.0~2.4	0.2~0.7

Sample	Abbreviation	Hole size (μm)	Diamond (V%)	Binder phase
Element Six Syndax-3	DiaSiC	22	80	Si, SiC (20%)
Syndite	DiaCo	25	95	Co (5%)
AMB 90	PcBN-90%	10	90	AlN, AlB_2 (10%)
DBA 80	PcBN-80%	6	80	TiN, Al_2O_3 (20%)

9.2 The Tweeter Diaphragm

Diamond's atomic bonding has only one type (i.e., sp^3) and its lattice structure extremely simple. Consequently, diamond's sound is very "pure" with the highest peak to noise ratio. In addition, diamond's high mechanical strength can allow it to be very thin. The small volume coupled with its low density has made diamond diaphragm so light that it can be accelerated by the electromagnetic signals with very little energy.

In addition, diamond's high hardness makes the diaphragm last for a very long time, its low friction means less heat is generated during vibration. The high thermal conduction will also keep it cold and hence little thermal stress is created.

The synergism of pure sound, lightness, coldness, and longevity makes diamond film the ultimate tweeter diaphragm for speakers.

The human hearing frequency ranges from 40 Hz to 16 kHz. Conventional tweeter diaphragms are made of alumina or titanium that can vibrate at a frequency of up to 20 kHz. As early as in 1987 Sumitomo Electric and Victor pioneered the coating of DLC on alumina (Fig. 9.2). The results were the extension of frequency to about 35 kHz, which is audible by dogs.

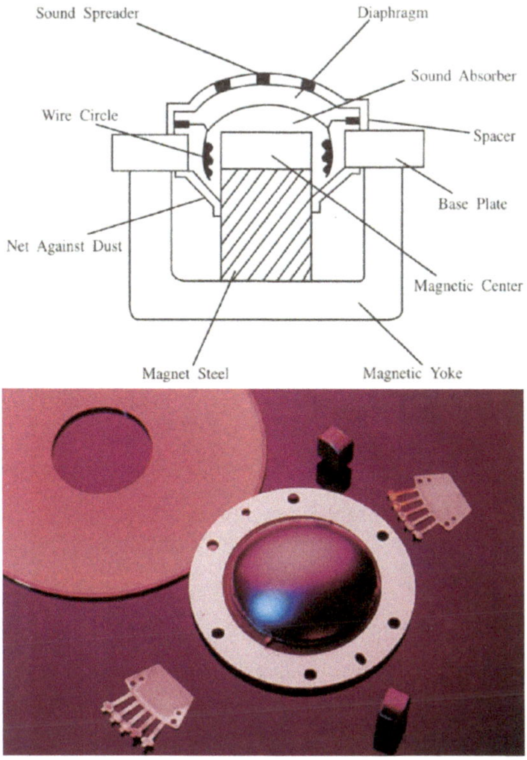

Figure 9.2 The structure of a tweeter (top) and a DLC-coated diaphragm (bottom) (Diamonex brochure).

Although the DLC coating is effective to extend the frequency range of reproduced sound, but the ultimate diaphragm is made of diamond itself. Using a reverse casting technology (e.g., 2003 UK Patent GB2,366,804, Cast diamond tools and their formation by chemical vapor deposition), freestanding diaphragms have been made without further requirements of mechanical finishing.

Figure 9.3 Tweeter diaphragm made of a free-standing diamond film deposited by a reverse casting technology invented by the author (Kinik product).

The free-standing diamond film tweeter may extend the frequency range beyond that achieved by DLC-coated ceramics and reach 80 kHz, which is far beyond what human ears can hear (Fig. 9.4). However, the ability of the diamond diaphragm to reproduce sound with such a high frequency means that the human ears will have the best sound at lower frequencies.

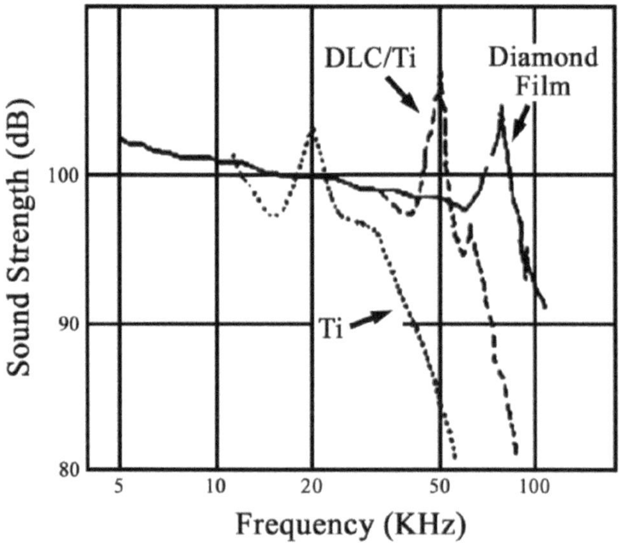

Figure 9.4 The frequency ranges of tweeter diaphragm made of different materials. The final surge of intensity near the high-frequency end before its decline is due to resonance.

9.3 SAW Filters

With the advancement of computer and Internet technologies, human civilization has entered the information age with gluts of information coming from all places. However, the transmission of these vast amounts of information still relies on connected wires, cables, and optical fibers, which is very cumbersome. Hence, new communication techniques (e.g., Bluetooth, WLAN) are now being implemented to phase out all connecting wires, beginning in office and at home. These wireless communications require specific frequencies, so filters are needed to gate the communication channels.

With the computer becoming ubiquitous, the future computation will become amorphous. In this regard, we do not come to computers to retrieve information, but the information will come to us wherever we go. Thus, the wireless cell phones will also function as computer, Internet, television, multimedia, etc. This means that most data in the future will be transmitted in air, and most likely they will be relayed by satellites.

In the language of wireless communication, frequency means information. Current cell phones receive and transfer mostly vocal messages that can be accommodated by low frequencies (e.g., 2.4 GHz at 10 Mbps), but with the requirement to transmit pictures or movies, much higher frequencies (e.g., 5.8 GHz at 50 Mbps) will be needed to speed up the transmission. Hence, the wave filters must increase their frequency.

Although there are several ways to filter frequency (e.g., resistor–capacitor combination, multiple layer ceramics), but the simplest design is surface acoustic wave (SAW) filter. SAW filter utilizes a thin layer of piezoelectric ceramic material that allows only the frequency of resonance to pass. The construction is simple and the device is small. Moreover, it is fully compatible with the existing thin film technology widely employed to make semiconductor devices. As a result, SAW filters are used extensively in many wireless electronics (Fig. 9.5).

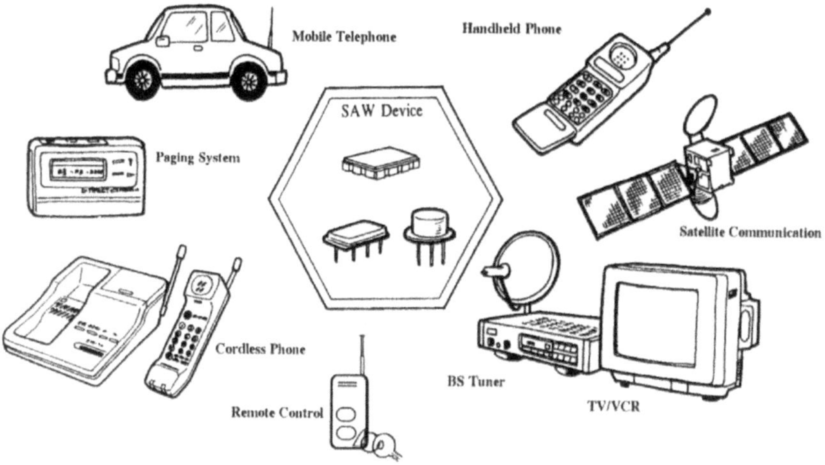

Figure 9.5 The omnipresence of SAW filters in electronic devices.

Common piezoelectric ceramics used in SAW filters today are quartz, ZnO, $LiTaO_3$(LT), $LiNbO_3$(LN). For example, there are at least four such SAW filters installed in each cell phone. The total number of filters produced worldwide each year is now approaching the world population (7 billion).

The SAW filter is operated by converting an electromagnetic signal into mechanical vibrations. This vibration must be in tune with the frequency preset otherwise it will be filtered out. The frequency (f) allowed to pass is determined by SAW speed (V) of the piezoelectric material and the wavelength (λ) as follows:

$$V = f\lambda, \text{ or } f = V/\lambda$$

The wavelength of the surface vibration is controlled by a pair of interdigital tranducers (IDT). The IDT set is typically made by depositing a thin (e.g., 50 nm) layer of aluminum (e.g., by PVD sputtering) and subsequently etching (e.g., by photolithography) it away to form a specific pattern. The pair of IDT will alternate their polarity upon intercepting the electromagnetic wave. The wavelength is determined by the periodicity of the pattern for each IDT, or half the periodicity of both IDTs.

Since the SAW speed is fixed for each piezoelectric ceramic, the frequency filtered is proportional to the spacing of IDT, so the closer the spacing, the higher the frequency. However, the spacing of IDT is limited due to the mutual interference of signals, the insulating property of the ceramic, and the power of device. Thus, with the narrowing of the spacing between IDT, the signal will be distorted and electrical current will leak. Eventually, the temperature becomes too high, so the fragile aluminum IDT melts down.

Since there is a limit to increasing the frequency by narrowing the spacing of IDT, the alternative is to use a piezoelectric material that has a much higher SAW speed. In order to be piezoelectric, the deformation of the crystal lattice must allow shear movement and there must be ionic atoms that show different electrical potentials with respect to various degrees of shear movement.

As diamond has the highest SAW speed, it would be ideal to be used as SAW filter. However, diamond's lattice is isometric so its elastic deformation is symmetrical. Consequently, there is no shearing component for atomic movements. Moreover, carbon's bonding is covalent that does not respond to external electromagnetic signal. Hence, diamond is not a piezoelectric material.

In order to make use of diamond as a sound transmitter for making SAW filter, the surface of diamond must be coated with a thin layer (e.g., <1 μm) of a piezoelectric ceramic.

Sumitomo Electric pioneered the design of diamond SAW filter by using a sputtered ZnO layer as the piezoelectric ceramic (Fig. 9.6). Because ZnO is only used to convert electromagnetic signal into mechanical vibrations, it can be very thin (e.g., <1 micron). Moreover, surface wave cannot penetrate much deeper than about 5 wavelengths, so the diamond thickness may not be much (e.g., <20 microns).

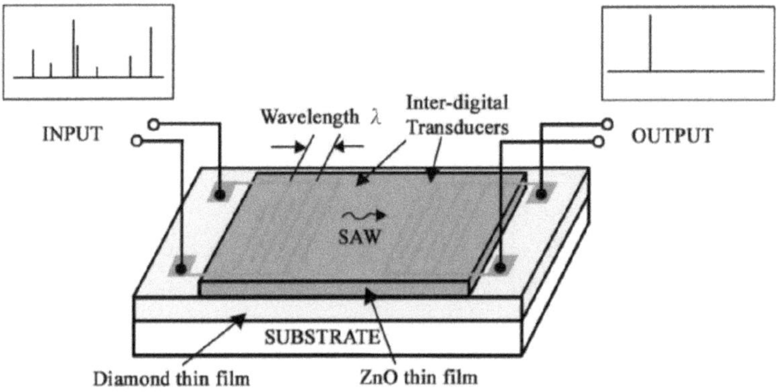

Figure 9.6 Diamond/ZnO SAW filter as designed by Sumitomo Electric.

A process of making diamond/ZnO SAW filter design was developed with the final product dimensions as shown in Fig. 9.7. The process starts with using a 100 mm boron-doped (P type) silicon wafer in the (100) orientation. The resistivity of the wafer is 1–10 ohm-cm. The wafer was deposited with a diamond film by hot filament CVD with a gas mixture of 1% CH_4 diluted in H_2. The atmosphere had a pressure set at 80 torr. The CVD diamond film was polished to a mirror finish (Ra = 3 nm). Subsequently, ZnO was RF (150 W) sputtered on the finished diamond substrate heated at 200°C. The target was zinc that was sputtered in an atmosphere with 60% Ar and 40% O_2. The pressure was set at 0.02 torr. Later the sputtered ZnO was deposited with aluminum by DC sputtering. The deposited layer was etched after photolithography to form IDT. Finally, the SAW filter was packaged in SiO_2 by depositing with LPCVD.

The important attributes of SAW filters are high coupling coefficient, which allows effective conversion of electromagnetic signals into mechanical vibrations and vice versa; low insertion loss, which minimizes the energy loss due to such conversions; small phase slope, which defines a sharp borderline between passing signal and filtered noise; low temperature-dependent performance, which avoids the shift of frequency due to temperature changes. The diamond/ZnO SAW filter has met all these criteria as shown in Table 9.2.

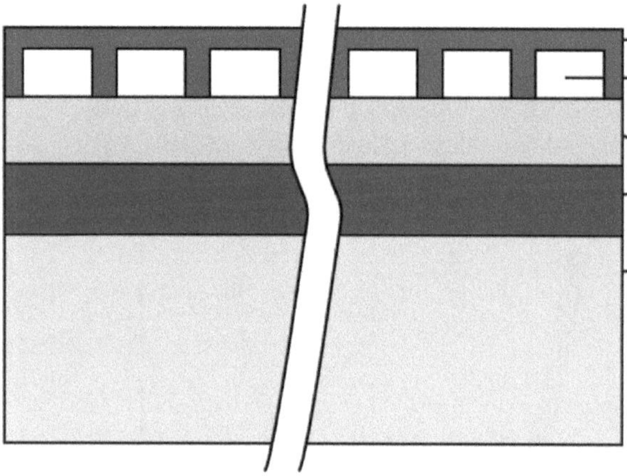

Figure 9.7 The simulated design of diamond/ZnO SAW filter (not to scale).

Table 9.2 Comparison of SAW filter performance

Attribute	Diamond/ZnO*	Quartz
Insertion loss (dB)	8	14
Size (mm^3)	25 (3.8 × 3.8 × 1.7)	120
T Coefficient (PPM)	120 (–40 to 85°C)	170
Phase slope (deg/kHz)	-0.04	-0.07

*Sumitomo Electric's SumiWave® product (2.488, 2.666 GHz).

Sumitomo Electric introduced diamond/ZnO SAW filter in 2002 with 3 inch wafer and diced size of 3.8 mm (Fig. 9.8).

Because of the coupling of diamond with a much higher SAW speed, the frequency range of diamond/ZnO is much higher than that for ceramic filters (Fig. 9.9).

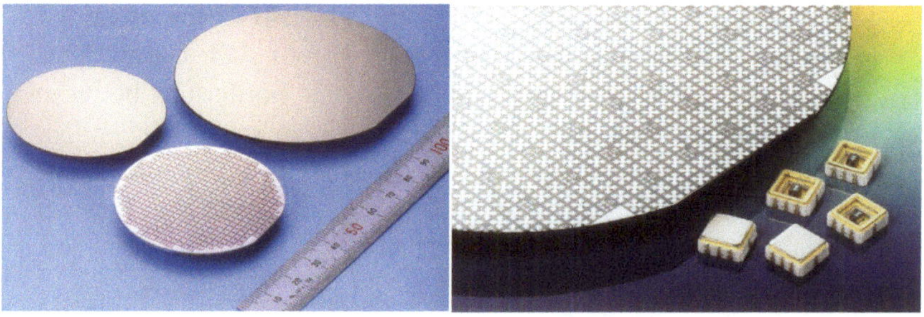

Figure 9.8 The diamond/ZnO SAW filter introduced by Sumitomo Electric (SumiWave® products). The wafers are 3 inch (76 +/−0.3 mm) with a total thickness of 0.8 mm. Diamond thickness is >12 microns, Wafer bow <45 microns, and ZnO thickness uniformity <4% (left).

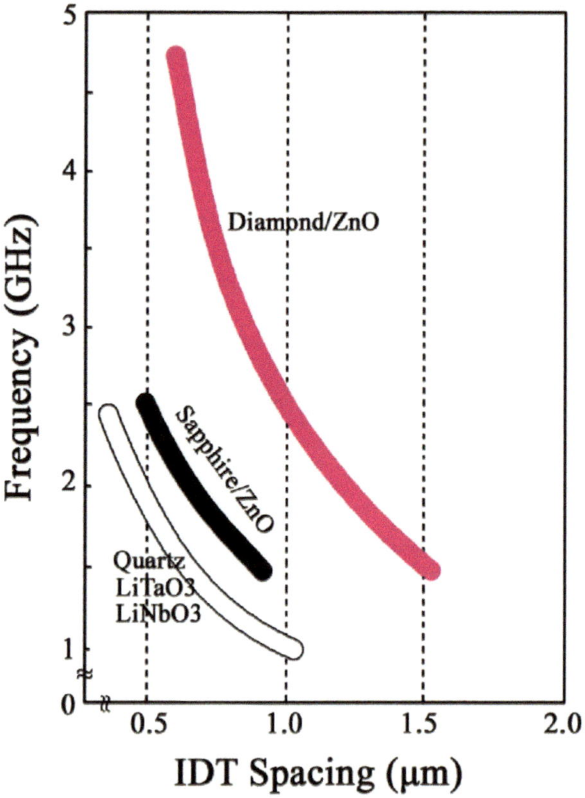

Figure 9.9 The working range of frequency of diamond/ZnO SAW filters in comparison with conventional designs.

The diamond/ZnO SAW filter can define a well-filtered signal (Fig. 9.10).

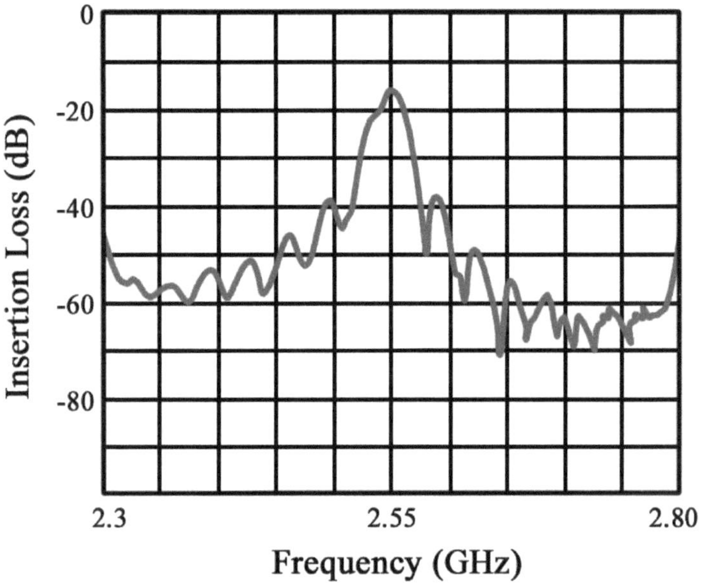

Figure 9.10 The frequency profile passed a diamond/ZnO SAW filter.

The high breakdown voltage of diamond can ensure a more reliable performance over a much longer time frame of diamond/ZnO SAW filter (Fig. 9.11).

One of the side benefits of using diamond as the booster of elastic wave is that diamond can also serve as a heat spreader. This is critical as the power of SAW filter will increase with the increase of communication distance. Without an effective heat spreader as the current designs of ceramic SAW filters, the IDT will soon be burnt out. In contrast, diamond/ZnO SAW filter can sustain a much higher power without overheating due to the fast dissipation of heat by diamond (Fig. 9.12).

Another merit of using diamond/ZnO is their low temperature coefficient of frequency shift (Fig. 9.13). This is important as travelers want to be connected by cell phones or other telecommunication devices from equator to poles, which span a large change of temperatures.

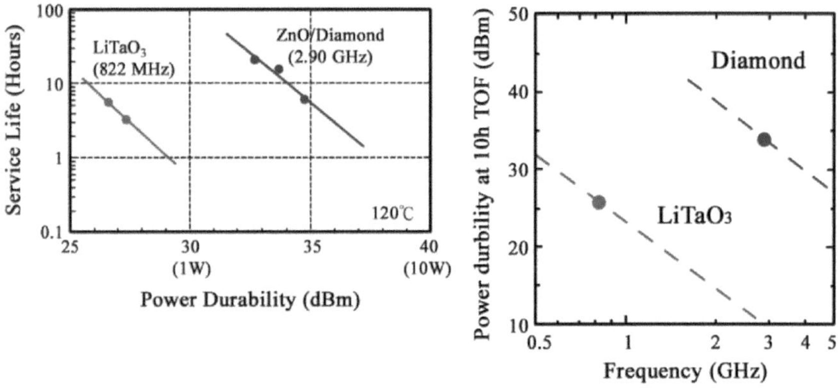

Figure 9.11 The reliability of diamond/ZnO SAW filter as a function of power (left). Frequency dependence of power durability with accelerated test at 120°C during 10 h TOF (dBm) (right) (Sumitomo Electric brochure).

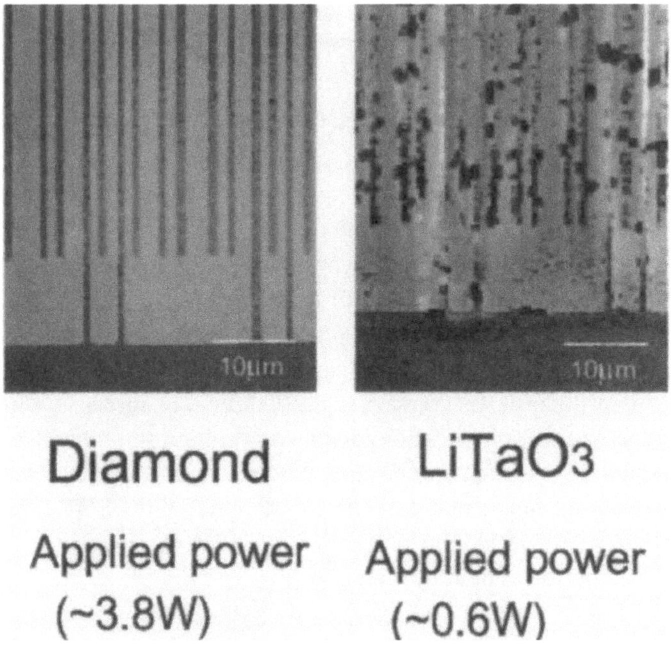

Figure 9.12 Diamond/ZnO SAW filter can support a high-power operation (left), but conventional ceramic SAW filter will break down even if run with modest power (right).

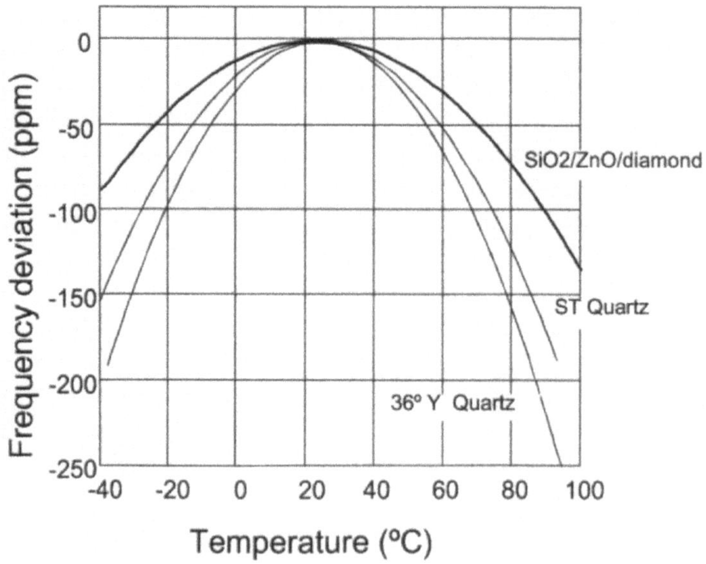

Figure 9.13 The sensitivity shit of communication frequency with temperature. The total shift of frequency from –40 to 85°C is 90 PPM for SiO_2/ZnO/diamond SAW filter that can transmit wave at 9000 m/s; 148 PPM for ST quartz, at 3160 m/s; and 195 PPM for 36° Y quartz, at 5100 m/s.

9.4 Advanced Saw Filters

Although diamond/ZnO SAW filters can outperform conventional ceramic SAW filters in frequency, power, life, and temperature deviation, but there are limitations that can be further improved. For example, ZnO has a very low coupling coefficient (1%), so the acoustic wave generated at the diamond interface is very weak. This problem is compounded by the fact that ZnO is a polycrystalline film, although the crystals have a preferred orientation with the hexagonal planes parallel to diamond's surface, the orientation is not perfect. Moreover, the transfer of mechanical energy to diamond at the interface is also weak. This is because strong chemical bonds are not formed between ZnO and diamond, so the energy transfer must rely on the weak van der Waals force.

However, the largest difficult of using diamond/ZnO SAW filter lies in the polishing of CVD diamond film deposited on the silicon wafer. Although CVD diamond deposition is expensive, the polishing of extremely hard diamond is costlier. Even if polishing can be accomplished with brutal force, the surface finish is not super smooth (Ra = 3 nm). The surface roughness can scatter the surface wave and this damping effect is more pronounced with high wave speed propagated along the diamond/ZnO interface.

A new way of making diamond SAW filter is to grow diamond film directly on $LiNbO_3$(LN) surface, which has a super smooth finish (Ra = 3 Å). LN has a much higher coupling coefficient (6–9%), so the insertion loss is minimized. In addition, it is a single crystal, so the grain boundary effect no longer exists. The availability of such a diamond/LN SAW filter will drastically improve the efficiency of wireless communications (Fig. 9.14).

In addition to use polycrystalline diamond as the sound transmitter, single crystal diamond can have a higher sound velocity at the same time with much more coherent vibrations. In other words, the signal-to-noise ratio of filtered wave can be much higher. Thus, diamond SAW filter using (111) plane for sound transmission can produce most clear sounds at highest frequency (Fig. 9.15).

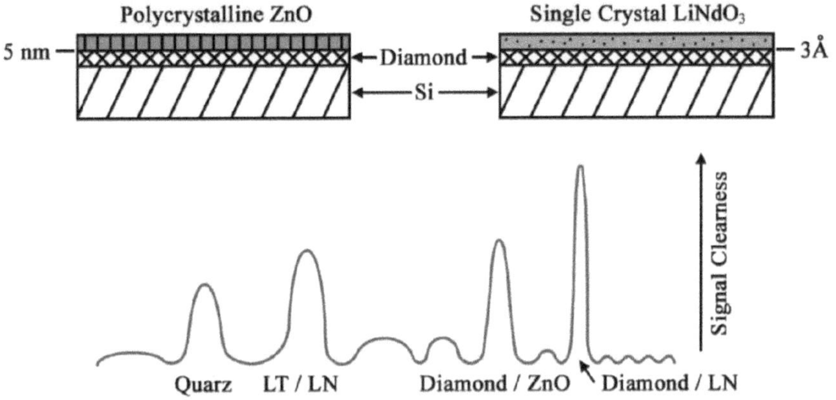

Figure 9.14 The new design of diamond/LN SAW filter can improve the signal-to-noise ratio of the filtered message at very high frequencies.

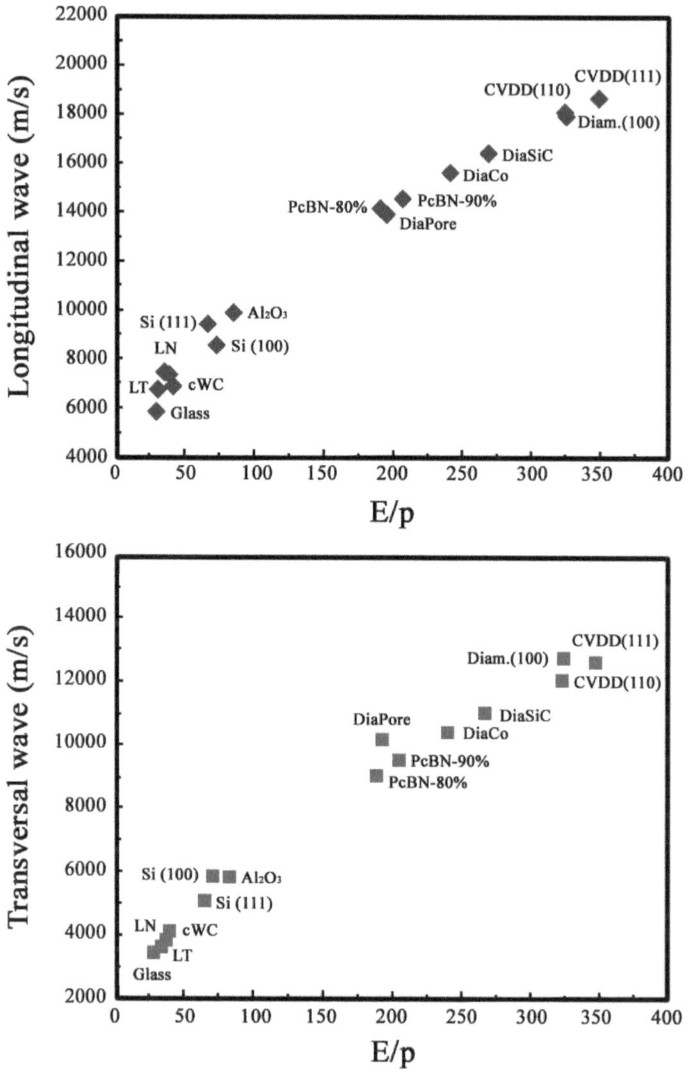

Figure 9.15 The velocities of longitudinal (top) and transverse (bottom) sound waves for various diamond materials. Note that single crystal diamond with either (111) or (100) orientation has the highest sound velocity of all materials.

The elastic properties of various diamond materials are compared in Table 9.3.

Table 9.3 The elastic properties of various diamond materials

Sample	E (Gpa)	B (Gpa)	ν
Diam. (111)	—	—	—
Diam. (100)	1143	372	0.013
CVDD (111)	1223	481	0.07
CVDD (220)	1136	470	0.09
DiaSiC	904	367	0.08
DiaCo	932	388	0.09
DiaPore	711	206	0.07
PcBN-90%	704	311	0.12
PcBN-80%	687	331	0.15
Al_2O_3	330	208	0.24
Si (111)	157	126	0.29
Si (100)	170	64	0.06
cWC	615	360	0.22
LN (Y-128°)	166	176	0.34
LN (Y-64°)	181	159	0.31
LN (Y-42°)	234	223	0.32
Glass	73	45	0.23

Chapter 10

Diamond Semiconductors

Diamond is the ultimate material for all future computations that include semiconductors.

10.1 The King of Semiconductors

Semiconductor is the enabler of computation and information, the newest era of human civilization. Diamond structure is the common base for semiconductors that include Si, GaAs, and SiC, the most advanced semiconductors used today. Diamond itself has most extreme properties for semiconductors of the future. Among them are fast movement of electrons and holes, highest thermal conductivity for thermal dissipation, lowest leaking current, and highest breakdown voltage for high power applications, all bottleneck-breaking necessities for making the ultimate semiconductors, and the fastest computers in the future.

The current use of Si as the base material for semiconductor was not chosen for its performance, but adapted because of

Handbook of Industrial Diamonds: Diamond Films and Carbon Coatings
James Sung

ISBN 978-981-4310-00-0 (Hardcover), 978-1-003-18045-6 (eBook)
www.jennystanford.com

convenience from the first Ge semiconductor. The branching into GaAs semiconductor is for a certain niche (e.g., high speed) applications that are with limited potentials. The moving toward SiC is an intermediate step following the natural path of most logical semiconducting materials, from Ge, Si to diamond (Fig. 10.1). These natural semiconductors are pure covalent compounds that differ from ionic semiconductors of III–V Groups (e.g., GaAs, GaN). The ionic semiconductors have a high leaking current and low breakdown voltage, so they are not suitable for making the future semiconductors of high power and high frequency. In contrast, diamond, the crown jewel of the Group IV elements, is located at the summit of pure covalent semiconductors. It will be the ultimate semiconductor for the advancements of computation power and speed.

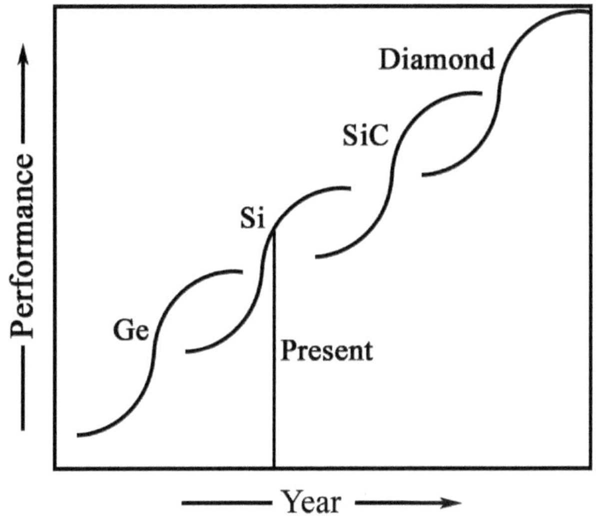

Figure 10.1 The incremental evolution of semiconductor's performance relies on Moore's Law of process optimization. The quantum revolution of semiconductor's performance depends on material's intrinsic capabilities.

Although the progress of Si semiconductor has been impressive following the geometrical increments of Moore's Law, its continuation of in advancing this fractal pattern will reach a limit in next decade. By then, it can no longer handle the high power input and high frequency output required for high-speed computation, nor can it dissipate the heat fast enough for high temperature applications and reliable performance. These problems can only be solved by diamond with its intrinsic capabilities designed for channeling power, frequency, voltage, temperature, and reliability, all critical for a semiconductor's performance (Fig. 10.2).

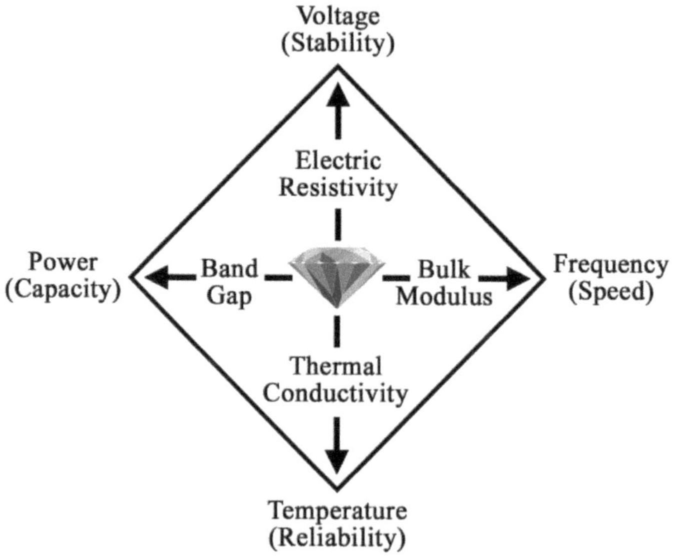

Figure 10.2 Diamond's royal properties and its semiconductor capabilities.

Current leading semiconductors (Si, GaAs, SiC) are no match to diamond as advanced semiconductor in many key attributes (Table 10.1). For example, as transistor for amplifying the power of radio frequency (R. F), the prime indicator for performance is the output power. This power is determined by the product of load impedance and the square of frequency. Based on this figure of merit, diamond is 8200 times of Si, 500 times of GaAs, and 8 times of SiC.

For comparing demanding applications of semiconductors in high power, frequency and temperature, certain combinations of material characters are required, these figures of merits all confirm diamond's king nature of semiconductors (Table 10.2).

Table 10.1 The comparison of semiconductor properties

Property	Diamond	b-SiC	GaAs	Si
Lattice constant	3.567	4.358	5.65	5.430
Thermal expansion coefficient	1.1	4.7	5.9	2.6
Density	3.515	3.216	5.27	2.328
Melting point	4000	2540	1238	1420
Band gap	5.45	3.0	1.43	1.1
Saturated electron speed	2.7	2.5	1.0	1.0
Charge mobility				
Electron	2200	400	8500	1500
Hole	1800	50	400	600
Critical electrical field	100	40	60	3
Dielectric constant	5.5	9.7	12.5	11.8
Electrical resistivity	10^{23}	150	10^8	10^3
Thermal conductivity	20	5	0.46	1.5
Absorption edge	1.2	0.4		1.4
Refractive index	2.42	2.65	3.4	3.5
Hardness	10000	3500	600	1000
Johnson's figure of merit	73856	10240	62.5	9.0
Keyes' figure of merit	444	90.3	6.3	13.8

Table 10.2 Figures of merit for demanding semiconductor applications

Figure of merit	Johnson	Keyes	Baliga
Equation*	$\propto E_b v_{sat}$	$\propto \lambda \left(\frac{v_{sat}}{\varepsilon}\right)^{1/2}$	$\propto \varepsilon \mu E_b^3$
Description	Power handling of high-frequency devices	Thermal dissipation of high-frequency devices	Power handling of low-frequency devices
Si	1	1	1
GaAs	11	0.45	28
SiO–6H	260	5.1	90
GaN	790	1.8	910
AlN	5,120	2.6	31,670
Diamond	6,220	32	11,700

*E_b, electric breakdown field; v_{sat}, saturated carrier velocity; λ, thermal conductivity; ε, permittivity; μ, carrier mobility.

In addition to be superior in many semiconductor attributes, diamond's thermal stability (e.g., >1000°C versus Si's <100°C) and thermal conductivity (e.g., >10 times of Si) are also much higher than all other semiconductors.

10.2 Semiconductor Properties of Diamond

Diamond's high resistivity and high breakdown voltage stem from its strong covalent bonds. In order to compensate for the decrease of the energy for bonding valence electrons, its anti-bonding conduction orbitals are pushed much higher. As a consequence, diamond has the highest band gap, i.e., the energy difference between conduction orbitials and bonding orbitals, of all covalent structures. This band gap (5.45 eV) is much higher than that of Si (1.2 eV) (Fig. 10.3). The high band gap makes diamond the best insulator material, so it can withstand the highest voltage difference without leaking electrical current.

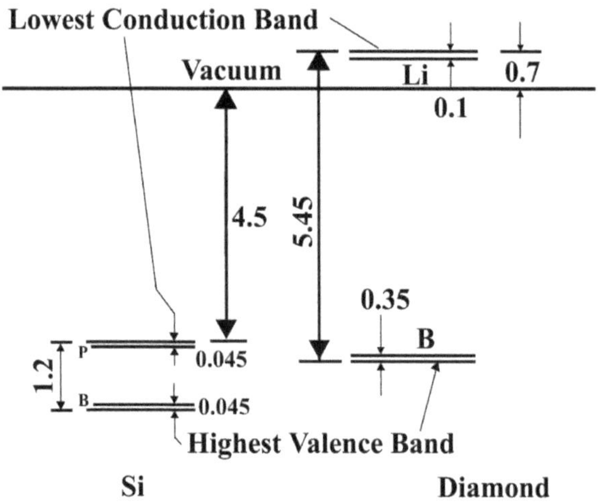

Figure 10.3 Electron orbital energies (eV) of Si and diamond. The energies of vacuum and dopants are labeled.

The high band gap allows diamond to minimize the leaking current due to its lowest concentration of intrinsic carrier. This is the most important attribute for high power and high frequency semiconductors. The low intrinsic carrier concentration can allow diamond to maintain its integrity at the highest operation temperature, so it will not lose its semiconductor property and become a metallic conductor at high temperature (Fig. 10.4).

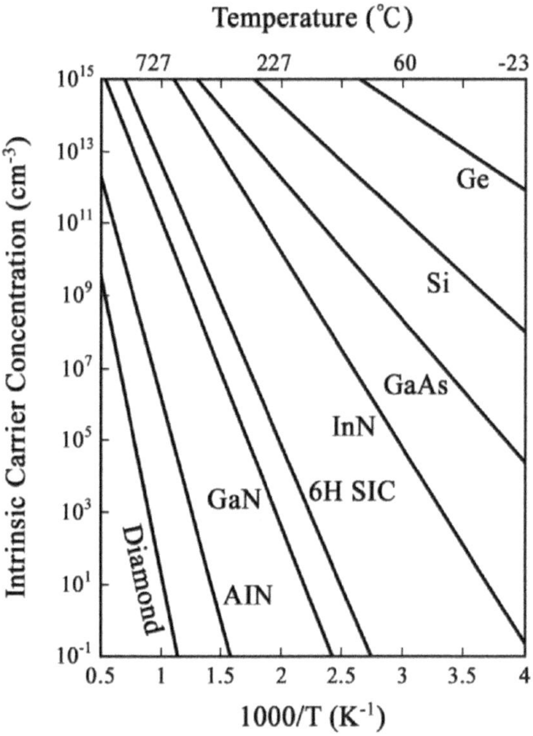

Figure 10.4 Diamond's intrinsic carrier concentration is much lower than all other semiconductors. Intrinsic carrier concentration determines the leaky current density and the threshold temperature for semiconductor to become metallic conductor. Diamond has the lowest leaky current density, so it can stand the highest temperature without losing its semiconductor status. Then intrinsic carrier concentration (n_i) is self-generated in a perfect crystal lattice due to temperature perturbation. The perturbation excites electrons in the valence band, moves them to conduction band, and leaves behind holes in the valence band. So $n_i = (N_e N_v)^{1/2}\exp(-E_g/kT)$, where N_e and N_v are electron density in conduction band and hole density in valence band, respectively. E_g is the band gap. K is Boltzmann's constant. T is the absolute temperature.

Because diamond semiconductor is virtually free from tunneled electrons agitated by heat, it can operate at much higher limits of voltage and frequency (Fig. 10.5). Moreover, diamond semiconductor also has the highest limits of the maximum power and the frequency of operation (Fig. 10.5).

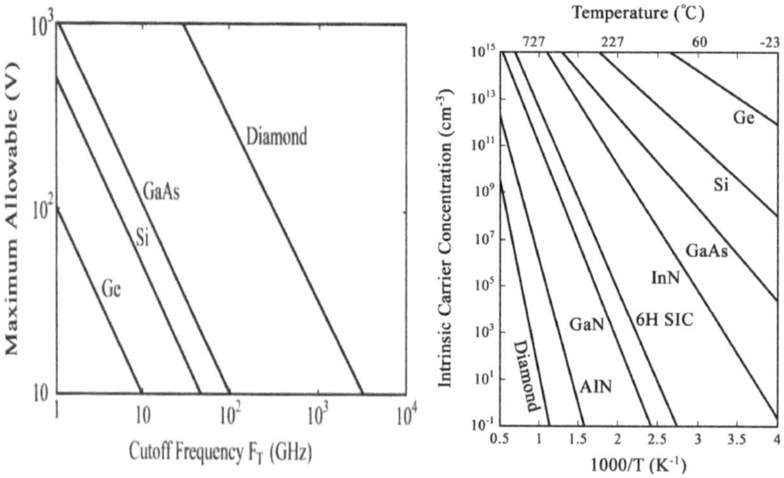

Figure 10.5 The inverse relationship between maximum allowable voltage and cutoff frequency for semiconductors (left). Note that diamond's tolerance limits are tens or hundreds times higher than other semiconductors. The inverse relationship between output power and radio frequency for transistors is shown (right). Note that diamond's tolerance limits are also tens or hundreds times higher than other semiconductors.

Not only diamond can tolerate a higher electrical field without losing its integrity, but also it can perform better under high electrical field. For example, at high electrical field (e.g., >50,000 V/cm), diamond's saturation electron velocity is twice that of other semiconductors (Fig. 10.6).

Although diamond's semiconductor properties far exceed any other semiconductors, but because it is a late comer, so unless the mature technology failed to optimize the performance of existing semiconductors, diamond will not be looked at for a solution.

As diamond semiconductor is still in the developmental stage, both its learning curve and high cost need to work out before commercial applications can take place. The first products may have to be in the niche areas where current semiconductors failed to address, such as microwave transistors, laser diodes, and high frequency communication devices using optical fibers or wireless radios.

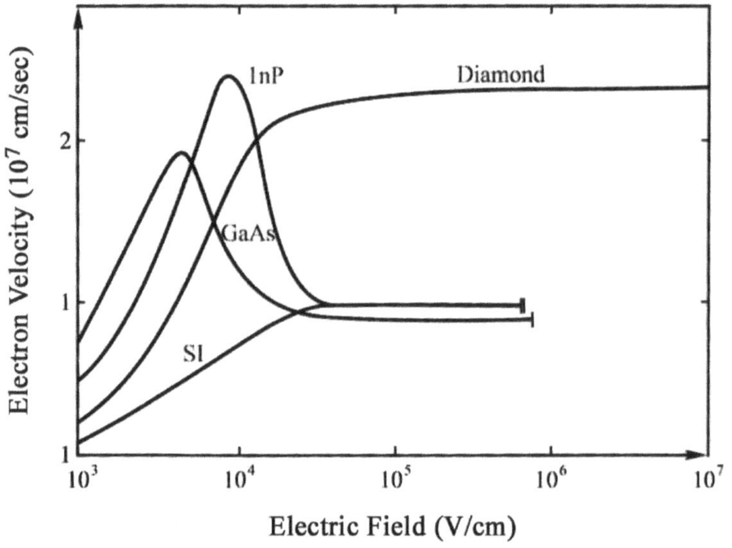

Figure 10.6 The electron velocity as a function of applied electrical field.

Another strength of diamond is that its high bond energy can tolerate the bombardment of high dose of energetic particles without damaging its crystal lattice. This radiation hardness is lacking for all other semiconductors that possess weak bonds of their atomic lattice. Hence, diamond semiconductor is ideal for using as radiation detectors.

As for the widespread applications of diamond semiconductors, one must wait for the derailment of Moore's law, which has been followed by the advancement of silicon technology, and the success of finding the right doping technology for making diamond semiconductors. Both the ending of Moore's Law for the advancement of silicon technology and the breakthrough for making diamond semiconductors may take more than a decade. By then, the transition of Si or SiC to diamond will become inevitable as will the arrival of diamond age as the climax of human civilization.

10.3 The Doping of Diamond

In order to be a transistor, a positively doped diamond and a negatively doped diamond must be joined together. In making the doped semiconductor, diamond must be free from defects that include grain boundaries. Hence, single-crystal diamonds of high purity are needed. Although gem diamonds may be available, they are expensive. Moreover, they are not compatible to the thin film technology that is needed to make layers of circuitry separated by insulators.

Although CVD diamond films are now routinely deposited, most such films are polycrystalline that contain large concentrations of various types of defects, in particular, grain boundaries. These defects can slow down or even stop electrons or holes. Hence, polycrystalline CVD diamond films have very low electron and hole mobilities. However, with proper orientation of nuclei (e.g., by hetero-epitaxial growth on orientated silicon substrate) and applying slow growth rates (e.g., by using low concentration of methane gas), diamond crystals with substantially same orientation can be grown (Fig. 10.7). The electron and hole mobilities of a highly textured diamond film may approach single crystal properties.

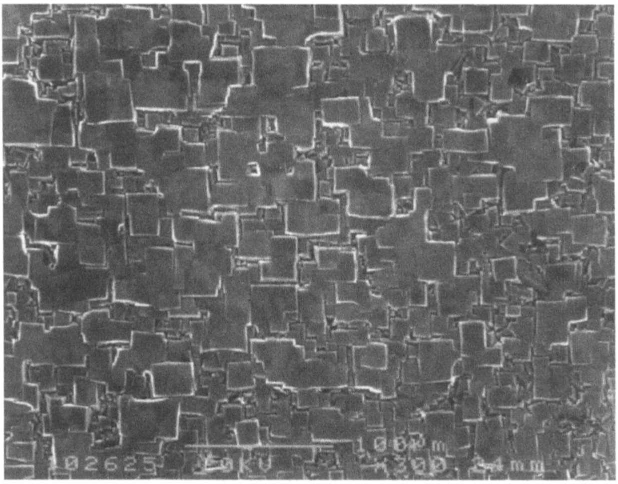

Figure 10.7 The (100)-oriented polycrystalline diamond film deposited on (100) surface of a silicon substrate (data of Wu, J. Y.).

Once the diamond lattice is perfected, suitable dopants of P or N type must be found. Because diamond atoms are very small, they may only be substituted by atoms Period 2 elements (e.g., Li, B, N). B-doped diamond show P-type acceptor property of semiconductor. However, the activation energy of boron atoms is high (0.35 eV) compared to that (0.045 eV), so their mobilities are quite low at low temperature. But because the resistivity of diamond lattice is 20 orders higher than that in silicon, many orders more boron atoms can be doped in diamond lattice without electrical shortage. As a result, the hole concentration in diamond can be many orders higher than that in silicon semiconductor. In addition, as diamond is highly thermally stable, the hole mobility can be further increased when the transistor is used at high temperature. Because of the availability of B-doped diamond, diamond transistors made of different levels of boron doping may be fabricated (Fig. 10.8).

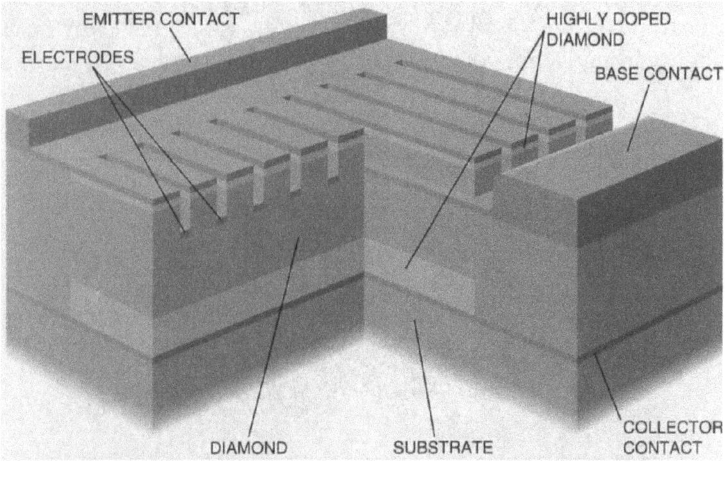

Figure 10.8 A diamond transistor that contains high and low boron atom concentrations.

The high concentration and mobility of electron holes can greatly increase the power of the diamond transistor. The high the power of the transistor, the shorter the switching time of the operation, and hence, the faster the computation speed (Fig. 10.9). Hence, a diamond transistor will be much faster than conventional semiconductors.

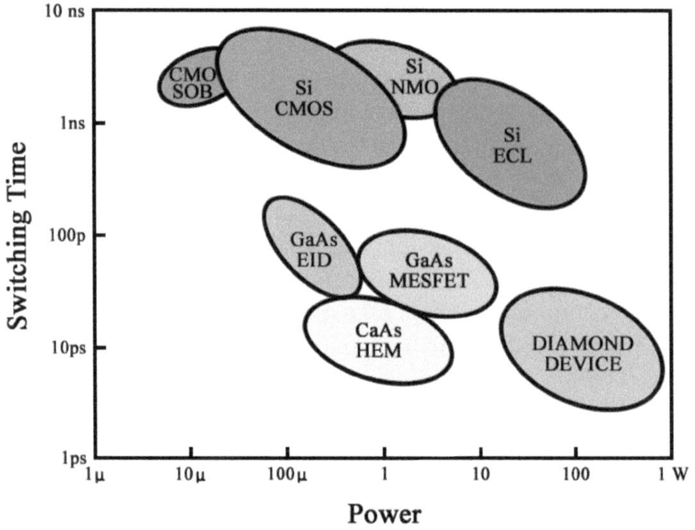

Figure 10.9 The decrease of switching time of semiconductors by increasing the power. Note diamond transistors have the shortest switching time and fastest speed.

The most difficult problem of making diamond transistor is to find a suitable N-type dopant. Nitrogen atoms have one more electron to donate, so they can be the N-type dopant. Nitrogen atoms are also small, so they can substitute carbon atoms in diamond lattice. Unfortunately, the extra electron is bound to nitrogen atom too tightly so the activation energy (1.7 eV) is prohibitively high. As a result, nitrogen atoms will not donate their extra electrons unless the transistor is used at a very high temperature.

Li atoms may also donate their extra atoms. Although Li atoms are slightly larger than carbon atoms, they can also occupy carbon atoms' position in diamond lattice. The activation energy (0.1 eV) of Li atoms is moderate so they can donate the extra electrons with ease. In addition to Li, P atoms are also capable of delivering electrons with high concentration.

With the promise to make diamond semiconductors that is at least 20 times faster than the fastest silicon semiconductor, three Japanese companies (Toshiba, Kobe Steel, and Sumitomo Electric) have planned to develop the first generation diamond chips by year 2008.

Not only a diamond semiconductor can be faster than a silicon semiconductor on planar design, but also it may become a three-dimensional semiconductor. Current semiconductors can only be fitted on planar design because their thermal conductivities are low. With its highest thermal conductivity, diamond can be the only semiconductor capable of dissipating its own heat, so its circuitry can be connected in three dimensions. The volumetric computation is much faster than planar computation. Moreover, as the circuitry is no longer in a series connection, massive parallel computation may take place, so a single 3D CPU can outperform hundreds of supercomputers connected together in series.

10.4 Diamond Vacuum Tubes

Solid-state transistors are much smaller electronic switches than old vacuum tubes, so when they made the first debut in the 1960s, transistors soon replaced vacuum tubes across the board. However, the resistance in vacuum is much lower than that in solid circuitry, so when electrical current is large, solid-state devices may burn out. Thus, for very high power (e.g., megawatts) and frequency (e.g., 1000 GHz) applications (e.g., microwave amplifier, infrared generator), the voltage and temperature will exceed the tolerance limits of current semiconductors. In this case, the vacuum tubes that can accommodate large current without overheating still dominate (Fig. 10.10).

However, diamond transistors are capable of carrying larger electrical current and dissipating much more heat than non-diamond transistors. Moreover, diamond's high radiation hardness and low dielectric constant can minimize the induction loss. Hence, diamond transistors can augment the domain of solid-state devices.

In addition, diamond semiconductors can be used as laser diodes that can emit infrared or visible light with tremendous energy. With its high thermal conductivity and low thermal expansion, diamond is the most stable reflecting mirror for synchronizing its own laser energy. Furthermore, diamond's unsurpassed Young's modulus and low density will also make the reflection highly consistent (Table 10.3). Not only can diamond generate synchronized light internally, but also it is the only laser diode that can dissipate heat itself.

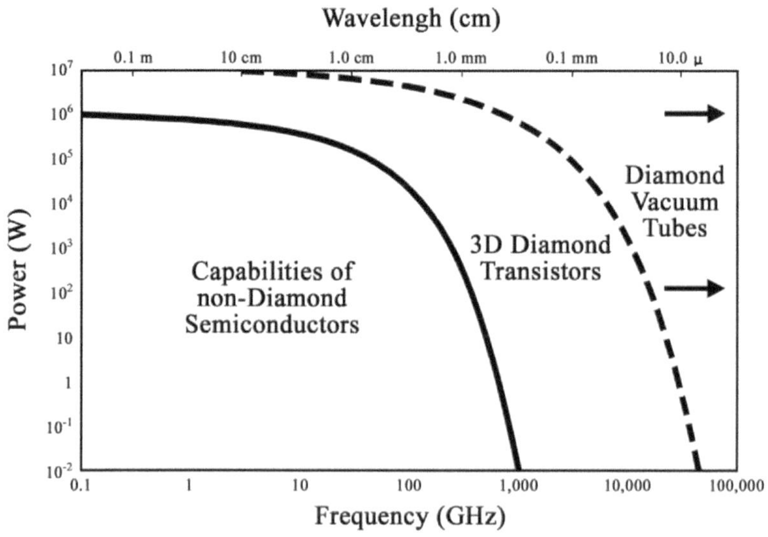

Figure 10.10 The capabilities of electrical devices. Note that non-diamond solid-state transistors are limited on power and frequency. Although 3D diamond transistors can effectively expand the domain of solid-state devices, diamondoid vacuum tubes will be the most powerful electronic gates ever invented.

Although there are multiple layers of interconnected circuitry for current semiconductors, but in essence, the logic layout is two-dimensional (2D), so the integrated circuitry is essentially planar. Despite this, the 2D IC has already run to the impasse of heat dissipation. In order to further densify the computing transistors and minimize the passive circuitry, the 3D IC must be envisaged. As diamond is the only semiconductor that can effectively dissipate its own heat, it can be made into the all powerful 3D chip for future ultra computers. Still these solid-state diamond devices will still be limited in power and frequency.

In addition to being the best computer chip, diamond is also the best cold cathode. Moreover, amorphous diamond can emit large current densities at very low voltage. Hence, diamondoids or diamond family materials can also expand the efficiency of vacuum tubes. Diamondoid vacuum tubes can be used in future ultra fast switches, high-energy particle source, and high-density electron generator.

Table 10.3 The figure of merits (FOM) of reflecting materials

Material	Thermal conductivity (κ) (w/mK)	Thermal expansion coefficient (α) ($\times10^{-6}$/K)	FOM (κ/α) ($\times10^{6}$W/m)	Young's Modulus (E) (GPa)	Density (ρ) (g/cm^3)	FOM (E/ρ) ($\times10^{6}$N/kg/m)
CVD Diamond	900	0.8	1125	1160	3.52	329
CVD SiC	200	2.4	83	466	3.21	52
W	170	4.6	37		19.1	
Zerodur	1.64	0.05	33	91	2.53	36
Mo	143	4.9	29	420	10.2	41
Si	100	4.0	25		2.34	
Cu	395	16.5	24	110	8.96	12
Ir	160	6.7	24		22.5	
Ag	430	19.5	22	83	10.5	8
Au	310	14.5	21	78	19.3	4
Os	80	4.8	18		22.5	
Be	147	11.6	13	276	1.85	149
Al	220	23.6	10	72	2.7	27

10.5 Diamond Optical Computers

Electrical circuitry has resistance, so during the computation of any circuitry, inevitably a proportion of energy is wasted as heat. However, if the circuitry is connected by optical fibers, this optical circuitry can put more energy into computation. In addition, the speed of light for the transmission of optical circuitry is also much faster than speed of electricity. Hence, optical transistors and circuitry can be much more powerful than the most advanced electrical devices including both solid-state transistors and vacuum tube switches.

Diamond is the best optical fiber because of its low absorption. Moreover, its high refractive index will assure that the light energy during transmission is trapped by total reflection from the wall of the diamond fiber. So diamond fibers will be the ideal material for connecting optical circuitry. Moreover, the interference pattern of diffraction and the bending of light by reflection and refraction can be used to construct an optical logic circuitry for computation. Such an optical computer will be much faster than any electronic computer. In addition to being the best semiconductor, diamond is, of course, the most superior optical computer possibly constructed. The diamond optical computer connected by diamond optical circuitry would make the superfast computation almost effortless (Fig. 10.11).

Figure 10.11 The illustration of optical computing with light.

In addition, diamond's lattice contains the smallest holes of all materials; hence, it can be the tightest sieve possible. As light wavc is also made of particles, diamond can be used as the tightest light strainer or optical sieve. For example, the hole size of diamond's (111) plane is only about 2 Å across, so it allows gamma ray to tunnel through and leave lower energy light behind. The energy of tunneled gamma ray photons is higher than 10^4 eV (Fig. 10.12).

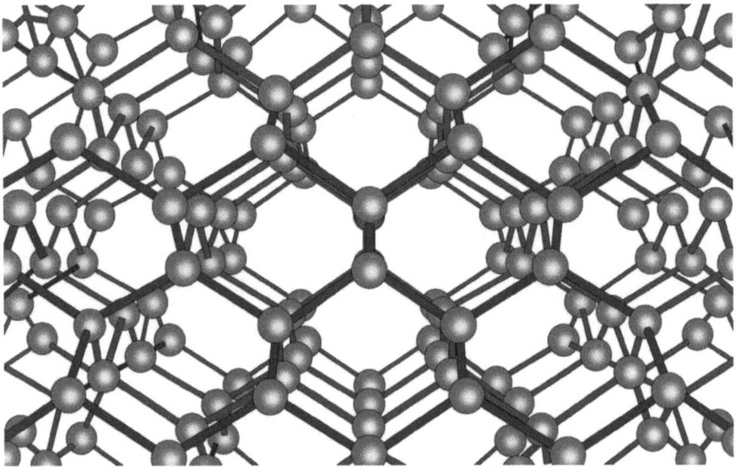

Figure 10.12 Diamond lattice may form the tightest optical sieve for straining photons of high energy.

If a high dose of such powerful photons is synchronized between a pair of diamond mirrors and then released to beam at a remote target, it can destroy it even from a large distance (e.g., for destructing a satellite). Hence, diamond can be made as the most powerful laser weapon in the future star war scenario such as that envisaged during the Reagan administration. Although this strategic defense initiative (SDI) was never left the drawing board, its potential for obsolete intercontinental ballistic missiles forced the former Soviet Union to deploy their own version of laser weaponry. This accelerated arms race became the last straw that collapsed the Soviet Union's already over-extended economy and brought its eventual demise.

The fastest computer available in 2003 was NEC's Earth Simulator, which had five times capability than the most advanced U.S. computers (e.g., HP's ASCI (Accelerated Strategic Computer Initiative) Q or IBM's ASCI White). The Earth Simulator can calculate 36 tera (3.6×10^{12}) flops (floating points per second). Each flops covers one arithmetic operation (plus, minus, multiply or division). Although this speed is awesome, it is dwarfed by what may be achieved by the calculation by a diamond optical computer. The energetic photons strained by a diamond sieve can vibrate at a frequency faster than 10^{18}. Such a frequency is thousands times faster than the above-mentioned Earth Simulator.

10.6 Diamond Biochips

All the above computations are based on series steps so the subsequent calculations are performed based on the data obtained in prior operation. The serial computers are known as Turing machine, named after Alan Turing, the inventor of digital computers.

Although serial computers may be arranged to run in parallel (e.g., with program subroutines or with different micro-processors), there are no transistors that can calculate based on the first arrival of multiple possible inputs. If the calculation can be performed in parallel by all operation units, then the speed of calculation can be greatly increased. This is particularly true if the operation units are numerous and the execution steps are few.

One way to increase greatly the operation units is to use functional molecules, particularly, DNA or protein. The smallest feature size of a silicon transistor was about 100 nm in 2003. In contrast, a functional molecule can be 100 times smaller, so the operation units may be packed 1 million times denser in three dimensions than that on a silicon chip.

Moreover, the functional molecules may be suspended in a liquid medium, so the calculation can be performed by simple chemical reaction such as redox (Fig. 10.13). Such an amorphous computer that contained DNA fragments was used to solve tough mathematical problems (e.g., the trip of a salesman to visit every city once without repeating any of the path) that may take years or centuries to solve.

Alternatively, different segments of DNA may be unzipped and they are attached to a diamond DNA chip and submerged in a solution. The sequence of nucleotides for these unpaired DNA encodes all possible solutions for a set of equations. Subsequently, single strands that encode the first equation are poured onto these unpaired DNA. They will search for the answers and twist close those that satisfy the conditions. After washing away of the unmatched lone strands, an enzyme can be spread over the attached DNA. The enzyme can destroy any and all unpaired DNA that do not satisfy the first equation. After removing the twisted strand by heating, the answers that satisfy the first equation will remain on the diamond chip. The process can be repeated to find the solution that satisfies the other equations. The surviving ones that have met all conditions of all equations can be encoded to provide the final solution (Fig. 10.14).

Diamond can be the best DNA chip because it contains the densest dangling electrons on the surface for bonding DNA strands. Moreover, its surface can be the flattest. It is also the most stable substrate that remains intact after immersing to various solutions and subject to numerous heating cycles. Hence, it can provide the most reliable answers to DNA computing.

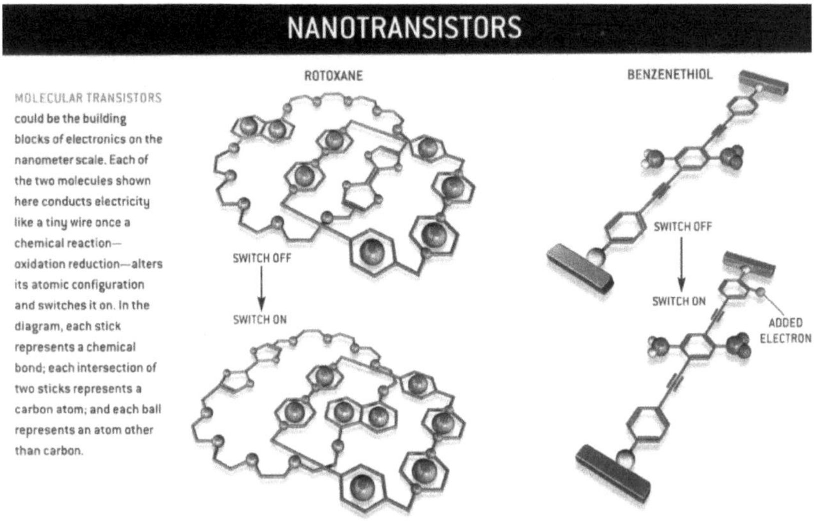

Figure 10.13 The molecular transistors that may be switched by either oxidation or reduction. In the diagram, carbon atoms are omitted between any two links that represent bonds.

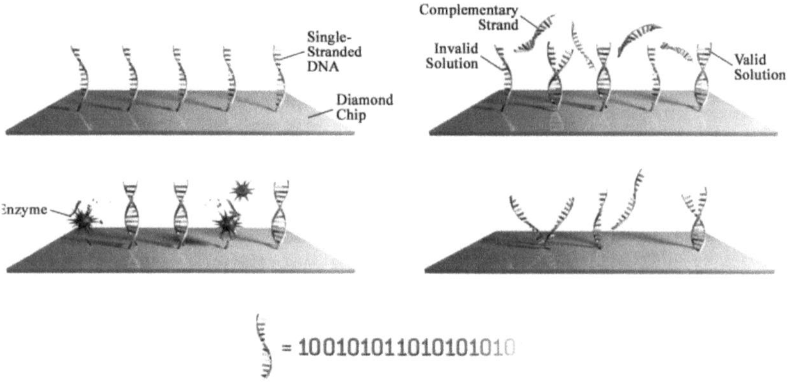

Figure 10.14 Diamond DNA chip can be used to solve efficiently complicated simultaneous equations. Unzipped DNA strands are attached to diamond DNA chip. The sequence of their nucleotides encodes all possible solutions to the simultaneous equations (top, left). The copies of single strand DNA that encode the first equation are poured onto the above chip. They will combine those unpaired DNA that satisfy the requirement of the first equation (top, right). Any unmatched lone DNA strand will be removed by enzyme (bottom, left). The combined single strands are then removed by heating (bottom, right). Subsequently, the single strands that encode the second equation are poured on the diamond chip that contain unzipped DNA that satisfies the first equation. The process is repeated until all equations are satisfied. The surviving DNA remaining on the diamond chip is read to find the answer that satisfies all simultaneous equations.

Another way to use diamond as a DNA chip is to store DNA code in two dimensions. All biological DNA record messages in one-dimensional sequence of four nucleotides. These four nucleotides can be attached to a diamond chip in such a way that their pattern shows all possible proteins that may be expressed by DNA. The human DNA contains about 3 billion nucleotides that may stretch to reach almost a person's height (about 160 cm). Most (>95%) of these nucleotides belong to dormant genes that do not express protein. If the nucleotides of working genes are grouped to form a two-dimensional pattern, they will occupy an area of only 10 microns across.

For a cell to make a particular protein, different parts of DNA must be abstracted and combined together to provide the right sequence of nucleotides. This is a slow process that can incur mistakes. The above DNA chip may provide an expedient and reliable way to express protein. Moreover, it can also be used to assemble protein by employing transverse RNA that carries a triplet of bases to encode an amino acid. In addition, DNA of different designs may be assembled by using such a DNA chip. Thus, diamond DNA chip may be used to research new biological species and even to manufacture such a species.

10.7 Diamond Quantum Computer

Turing machines can calculate data in one dimension, i.e., in sequence. Molecular computer can calculate data in three-dimensional space. But quantum computers can calculate data in parallel universes. So in terms of speed, quantum computers are the ultimate machine for solving all problems. In fact, our brain is a partial quantum computer that can perform parallel computation and reach the conclusion by reassembling the data. This is why the Russian chess king can hold the fastest super computers at bay for decades before 1996 when IBM's Deep Blue finally outsmarted the human brain by using the brutal force with many times more computing transistors than neurons in human brain.

Quantum computing is based on a simple fact that when a quantum (e.g., an electron or an atom) is in total isolation, it can evolve in superposition, i.e., it can evolve in all possible ways. So it can branch off in numerous histories until the time it is brought back to reality. Hence, while it is branching off, it is occupying different spaces in different universes. As soon as it is interfered with any particle in our universe, all the outcomes vanish, except the one that is coherent to everything else existed in our universe.

The power of quantum computing lies in the ability of the quantum to split itself in every Planck time (10^{-43} second), the shortest duration that each quantum state can survive. This is an enormous short period of time, so a few correlated quanta that are totally isolated from our universe can traverse astronomical possibilities and reach a conclusion in a very short period of time. For example, if 40 quanta can be tied together in a total isolation, it can be a faster computer than even the fastest ultracomputer that has existed (e.g., Earth Simulator). With a quantum computer that contains 100 quanta, all national security codes that are based on the products of two very large prime numbers can be broken in a matter of minutes.

The quantum computer that contains two quanta has been experimentally tested and it confirmed the validity of the power of quantum computing. However, to make quantum computer with higher numbers of quanta, it would require the isolation of the system to unusual degree. Normally, the quanta envisaged are either electrons or atoms. The spin state of an electron or atom (up and down) can be used to represent 1 and 0 of the binary code. The spin state can be manipulated by using a laser so the quantum switch can be turned on or off by control. However, it is very difficult to isolate a group of electrons or atoms without being interfered by the particles outside the system so a quantum computer of many quanta cannot be built.

Diamond, with its high rigidity and thermal stability, can provide an isolated system of high stability at near absolute zero temperature. For example, the dangling electrons on the (111) surface of diamond would be ideal quanta for flipping on and off by irradiating with a focused laser beam. Alternatively, the hydrogen atoms that attach to the (111) surface may also be used as the computing quanta.

In addition, the color centers of N-V vacancies in a perfect diamond lattice can be used as isolated computing system. These vacancies are next to three carbon atoms and one substitutional nitrogen atom. The N-V quanta can be controlled by laser beam with a wavelength of 490–640 nm.

The N-V vacancies can be created by irradiating perfect diamond lattice with high-energy (e.g., 2 MeV) electron beam and subsequently annealed (e.g., 900°C) in helium atmosphere.

In summary, diamond holds the key to be the ultimate computing material for future semiconductor transistors, cold cathode vacuum tubes, optical sieves, DNA chips, and isolated media. So it will drive most advanced Turing machines, vacuum devices, optical computers, chemical computers, and quantum computers. Diamond is forever. It will lay down the foundation for the future culmination of human civilization and the arrival of the eternal diamond age.

Chapter 11

Diamond Electrodes

Diamond can be the most sensitive detector with high reliability.

11.1 Boron-Doped Diamond

Diamond is bonded by strong covalent bonds in high symmetry; hence, it does not have electrons to move around. In fact, pure diamond (Type IIa) is the best insulator of all materials with a resistivity as high as 10^{16} ohm-cm. However, with the addition of nitrogen up to 1000 PPM (e.g., Type Ia), the resistivity may decrease by a thousand folds. Although boron doping (up to 100 PPM) is rare in natural diamond (Type IIb), it can be heavily doped in either high-pressure diamond grit or CVD diamond film. The resistivity can be as low as 1 ohm-cm, which is comparable to a metal in electrical conductivity (Fig. 11.1).

Boron-doped diamond (BDD) can be conveniently deposited by a CVD method onto a metallic substrate such as graphite, silicon, titanium, niobium, tantalum, molybdenum, or tungsten.

Handbook of Industrial Diamonds: Diamond Films and Carbon Coatings
James Sung

ISBN 978-981-4310-00-0 (Hardcover), 978-1-003-18045-6 (eBook)
www.jennystanford.com

A typical CVD process uses sp^3 bonding methane (e.g., 1%) as the carbon source. The carbon source is mixed with overabundance of hydrogen gas and together they are decomposed at high temperature. The gas may be heated by several means such as by hot filaments, microwave plasma, or arc jet. The dissociated carbon atoms are catalyzed by the surrounding hydrogen atoms to maintain their sp^3 bonds. The connection of these sp^3 bond carbon atoms on the substrate becomes diamond.

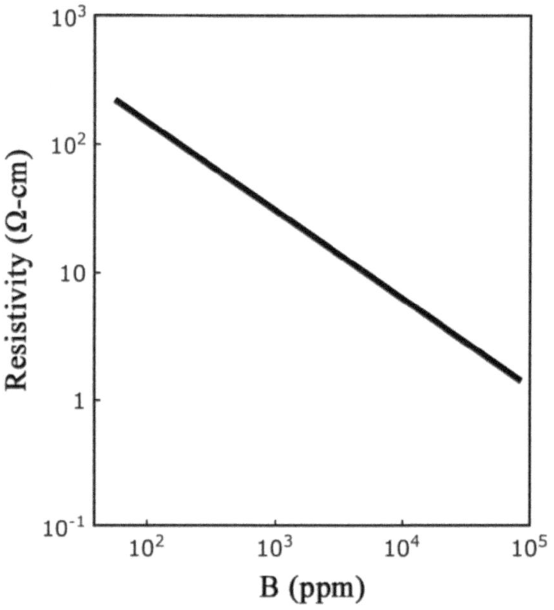

Figure 11.1 The exponential decrease of resistivity for diamond by increasing doping of boron.

During the CVD process, if the gas mixture contains boron, then BDD will be formed. Boron may be introduced by incorporating B2H6 (e.g., 10 PPM). Alternatively, one of the gases may pass through warm liquid of trimethyl boride ((CH3)3B) or boric acid (H3BO3) so low dose (e.g., 100 PPM) of boron may be incorporated in the gas mixture. Still, boron will be enriched substantially in the deposited diamond film. If the boron doping level in BDD exceeds about 1000 PPM, its resistivity will be comparable to the refractory metal substrate.

11.2 Annealing of Diamond Film

Hydrogen absorbent in diamond film may increase the conductivity. Hence, by eliminating hydrogen contamination, the resistivity of diamond film may increase substantially. For example, annealing diamond films in air up to 800°C may increase the resistivity by one million times.

11.3 Overpotential of BDD Electrode

The deposited BDD along with the metal substrate can be used as an electrode. The BDD electrode has many superior properties that excel conventional metal electrodes (e.g., nickel, platinum) or even glassy carbon (GC) electrodes (Fig. 11.1). One of the major advantages is BDD's extreme inertness toward chemical reactions. The surface of BDD will remain intact even the electrode is used in electrolysis of highly reactive liquid.

Another advantage of using BDD is that it has an overpotential in water solution that is significantly larger than all other electrodes. When water is dissociated, it will form H^+, which will combine to form H_2 gas at the cathode, and OH^- ions, which will combine to form O_2 gas at the anode. Normally, these ions will polarize the electrodes so the applied voltage will drop. But BDD will prevent such polarization, so it will take an additional negative bias to evolve hydrogen at cathode, and an additional positive bias at anode to evolve oxygen at anode. As a consequence, an overpotential can be maintained during the entire duration of electrolysis. This is an enormous advantage for performing electrolysis. For example, there are certain ions that are either difficult to be reduced at cathode (e.g., alkali elements) or their concentration is so low (e.g., Pb) that they cannot be detected with lower voltage. By applying the overvoltage, water will not be dissociated so these hard-to-be-reduced ions can be separated and deposit on either cathode or anode.

The broad working potential window in aqueous electrolytes allows the production of strong oxidizing agents (e.g., hydrogen peroxide, ozone, chlorine). For example, BDD has been used to electrolytic dissociation of concentrated hydro sulfuric acid (H_2SO_4)

and produce peroxodisulfuric acid $H_2S_2O_8$, a strong oxidizing agent that cannot be obtained by using normal metal electrodes. Peroxodisulfuric acid is used as corrosive agent for etching copper foil coated on printed circuit board. BDD has also been used to dissociate the molten salt of KF·2HF and produce highly corrosive fluorine gas. These extreme applications cannot be accomplished by using conventional electrodes that lacks the overpotential for dissociation. Worse still, they will soon be corroded away.

The combination of high stability and high sensitivity has made BDD highly valuable in many electro-chemical applications. In particular, its high signal-to-noise ratio can make the detection discrete. This discreteness is useful for analyzing certain biological or medical substances. For example, the critical chemical dopanmine (DA) is the messenger between neurons. If it is out of balance, brain disorder will ensue. The oxidation potential of DA is very close to its associated ascorbic acid (AA), so they cannot be separated by ordinary electrolysis. Normally, CVD diamond film is deposited under hydrogen atmosphere, so the carbon atoms on diamond's surface are hydrogen terminated. Unless it is heated above about 800°C, the absorbed hydrogen will stay indefinitely. The presence of hydrogen on the surface will make diamond hydrophobic. It also raises the overpotential of BDD. However, if BDD is treated in oxygen plasma (e.g., using RF as the energy source), the surface hydrogen will be replaced by oxygen so diamond will become oxygen terminated. In this case, diamond becomes hydrophilic and the surface resistance in water solution will decrease. In this case, the oxidation potential of DA will be higher than AA so they can be separated by using BDD as anode.

11.4 Trace Element Analysis

In addition to separation of DA from AA, BDD has also been used to analyze urine acid, amines, sulfa drugs and chlorophenois and other medical-related substances. In addition, BDD is also used to detect trace of carbon hydrates (e.g., sugar), proteins, or even amino acid.

Another useful feature of BDD is to use it to detect trace amounts of metallic constituents. The resolution of such analysis is so high that it can differentiate same-composition materials manufactured by different suppliers. In essence, BDD can be used to quickly analyze samples and establish the "finer print" of the characteristic pattern of the composition. Such a sensitive detection is valuable for identifying polluting source of environments or forensic evidence of crimes.

Because of BDD's chemical stability, the electrical current pass through BDD electrode has a very high signal-to-noise ratio. Due to this high resolution of electrical signal, BDD electrodes can be used to detect trace amount of ions in the solution. For example, it was demonstrated that BDD was capable of detecting lead at a concentration as low as 4×10^{-9} M (A. Manivannan, et al., 1999, *Electrochemical and Solid State Letters*, **2**(9), 455–456). Hence, BDD electrode can be used to identify pollutants such as lead, cadmium, and other heavy metals.

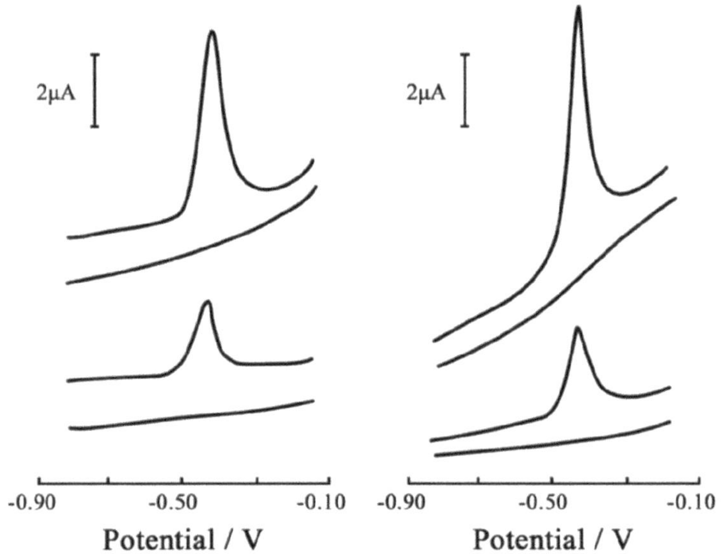

Figure 11.2 Voltammograms of lead at anode for BDD (top profile) and GC (bottom profile). For each profile, the lower smooth curve shows the background electrolyte with reference as Ag/AgCl. The electrolyte is 0.1 M/l KCl with 41 μM/L $Hg(NO_3)_2$ (left diagram); and 0.1 M/L KNO_3 with 41 μM/L $Hg(NO_3)_2$. C(Pb^{++}) = 100 μg/L. The scan rate was 100 mV/sec.

11.5 Fingerprinting Analysis

Another use of BDD is for quick analysis of chemical compositions. The technique relies on BDD's extreme hardness, corrosion resistance, and electrical conductance. The BDD formed contains rugged surface with faceted diamond (e.g., octahedral faces) grains of tens microns in size. This surface can serve as the best diamond file so it can be used to attach sample (e.g., an alloy) by simply rubbing against it. The sample attached BDD is then dipped into the electrolyte and used as an anode. The voltage–current profile obtained during the electrolysis can be used as the "finger print" for the determination of composition as well as its source. This method of analysis is known as abrasive differential pulse stripping volumetry or AbsDPSV. The sample testing can be in a matter of minutes and almost any material can be analyzed, so this technique is very handy and versatile (Fig. 11.3).

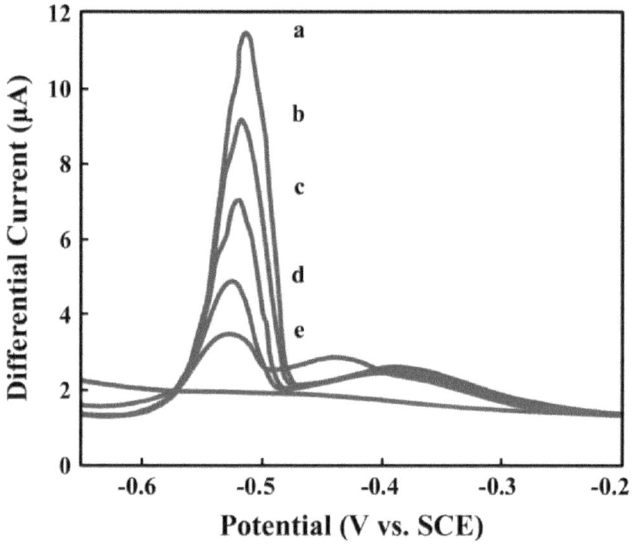

Figure 11.3 AbsDPSV analysis of $Pb(NO_3)_2$ in 0.1 M KCl (pH = 1). The overpotential of BDD was −1.0 V (vs. SCE). The time of analysis was 2 min. In the diagram, the concentration of lead decreases from 2×10^{-6} M (a) to 4×10^{-7} M (e). (A. Manivannan, et al., 1999, *Chemical Letters*, 851–852).

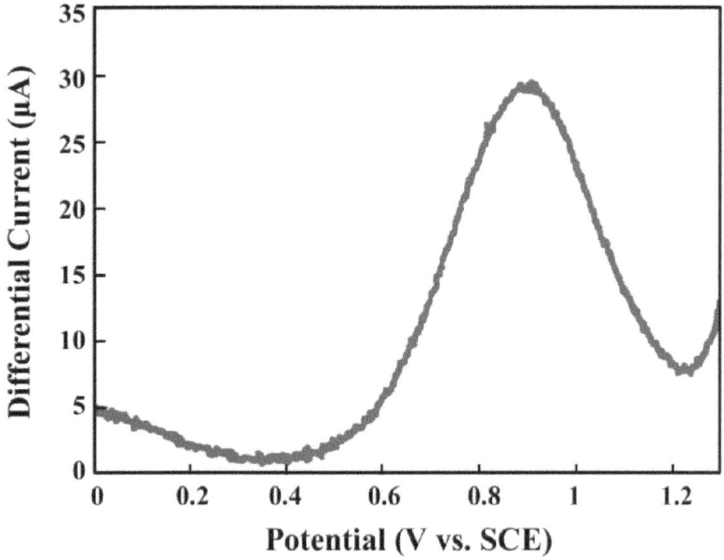

Figure 11.4 The profile of iron as analyzed by absDPSV.

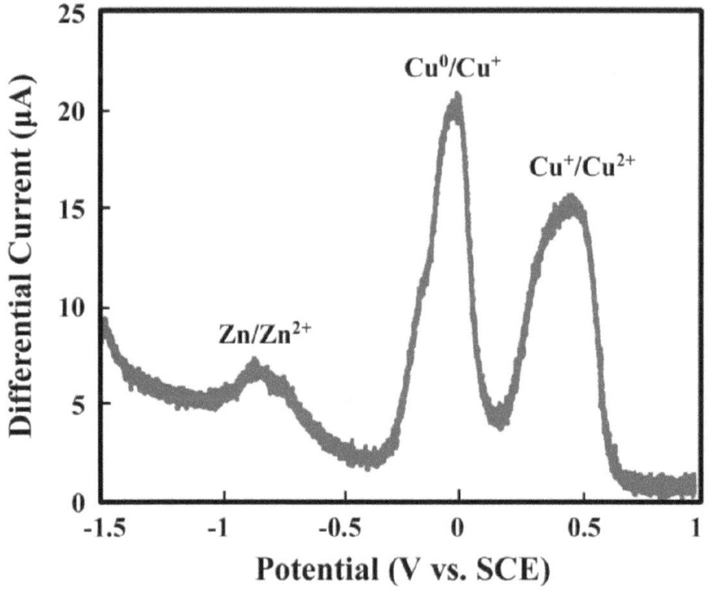

Figure 11.5 The profile of electrolysis showing the oxidation of copper in brass (65% Cu, 35% Zn).

11.6 Water Treatment

BDDs cannot even be dissolved in strong acids (e.g., aqua regia), or bases (e.g., KOH). Such chemical stability makes BDD highly desirable as electrodes working in very corrosive environments, such as by dissociating wastewater.

One way to treat wastewater is to oxidize its organic pollutants (e.g., oil, cooling fluids, toxic compounds). Compared to other electrode materials, BDD can produce hydroxyl radicals with a higher current efficiency. These hydroxyl radicals can effectively mineralize organic impurities n water. The result will be cleaned water that may be released for municipal use. During this wastewater treatment, calcium or other oxidizing agent may build up at the cathode; however, as BDD is stable for both cathode and anode, the polarities of these two electrodes can be reversed to remove the attachment of contaminants so the electrodes can maintain clean.

In addition to being chemically stable, BDD is also highly thermally stable. In fact, it is much more thermally stable than CVD diamond film. The reason is that although carbon atoms have four valence electrons. This is why they form tetrahedral diamond lattice. Although carbon atoms inside diamond lattice are bonded to four neighbors, their surface (e.g., octahedral faces) atoms have three neighbors at the most. As a consequence, there are always one or two valence electrons that are not bonded. These dangling electrons are highly reactive, which makes diamond unstable at high temperatures. For example, CVD diamond will oxidize in air above 600°C. But BDD contains boron atoms that carry only three valence electrons instead of four. Hence, there are no dangling electrons available for reaction with foreign atoms. Because of this added inertness, BDD can be more thermally stable than CVD diamond (Fig. 11.6). For example, it will not be oxidized in air unless the temperature is more than 800°C.

Because of BDD's chemical inertness and thermal stability, not only its electrode can survive the electrolysis in corrosive liquid, but also it can carry a very large electrical current density, up to 10 A/cm^2. Such a capability makes it a highly aggressive electrode that if applied to wastewater treatment may clean up a lot of wastewater quickly. For example, large plates (e.g., up to 1 m^2 by DiaChem electrode) with thin (e.g., up to 5 μm) coating of BDD may be used that may carry an electrical current as high as 100,000 A per electrode!

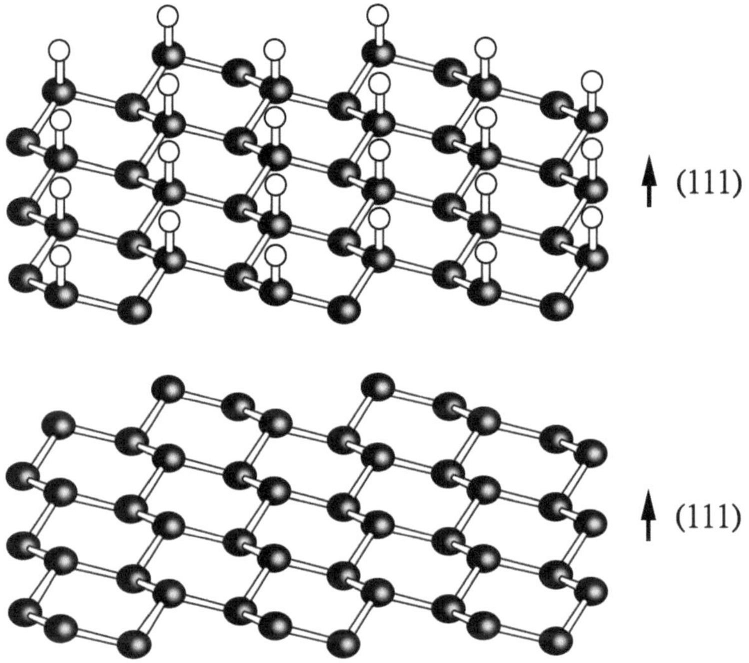

Figure 11.6 Diamond's (111) face contains dangling electrons that may be hydrogen terminated (open circle) (top), but BDD's same face has no dangling electrons so it is much more thermally stable than pure diamond.

11.7 Optical Transparent Electrode

Yet another application of BDD is to make use of its transparency for spectro-electrochemical measurements in the UV, visible, and IR regions of electromagnetic radiations. In this case, the boron doping may not be high (e.g., 100 PPM), or the thickness should be thin (e.g., a few microns). For freestanding optical transparent electrode (OTE), optical grade BDD with low amount of grain boundaries should be used. In order to minimizc the surface scattering of light, the diamond film's surface should be smooth with a roughness less than 30 nm. The substrate used must also be transparent in the region of interest. For example, undoped silicon substrate can be used for IR measurements; and quartz substrate for UV to visible range of spectrum. In the latter case, in order to avoid the scattering of light by grain boundaries, nanometer-sized diamond may be used to seed the BDD so the grain size will be smaller than the wavelength of the measured spectrum.

OTE has been used to characterize many organic redox systems (e.g., biomolecules of cytochrome c) with high sensitivity, reproducibility, and stability in the wide range of spectroscopy (Fig. 11.7).

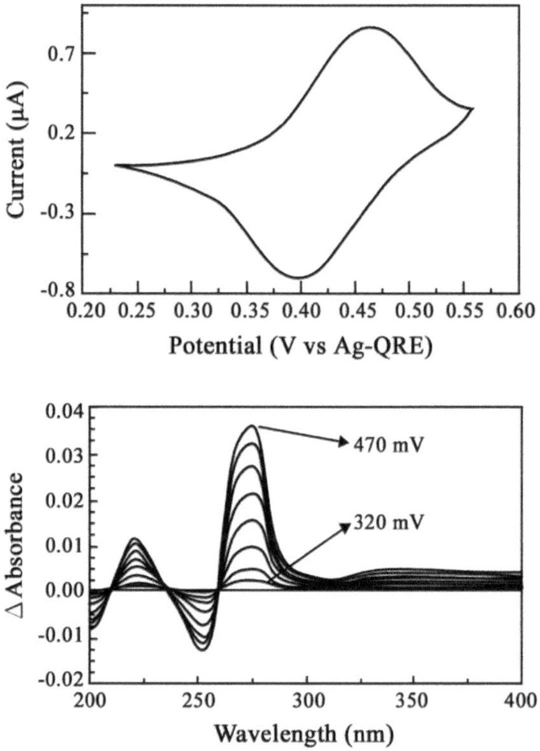

Figure 11.7 The redox phenomena of chloropromazine (CPZ). Background corrected, thin layer voltammetric current-voltage plot for diamond/quartz OTE in 0.1 mM CPZ + 10 mM H_2SO_4. Scan rate = 2 mV/sec (top). A series of UV-visible absorbance difference spectra for 0.1 mM CPZ + 10 mM $HClO_4$, as the potential was stepped from 0.32 to 0.47 V vs. Ag-QRE (bottom) (*Anal. Chem.*, 2002, 74, 5924).

Chapter 12

PVD of Diamond-Like Carbon

Diamond cannot be coated on ordinary materials, so we have diamond-like carbon instead.

12.1 Diamond-Like Carbon

Diamond is too extreme to be able to adhere to ordinary materials; hence, a less extreme material called diamond-like carbon (DLC) is needed as a coating material. Although DLC is not exactly diamond, it has certain attributes (e.g., wear resistance) that may approach diamond, but the other properties (e.g., thermal stability) are compromised.

In order for diamond to be less diamond, carbon atoms do not form the regular diamond lattice. Instead, each carbon atom may assume either diamond bonding (i.e., sp^3) or non-diamond bonding (e.g., sp^1 or sp^2). Alternatively, every carbon atom may possess both characters simultaneously. However, there are two possibilities that diamond and non-diamond bonding can

Handbook of Industrial Diamonds: Diamond Films and Carbon Coatings
James Sung

ISBN 978-981-4310-00-0 (Hardcover), 978-1-003-18045-6 (eBook)
www.jennystanford.com

coexist, either every atom has the same degree of mixing characters (e.g., every bucky ball atom has 72% of $sp^2\pi$ character and 28% of sp^3 character), or they differ one by one (e.g., every amorphous diamond atom is unique in bonding character). All the above possibilities of diamondoids are listed in Table 12.1.

Table 12.1 Carbon kingdom with diamondoid families

Bond type	Atoms involved	Structure	
sp^1	All Identical	Carbyne	=C=C=C=C=
			—C≡C—C≡C—
	Partial same	Polymer	
	All different	Amorphous Carbon	
sp^2	All identical	Graphite	
sp^3	All identical	Diamond	
sp^1+sp^2	All identical	Benzene	
	All different	Amorphous graphite	
	Partial same	Coal	
sp^2+sp^3	All identical	Bucky ball carbon nanotube	
	All different	Amorphous Carbon (a-D)	
	Partial same	Nanocyrstalline diamond	
sp^1+sp^2+sp^3	All different	Diamond-like carbon (a-C:H)	

Some of the above-mentioned mixing bonding characters of diamondoids are illustrated in Figs. 12.1 to 12.6.

In contrast with a bucky ball, which has a regular structure, DLC is highly chaotic with a high entropy (Fig. 12.2).

Although DLC may have chaotic relationship of their atoms, it often contains domains that show some regularity of structures. Some of such domains are already represented by clusters with diamond structure (Fig. 12.3).

Alternatively, the carbon may form various combinations of graphitic layers and diamond-like structures (Fig. 12.4).

If there are plenty of hydrogens, the DLC structure may contain domains of polymers (Fig. 12.5).

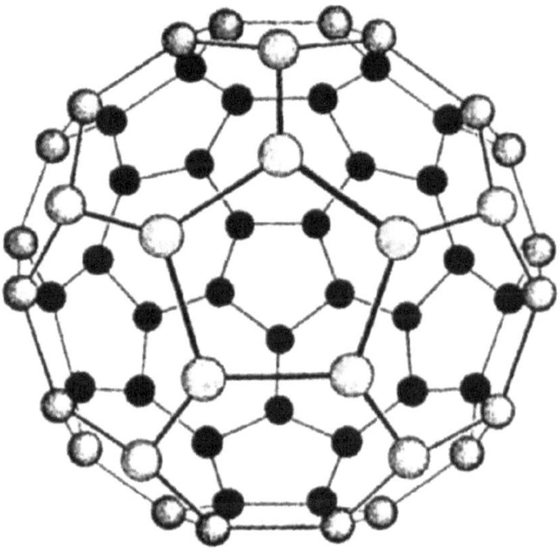

Figure 12.1 Bucky ball (C_{60}) contains all identical carbon atoms each possesses 72% of $sp^2\pi$ character and 28% of sp^3 character.

Figure 12.2 A 3D model of DLC showing the random distribution of carbon atoms with graphitic (sp^2) and diamond's (sp^3) bonds.

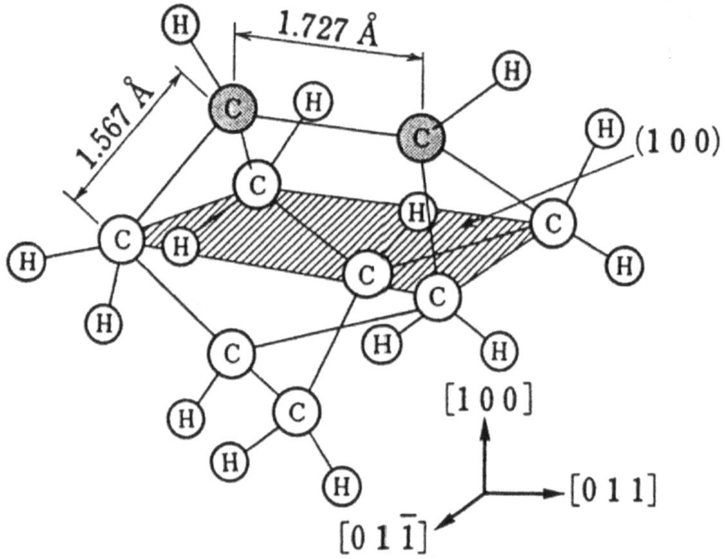

Figure 12.3 A diamondoid molecule of C_9H_{12} that reveals diamond's crystallographic orientations. Methane is hydrogen terminated single atom diamond; C_9H_{12} is hydrogen terminated 9 atoms diamond.

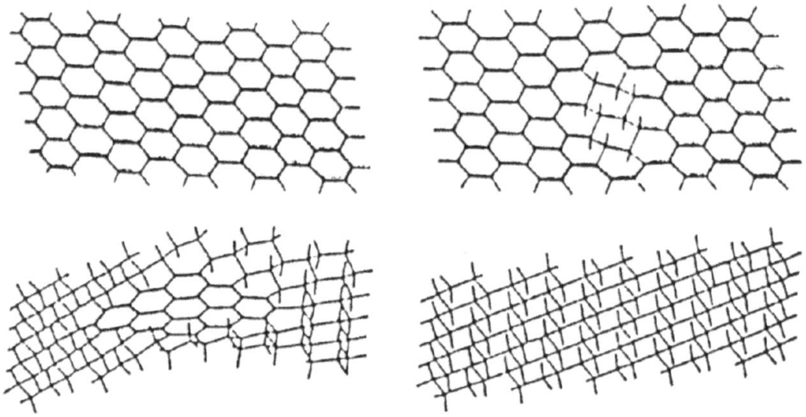

Figure 12.4 Carbon atoms may assume all graphitic bonds (top left), most graphitic bonds but with partial carbyne character (top right), mixed characters of graphite and diamond (bottom left), and all diamond bonds (bottom right).

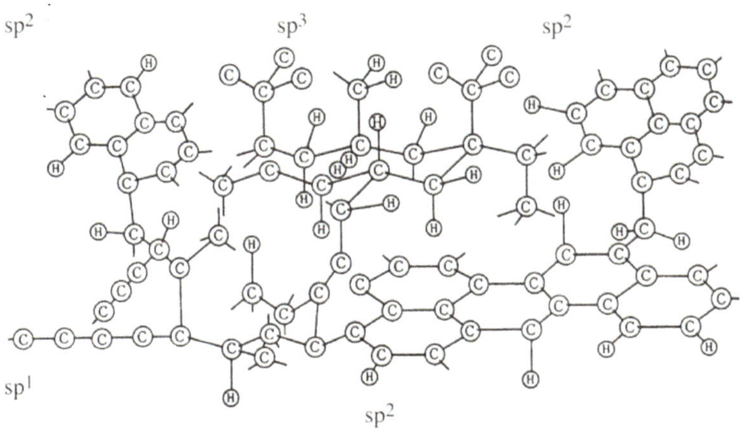

Figure 12.5 A hydrogenated DLC that contains domains of various polymers.

However, there is also DLC that shows no obvious structure but includes a random network of carbon atoms with other atoms (Fig. 12.6).

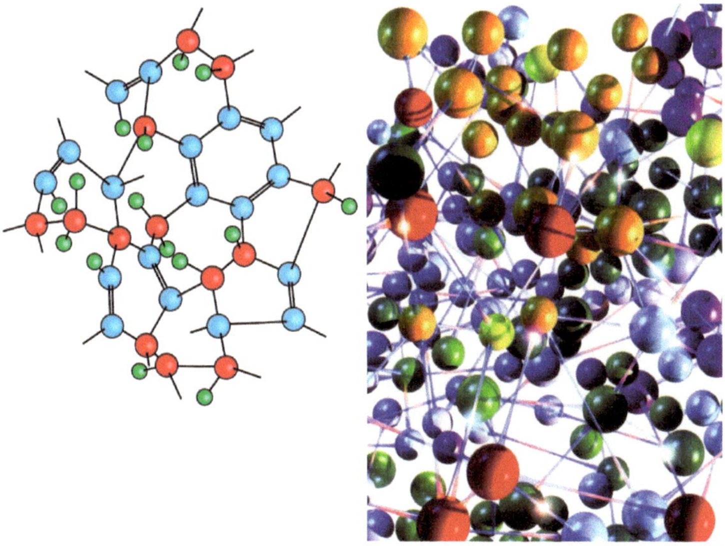

Figure 12.6 Hydrogenated DLC may contain randomly bonded atoms. Diamond atoms are cross hatched; graphitic atoms are in gray shade, hydrogen atoms are small open circles.

The purest form of DLC is amorphous diamond. It contains only carbon atoms that piled together with no order at all (Fig. 12.7). Such a structure has the highest entropy of all DLC materials.

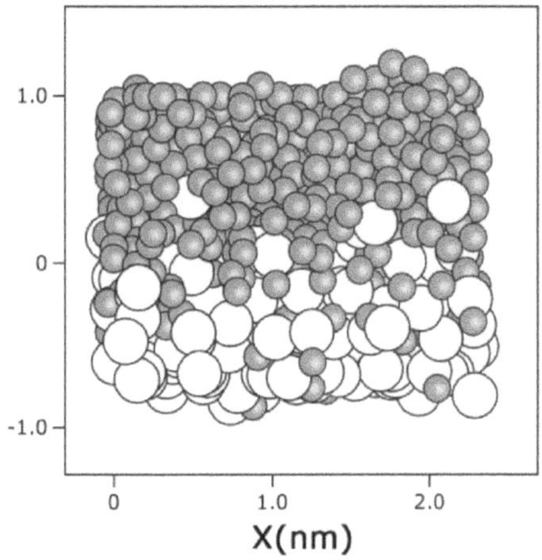

Figure 12.7 An atomic view of amorphous diamond coating on silicon substrate. The amorphous diamond could be deposited with a kinetic energy of 20 eV. The silicon is oriented with (001) face horizontal.

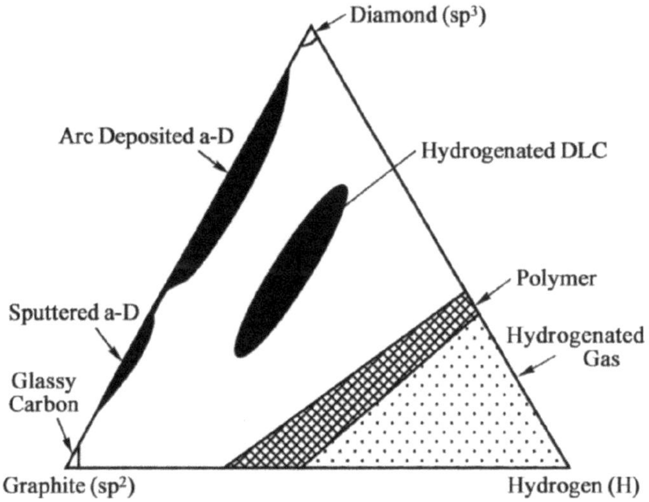

Figure 12.8 The ternary diagram of C–H system showing diamond, graphite, and polymer. DLC is the composite material of these three components in various proportions.

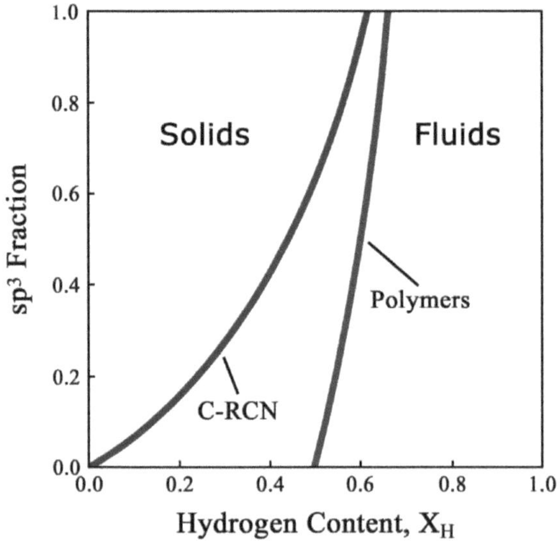

Figure 12.9 Both DLC and polymer are sp^3 dominant carbon materials, but their graphitic carbon to hydrogen ratios differ. DLC contains more graphitic carbon than hydrogen atoms; polymer is reversed.

12.2 DLC Properties

The properties of DLC may be compared with those of other carbon-containing materials as listed in Table 12.2.

Table 12.2 The properties of carbon-containing materials

Material	Sp^3 (%)	H (%)	Density	Hardness
Graphite	0	0	2.27	<0.01
Glassy carbon	0	0	1.3–1.6	2–3
Evaporate carbon	<5	<5	1.9–2.0	2–5
Sputtered DLC	5–10	<5	2.2–2.6	10
Amorphous diamond	>85	<5	2.9–3.4	80–100
Hard DLC	40–50	30–50	1.6–2.0	10–30
Soft DLC	50–80	50–60	0.9–1.2	<5
Polyethylene	100	66	0.9	0.01
Diamond	100	0	3.52	100

DLC may retain some diamond properties, but as it is a composite material, its diamond character grades as the composition and the structure vary (Table 12.3).

Table 12.3 Diamond and DLC properties

Property	Natural diamond	CVD diamond	DLC (a-c:H)	DLC (a-D)
Hardness	100	80–100	10–50	80–100
Density	3.5	3.2–3.4	1.7–2.2	3.0–3.2
Frictional coefficient	0.1	0.1	0.1	0.1
Roughness	—	0.1–3.0	Smooth	Smooth
Adhesion	—	Weak	Intermediate	Strong
Deposition temperature	—	> 600	20–325	20–150
Crystal structure	Crystalline	Crystalline	Amorphous	Amorphous

12.3 DLC and Silica

Amorphous DLC may incorporate other amorphous compositions, such as amorphous silica. In fact, DLC's C–H network may interpenetrated with silica's Si–O network (Fig. 12.10).

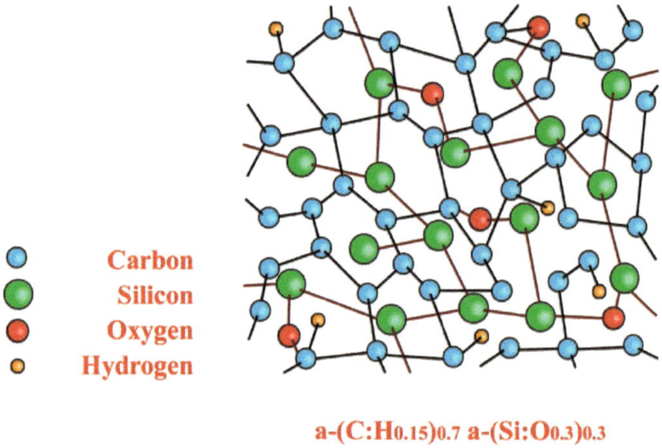

Figure 12.10 A C–H network that is interpenetrated by another Si–O network to form a silicate DLC (Dylyn brochure).

The DLC–silica composite possesses very low stress levels and frictional coefficient (Fig. 12.11) so they may be used as a solid lubricant. However, this composite structure is not very hard, so it is an intermediate material between DLC and polymer.

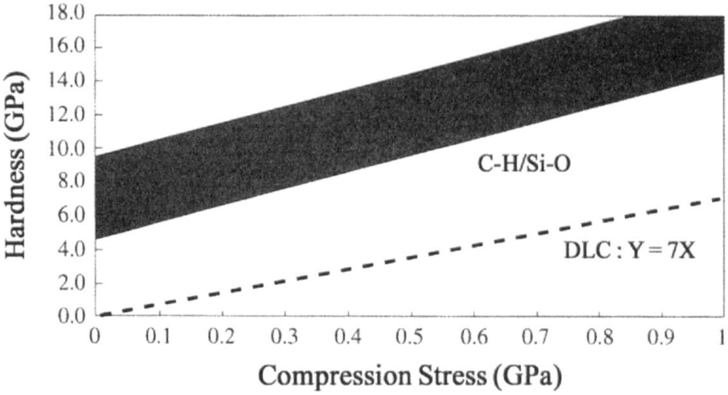

Figure 12.11 The incorporation of amorphous silica in DLC can reduce the compression stress substantially for a given hardness level (Dylyn brochure).

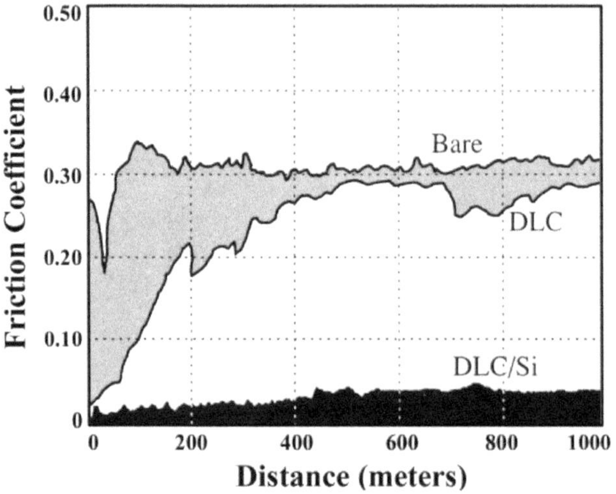

Figure 12.12 The low frictional coefficient of DLC may increase by the exposure to moisture, but C–H/Si–O composite has a steady lower frictional coefficient (Dylyn brochure). The test was performed by pin (alumina) on disk (1018) with no lubricant. The load was 100 grams with a speed of 20 cm/sec. The temperature was 25°C and the relative humidity was 32%.

DLC may also incorporate metals (e.g., tungsten) as either networks or inclusions (Fig. 12.13). If the amount of metal dopants is high, the DLC composite can be electrically conducting.

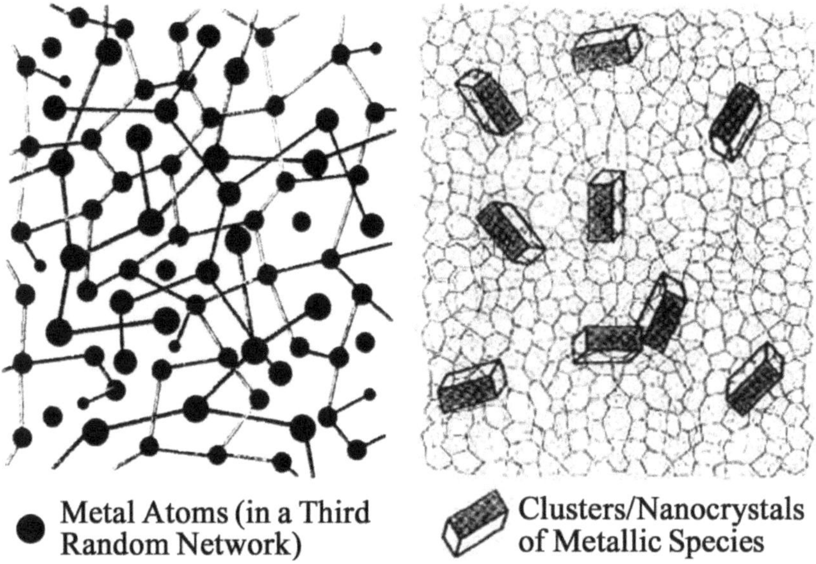

Figure 12.13 Small amount of doping metal atoms (large circles) may form a third network that interpenetrate within the intertwine C–H (small circles) and Si–O (middle circles) networks (left). A large amount of metal atoms may form nano-sized inclusions (right). Such a composite may be viewed as DLC cemented metal (Dylyn brochure).

Various DLC composites may also form nano-sized layers of superlattice (Fig. 12.14). Such a complicated structure may find use as exotic capacitor or semiconductor.

The diversification of DLC structures may be compared with silica structures with natural occurrence. Both of them are metastable and they may transform into more stable phases at high temperature.

For example, silica-saturated volcanic water may incorporate water and deposit as opal. It may also solidify to form water-free chalcedony. The latter, if deposited in banded structure is known as agate. The extension of agate in volcanic voids (geode) may crystallize to form quartz. The silica system may be compared with DLC composite as shown in Table 12.4.

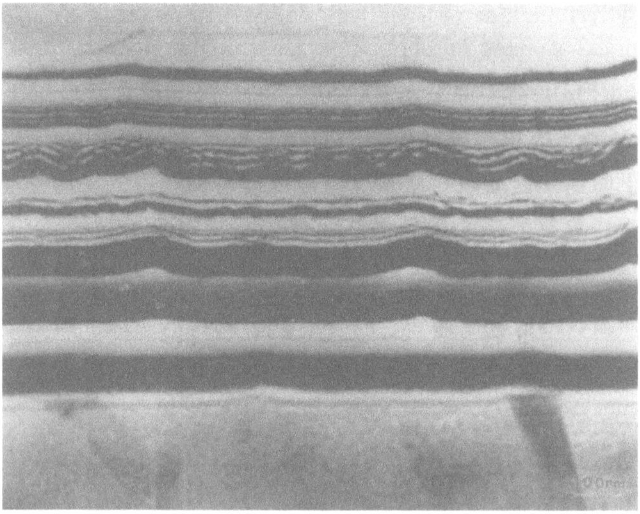

Figure 12.14 A 30-layer DLC–silica–metal superlattice that has a total thickness of only 0.6 micron. Some of the layers are only 1 nm thick. Certain layers are electrically conducting, and others are insulators.

Table 12.4 The comparison between DLC and silica systems

Name	DLC	Silica
Composition	C–H	Si–O
Fluid medium	Gas	Water
"Wet" composite	Hydrogenated DLC	Water containing opal
"Dry" material	Amorphous diamond	Chalcedony
Banded structure	Superlattice	Agate
Crystalline form	Diamond	Quartz
Reactor	CVD	Geode

12.4 Atomic Energy

DLC is formed at conditions grossly deviated from equilibrium. Specifically, the temperature is so low that there is no chance for atoms to move into equilibrium positions. Typically, DLC is deposited by physical vapor deposition (PVD) that does not involve chemical reactions. Hence, it lacks the thermal reactions (e.g., pyrolytic decomposition of methane molecules) as well as catalytic functions (e.g., touch-and-leave interaction of hydrogen atoms) that are typical in chemical vapor deposition of diamond films (Table 12.5).

Table 12.5 The contrast between PVD coating of DLC and CVD growth of diamond film

Process	PVD	CVD
Chemical reaction	No	Yes
Ambient temperature	Low	High
Equilibrium condition	No	Yes
Vacuum level	High	Low
Film thickness	Thin	Thick
Crystal structure	Amorphous	Crystalline

The PVD process is deposited atom by atom, or ion by ion. Atoms may be driven by internal (latent) energy; and ions, by external (field) energy. In the former case, the energy is typically very low (<1 eV), so the deposition rate is slow. Moreover, the motions of atoms are aimless (Brownian motion), so their adherence to the coating is by chance. Table 12.6 shows what carbon may gain such latent energy from either pressure or temperature.

Table 12.6 The latent energy of carbon atoms

Energy (eV)	Temperature (°C)	Pressure (GPa)
0.1	887	4
0.5	5529	22
1	11331	44
2	22936	88
3	34540	132
4	46145	176
5	57749	219

On the other hand, an ion may pick up the energy form an externally applied electrical field. The acceleration of ions by electrical field may impart individual ions much higher kinetic energy (e.g., up to 100 eV) than that normally achieved by heating. Moreover, the motion of ions driven by an external field can be guided to hit a substrate (e.g., with a negative bias) by trajectory. As the result of the aimed bombardment, coating rate can be much higher than that of wondering atoms.

If the energy of the ions are further increased (e.g., to 1000 eV), then instead of coating the substrate, these ions can actually sputter the coating material away, as in the case of ion beam etching. If the energy is further increased to up to 1 million eV, ions may simply penetrate deep into the substrate, a phenomenon known as ion implementation. Figure 12.15 shows the ranges of energies for various types of ions.

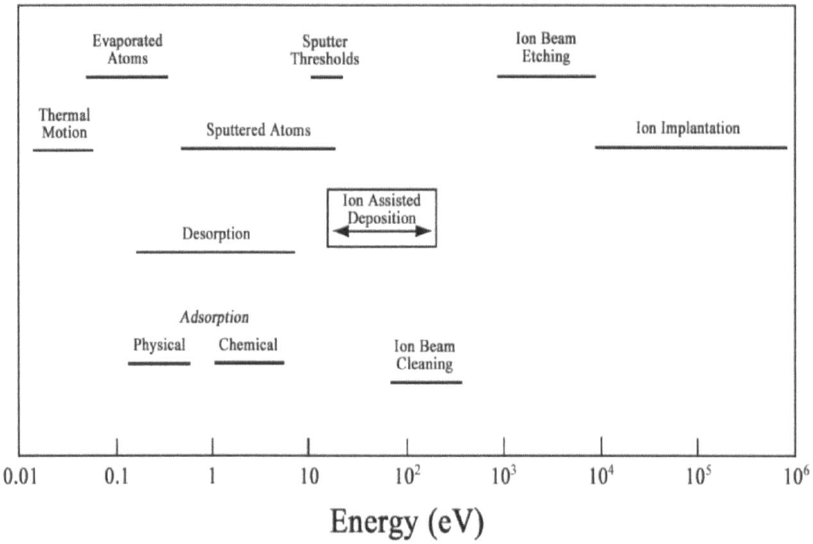

Figure 12.15 Energy of atom or ion and its effects on the substrate.

Although PVD processes are performed in partial vacuum, and at low temperatures, but in reality atoms being deposited are experiencing extreme pressures and temperatures. Most atoms are driven by an electrical field so they may move with variable kinetic energies. When they are smashed against a pile of atoms in the substrate, if the kinetic energy is too low, it will be bounced back by the repulsion of valence electrons. If the kinetic energy is moderate, it can transfer the kinetic energy to the bond energy and stick to the coating. During the impact, it may also transfer the energy instantaneously to a cluster of atoms. These atoms may experience transient pressures of tens of GPa and transient temperatures of thousands of degrees (Fig. 12.16). If the kinetic energy is excessive, a crater may be formed with atoms already deposited splashed out. For an atom with a kinetic energy of 1000 eV, the impact crater may affect an area of 20 μm across with a loss of about a million atoms. Thus, PVD process actually may be a violent process with continuous bombardments of high energy atoms.

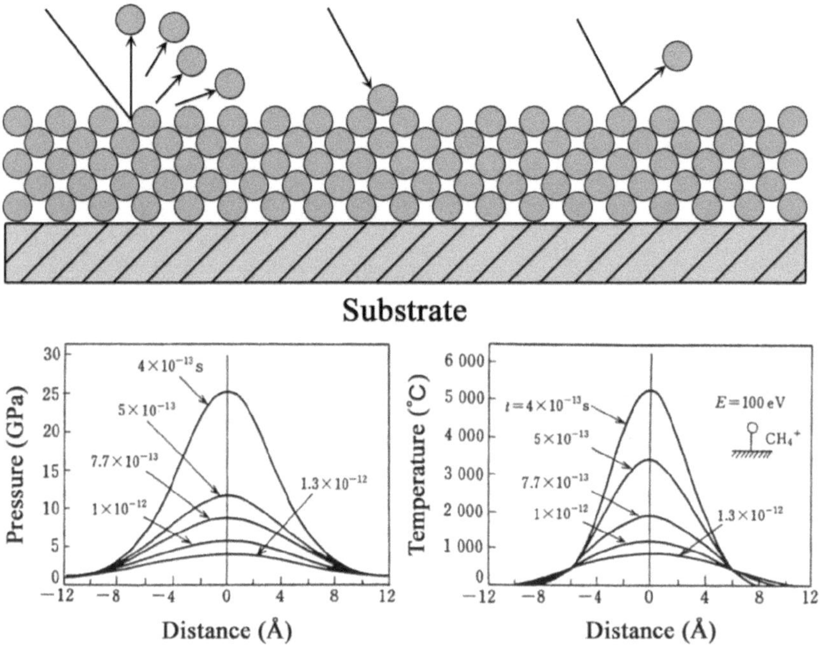

Figure 12.16 "The smashing of a carbon atom on the carbon substrate may have different results depending on the incident angle and the kinetic energy. A violent smashing may splash other carbon atoms away (top, left). The gentle impact may bounce back the carbon atom (top, right). Only when the impact is moderate can the kinetic energy be transferred to the bonding the atom to the substrate (middle). In this case, the substrate atoms are shocked to high pressure (bottom, left) and high temperature (bottom, right), but the pressure and temperature fall back to ambient conditions instantly.

12.5 Ion Bombardment

PVD of DLC coating is achieved by bombarding the substrate surface (e.g., Si) with individual ions of carbon atom (e.g., C^+) or carbon-containing molecules (e.g., CH_4^+). In order for the ions to stick to the DLC coating already deposited, the kinetic energy of these ions must at least exceed 3.7 eV required to form tetrahedral bonding of diamond structure. However, when the ion impinges on the DLC coating, its kinetic energy is spread to a group of impacted carbon atoms. As a consequence, most of the kinetic energy is dissipated in heat, i.e., vibration of surrounding carbon atoms, and only a fraction (e.g., 5%) is locked in as the bond energy. Hence, the impinging ion must carry an energy that is manyfold of that required to form tetrahedral bonds.

The efficiency of transferring kinetic energy into bond energy is dependent on the original kinetic energy. The higher the original energy, the more heat is generated during the impact, and the less energy is used in bonding. Hence, energetic beams tend to form graphitic (sp^2) bonds more than diamond (sp^3) bonds. The incidence angle may also affect the type of carbon bonds formed. A head-on collision at normal incidence will make tetrahedral bonds much easier than a glazing impact that tends to produce much heat by shaking a lot of atoms. Hence, the normal incidence of ions can form harder DLC coating than inclined incidence. The possible influence of kinetic energy and incidence angle of the ion beam on the fraction of tetrahedral bonding of deposited DLC is depicted in Fig. 12.17.

The fraction of tetrahedral bonding in a DLC may determine many useful properties, such as density, hardness, wear resistance, Young's modulus, surface smoothness, and even band gap. Hence, such properties are controllable by the kinetic energy of impinging carbon ions (Fig. 12.18).

The kinetic energy of ions is imparted by the negative bias applied at the substrate (e.g., eV is about 40% of biased voltage), so ultimately, this bias may be used to control the properties of DLC coating (Fig. 12.19). If the impinging ions are derived from hydrocarbon gas molecules (e.g., methane), then with the increasing bias, the hydrogen dominant transparent coating of polymer will give way to carbon dominant opaque film of DLC. With the further increase of the bias, DLC will eventually overheat to become graphitic carbon.

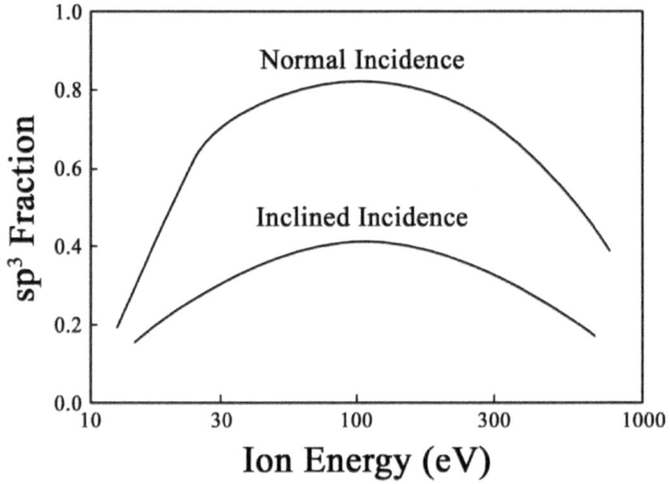

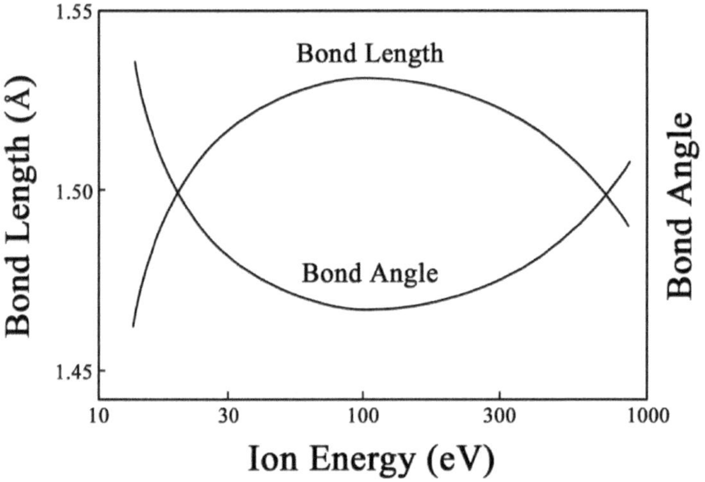

Figure 12.17 The highest fraction of tetrahedral bonding of DLC can be obtained by energizing ions around 100 eV with normal incidence (top). The variation of bond length and bond angle in tetrahedral bonding (bottom). Note that higher tetrahedral bonding ratio also implies more diamond like of the bonding.

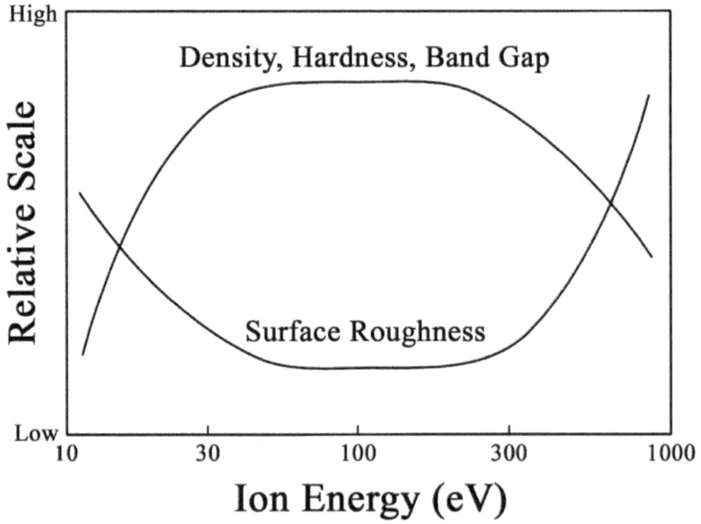

Figure 12.18 The kinetic energy of carbon ions may control the properties of DLC formed.

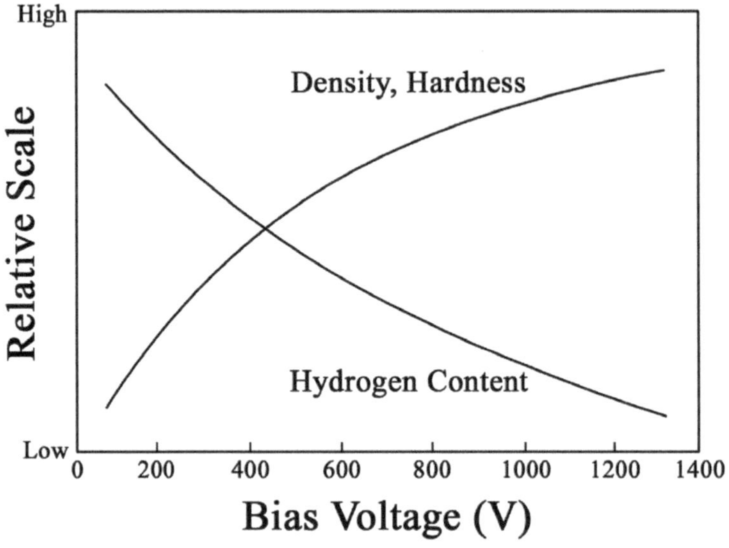

Figure 12.19 The higher negative bias at the substrate can increase the kinetic energy of methane ions. The result are decreasing hydrogen content and increasing diamond content of the DLC.

The tetrahedral bonds of carbon atoms in DLC are distorted in various degrees. The distortion energy will make tetrahedral carbon atoms highly metastable, so they can more readily than diamond atoms be converted to lower energy graphitic atoms. The higher the degree of distortion, the lower the temperature required for this conversion. The degree of distortion of tetrahedral bonds increases with the graphitic content in DLC; hence, the lower the amount of sp^3 character a DLC has, the lower the threshold temperature for it to be back converted to amorphous carbon. A DLC with a high sp^3 content (e.g., 80%) may resist back conversion up to 500°C that is much lower than the threshold temperature (e.g., >1500°C) for the graphitizaton of diamond. But a DLC with low sp^3 content (e.g., 30%) may back convert even at a temperature as low as <300°C.

During the coating process of DLC, the substrate will heat up due to the constant bombardment of ions. It is important to prevent the substrate from overheating; otherwise, tetrahedral bonds formed in DLC will be back converted to graphitic bonds (Fig. 12.20). The peak temperature of carbon bombardment is significantly higher than the average substrate temperature. In general, the substrate must be kept below 150°C; otherwise, sp^3 bonds of carbon atoms once formed may not be preserved.

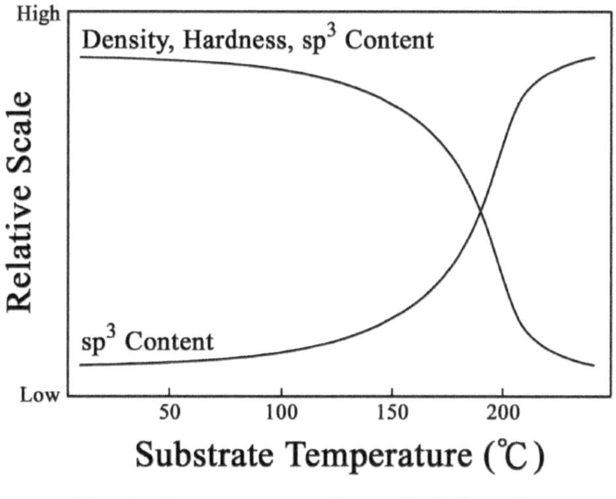

Figure 12.20 The back conversion of tetrahedral carbon atoms will accelerate above a threshold temperature of about 150°C.

In addition to the kinetic energy, incidence angle, and substrate temperature, other factors that may affect the properties of DLC films include the raw material of carbon, the vacuum pressure, and the plasma temperature. If the raw material contains only carbon, because the mass of ion is small, the kinetic energy required is also lower (e.g., 50 eV). Moreover, because there is not hydrogen atoms involved, the deposited film is denser, and it contains higher tetrahedral bonds (e.g., 85%). Such a DLC is amorphous diamond (a-D).

On the other hand, if the ion is derived from more massive molecules (e.g., benzene), the kinetic energy needed would be higher (e.g., 100 eV). But because the carbon ions contain multiple carbon atoms, the coating rate of DLC would be faster. However, because the a-C: H film contains hydrogen, it is not as hard.

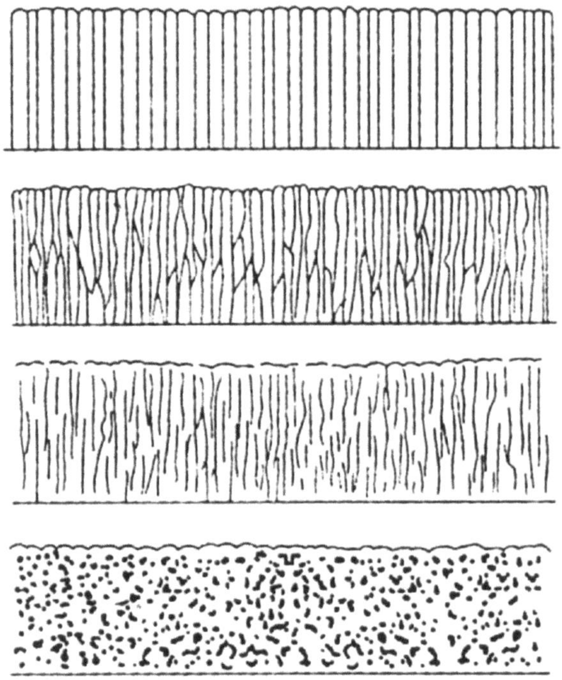

Figure 12.21 With the increasing substrate bias or the decreasing vacuum pressure, the kinetic energy of impinging ions will increase, so the DLC film will become denser and harder with structures as illustrated from bottom to top.

The effect of vacuum pressure is secondary. A higher vacuum pressure contains more gas molecules in the atmosphere. Because the mean free path is shorter, ions are more likely to collide with interfering molecules so their kinetic energies are reduced. Hence, the effect of increasing vacuum pressure is similar to reducing bias at the substrate (Fig. 12.21).

The plasma temperature may affect the concentration of ions and also their kinetic energy. A high plasma temperature will increase the content of tetrahedral bonding (Fig. 12.22). In fact, if this trend continues, the PVD process may grade into a CVD process with the possibility to form nanocrystalline diamond. If hydrogen atoms are also present, the nanocrystals will grow into micron-sized diamond so the film deposited is a CVD diamond film.

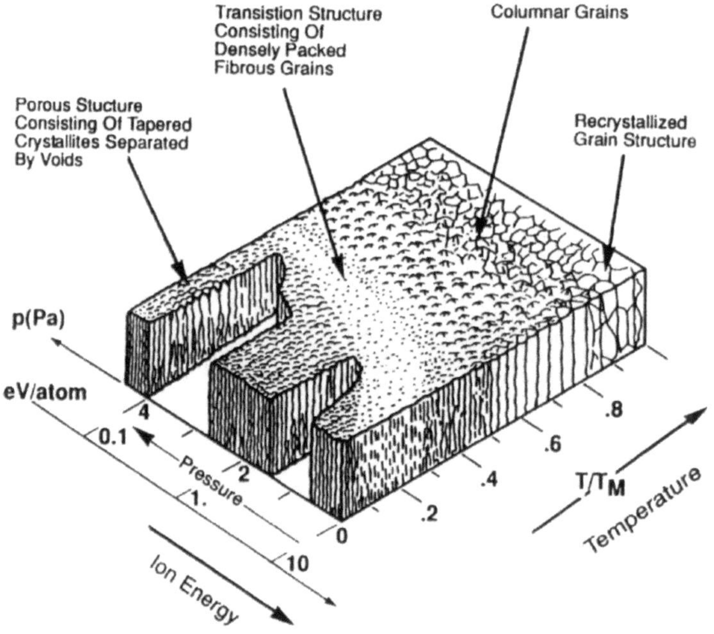

Figure 12.22 The microstructure of a PVD's DLC may grade into a CVD's diamond film with increasing plasma temperature.

12.6 Ion Generation

There are several ways to generate atoms, molecules, or ions of carbon materials and accelerate them toward the substrate to form DLC. In general, these gaseous items may acquire their energy by heating to form vapors as in the case of evaporation deposition. Alternatively, their energy may be imparted by bombarding with accelerated ions as in the case of sputtering coating. Another method is to shoot carbon-containing ions directly toward the substrate as in the case of arc deposition, also known as ion plating or ion-assisted deposition.

The simplest PVD coating method is evaporation deposition. However, carbon has the highest boiling point (e.g., >4000°C) of all materials, so it is not practical to vaporize it by heating. Moreover, the energy achievable by heating is very low (e.g., only 0.2 eV at 1770°C), although such low energy may be sufficient to coat certain non-carbon materials (e.g., a low-melting-point metal such as zinc), it would not be possible to form DLC. Hence, more energetic means must be used to coat DLC.

One way to provide energy is acceleration ions by applying an external electrical field. In order to do so, the carbon-containing gas must first be ionized. A convenient way to ionize atoms is by bombarding them with electrons. For example, by passing methane between a pair of electrodes biased with a high voltage, electrons may emit from cathode to anode through the passing methane molecules. As the result of the electron bombardments, many methane molecules are stripped with their electrons to form a plasma that contains methane cations and electrons (Fig. 12.23).

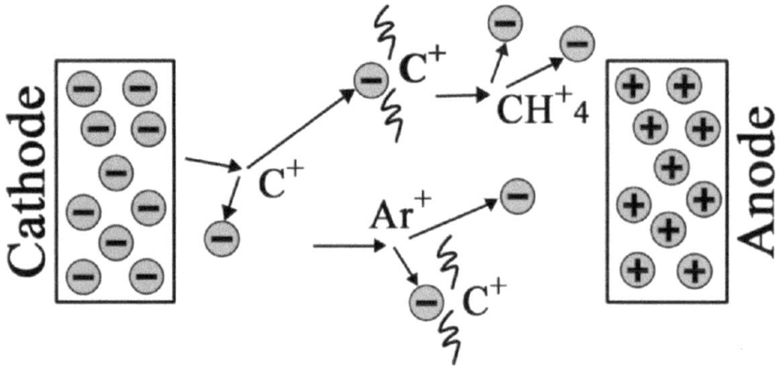

Figure 12.23 The formation of plasma by dissociating atoms and molecules with an electric arc generated between two electrodes. The figure also shows glow discharge by recombining cations and electrons to emit photons.

The plasma of methane gas may be formed inside an ion gun that allows the methane cations to be driven by a magnet toward the substrate that is negatively biased to attract methane cations (Fig. 12.24). The bombardment of methane cations on the substrate can form hydrogenated DLC (a-C:H).

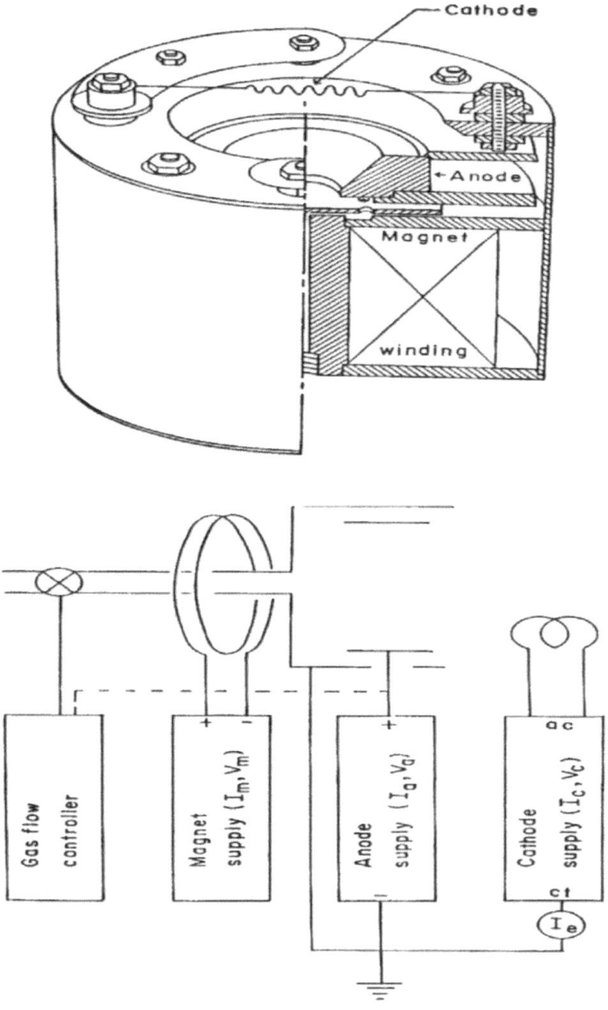

Figure 12.24 The methane cation generator (top) and its use as the ion gun of End Hall type (bottom). The top diagram shows the direction of magnet line that drives cations. The open circles denote methane molecules and the positive sign indicates methane cations.

The ion gun may also be equipped with a grid cathode in the front to further guide the shooting direction of methane cations (e.g., Kaufman design). Alternatively, the ionization may also be induced by applying radio frequency of electrical bias. These various designs of ion guns may drive methane cations to a kinetic energy of up to 1000 eV with a flux of ion current up to 10 A. The flow rate of the gas may reach 100 SCCM.

Alternatively, carbon atoms may also be vaporized and ionized by bombardment of electrons. While the electrons are constantly being knocked out of their host atoms to form cations, other electrons may join cations and release energy in the form of photons. This phenomenon is called glow discharge. The arc of glow discharge actually marks the path of electrons through the gas just like lightning displays the path of electrons through air.

When the vacuum is high, molecules and atoms are separated farther out; in this case, the resistance of electron passage is low. With the increase of gas pressure, the voltage must also increase to overcome the gas resistance. For example, it may take less than 1000 V to trigger an arc through a vacuum of less than 1 torr (about 1 thousandth of 1 atmosphere), but it may need tens thousands of volts for whipping an arc through air.

Once an arc is formed through a plasma, there are two distinct regions where the arc can be sustained: high voltage and low current, and low voltage and high current. Using the flow of water as an analogy of the flow of electricity, the former is a waterfall, and the latter, a river.

Normally, a high voltage–low current arc is generated with a high gas pressure, and low voltage–high current, with a low gas pressure. In the former case, the ions carry high kinetic energy, but their concentration is low. In the latter case, the reverse is true. These two conditions of electrical arc have their application in PVD coating of DLC. In general, the high voltage–low current arc is applied indirectly to coat a substrate with DLC by a process called sputtering. On the other hand, the low voltage-high current arc is used to spray carbon cations directly on a substrate to form amorphous diamond (Fig. 12.25).

The operation conditions for major PVD processes are summarized in Table 12.7.

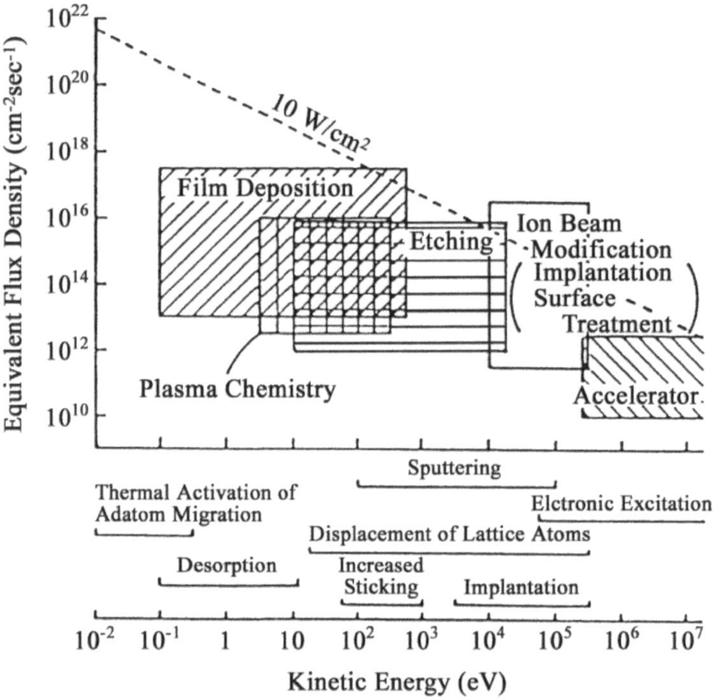

Figure 12.25 The applicability regions of various plasma processes. Note the inverse relationship between applied electric field and generated arc current.

Table 12.7 Typical coating parameters for PVD processes

Method	Pressure	Voltage	Current density	Ionized proportion	Kinetic energy
Evaporation coating	0	0	0.1–0.3	<1	10^{-3}
Sputtering deposition	>100	0.03–0.3	<1	0–10	10^{-3}
Arc deposition	<50	10^4–10^8	10–90	20–200	10^{-6}

12.7 Sputtering Coating of DLC

The high-energy, low-dose ions may not be suitable for coating DLC directly, but they may be used as the "impactor" to splash carbon atoms out of a graphite target. The splashed carbon atoms may deposit on a suitable substrate as DLC (Fig. 12.26).

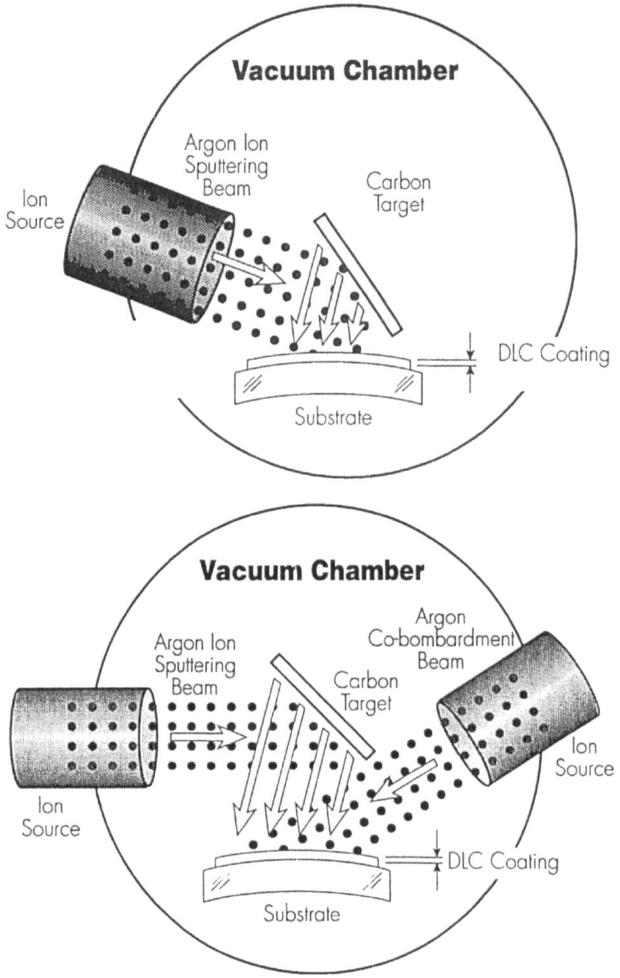

Figure 12.26 Argon ions are used to sputter carbon target to deposit DLC (top). Another argon source may be used to tuck tight the deposited coating (bottom).

Argon atoms are often used as the impactor for sputtering various materials. Argon is chosen because of its inertness with no possibility to contaminate the sputtered material by reaction. Moreover, the mass of argon's atoms is intermediate, so the impacted energy can be effectively transferred to sputtered atoms. Also, argon's atoms have low ionization potential so they are easy to form ions by bombarding with electrons.

During the sputtering, argon ions may be generated in an ion gun, or more commonly, they may be formed from local argon atoms that permeate the vacuum chamber. The argon atoms near the carbon target can be ionized in situ by a strong electrical field (e.g., by negative bias the target). The argon cations formed will be attracted toward the carbon target where they will sputter out carbon atoms or cations upon impact. However, during the bombardment of argon ions, a large amount of secondary electrons may be generated in the carbon target. These electrons may heat up the target and even form vaporized carbon atoms that will be wasted as they tend to form useless amorphous carbon.

In order to minimize the temperature of the carbon target, a set of magnets may be mounted behind it. These magnets can force secondary electrons to fly toward the center of the carbon target (Fig. 12.27). The center moving trajectory of secondary electrons can ionize ten times more argon atoms on the way so the yield of sputtering can be dramatically increased. Thus, by applying magnetron sputtering, DLC can be coated faster without overheating the carbon target.

The sputtered carbon atoms are typically with low kinetic energy. When they collide to form DLC, they form a loosely bonded carbon film with a low content (e.g., 10%) of tetrahedral bonding. Such a DLC film is not very useful. If another argon ion beam can be used to "hammer" the carbon atoms tight, a denser DLC coating may be formed. Such a DLC film may also contact a higher proportion of tetrahedral bonding (Fig. 12.28).

In addition to sputtering the carbon target with argon ions, powerful laser beams (e.g., 10^{12} w/cm^2 from an excimer laser) may also be used as the sputtering source (Fig. 12.29). Unlike the sputtering formed by ion bombardment that may contain mostly individual carbon atoms, the laser beam may splash out groups of carbon clusters (e.g., C_3 or C_4) that may contain very high kinetic energies (up to several thousand eV). In addition, molten nodules of carbon atoms may also form and they tend to mix in with deposited DLC. Because of these high-temperature phenomena, laser-sputtered DLC resembled to that deposited by an arc more so than ion-sputtered ones.

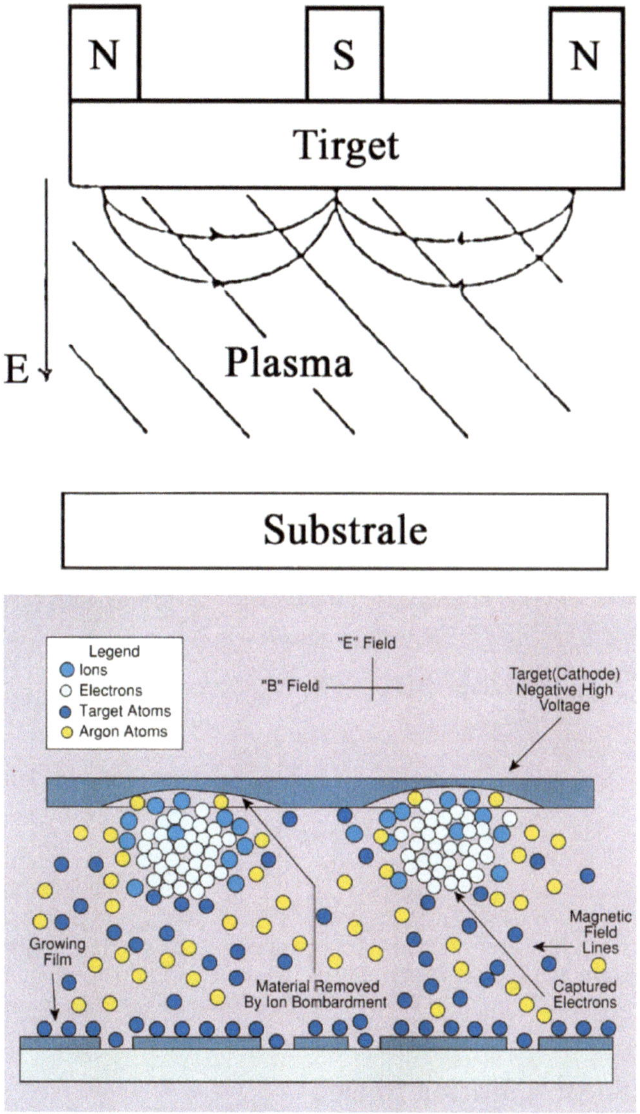

Figure 12.27 A magnet set can force secondary electrons to trajectory toward the center of the carbon target (top). This trajectory movement of secondary electrons can ionize 10-fold argon atoms for achieving much faster magnetron sputtering.

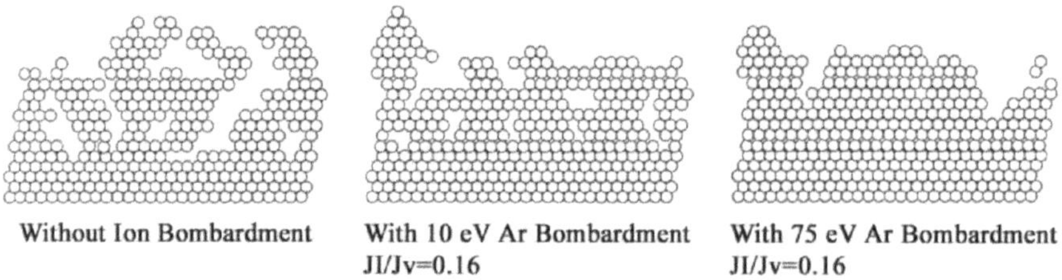

Figure 12.28 Loosely attached carbon atoms (left) can be smashed tight by bombarding with a low energy (10 eV) argon ion beam (middle). The film can be further dense packed by bombarding with a high-energy (75 eV) argon ion beam (right).

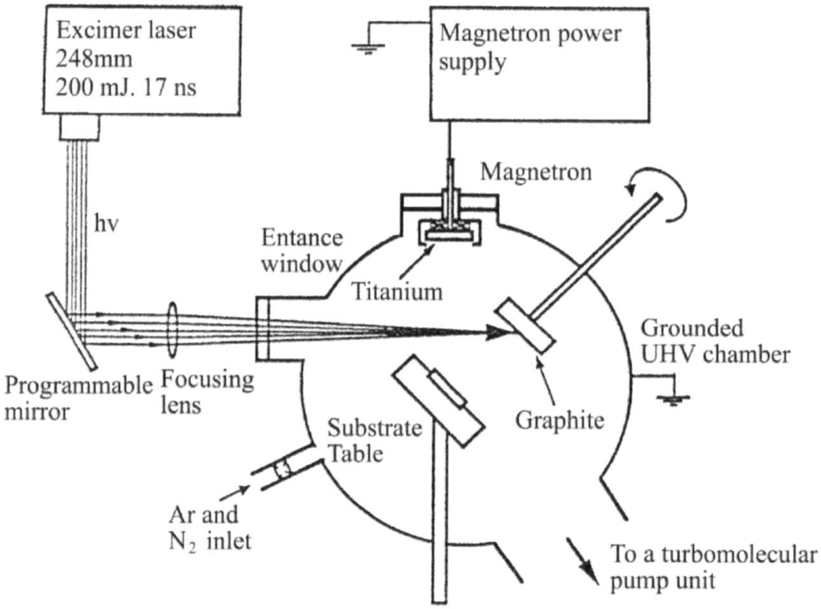

Figure 12.29 Laser sputtering deposition of DLC. The diagram also shows the ion sputtering unit for the possible addition of titanium in the deposited DLC.

12.8 Ion Plating of DLC

In contrast to the indirect deposition of DLC by sputtering, carbon-containing cations (e.g., CH_4^+) can be directly deposited on a negatively biased substrate (Fig. 12.30). In this case, carbon-containing cations form a low voltage high current arc in a high vacuum (e.g., 10^{-6} torr). Because the deposition is direct, the impinging ions have a higher energy than that sputtered by argon ions; hence, the DLC so formed may contain a higher content (e.g., 30%) of tetrahedral bonding. Moreover, because the flux of carbon-containing cations is high; hence, the deposition may also be faster than sputtered processes.

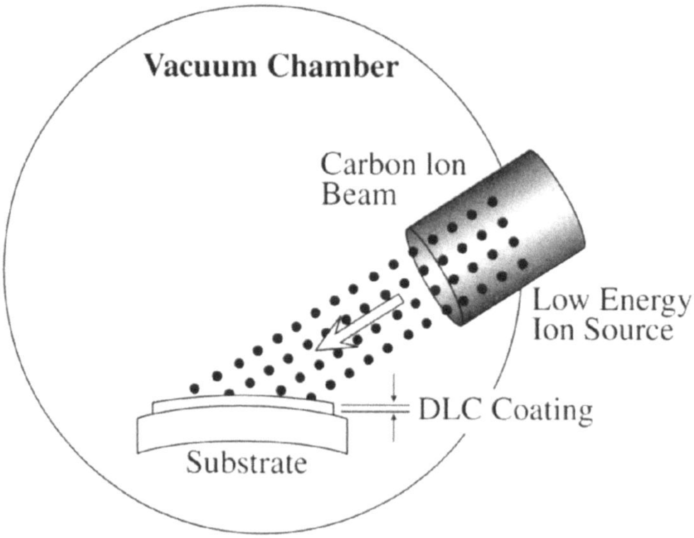

Figure 12.30 The arc deposition of DLC by bombardment of ion beam that contains methane cations.

Methane ion guns can deposit hydrogenated DLC on many substrates. The coating can be very thin but highly uniform (Fig. 12.31).

Methane ion guns have been used to coat extremely thin (e.g., <5 nm) hydrogenated DLC on magnetic head, magnetic disk, magnetic tape (Fig. 12.32). Such nanometers films may adhere with a strength exceeding 50 MPa. The film contains about 40% hydrogen. Its electrical resistivity is higher than 10^8 ohm-cm. These films are used primarily for protecting the magnetic data zone from wear or corrosion.

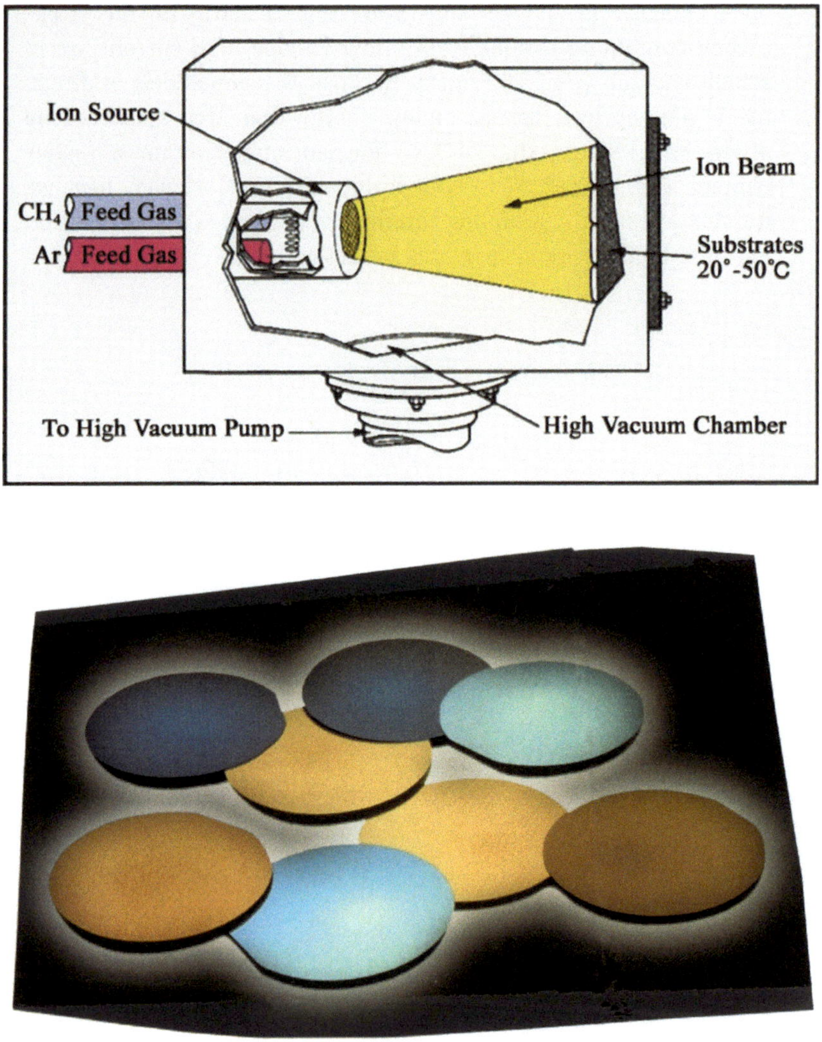

Figure 12.31 Methane cation gun is used to coat hydrogenated DLC (top diagram). Such DLC guns can be used to coat DLC on magnet heads, magnetic tapes, magneto resisters, hard drives, cassettes disks, to a thickness as thin as 5 nm. Thin (30–200 nm) coatings on silicon substrate may reveal different colors depending on thickness (bottom diagram). This color index can be used as thickness gauge for DLC coating.

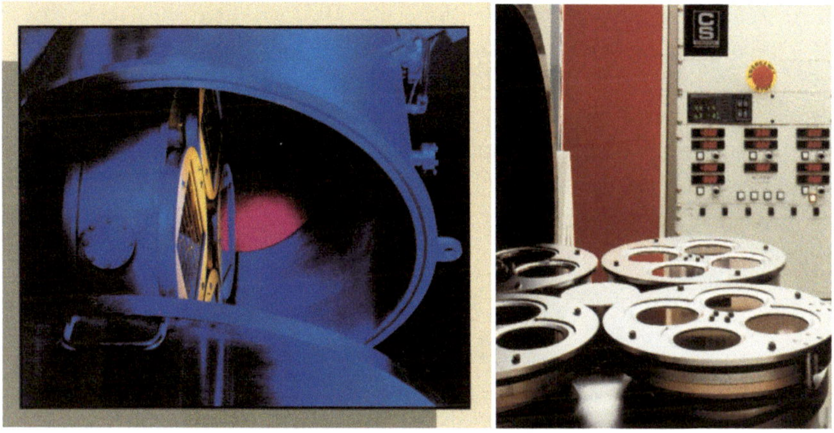

Figure 12.32 A production system for coating hydrogenated DLC (left) with rotating carousels for loading samples (right) to assure uniform coating (Commonwealth Scientific brochure).

The arc deposition of hydrogenated DLC may also be impacted by another argon ion gun, hence increasing the packing density as well as the content of tetrahedral bonding (Fig. 12.33).

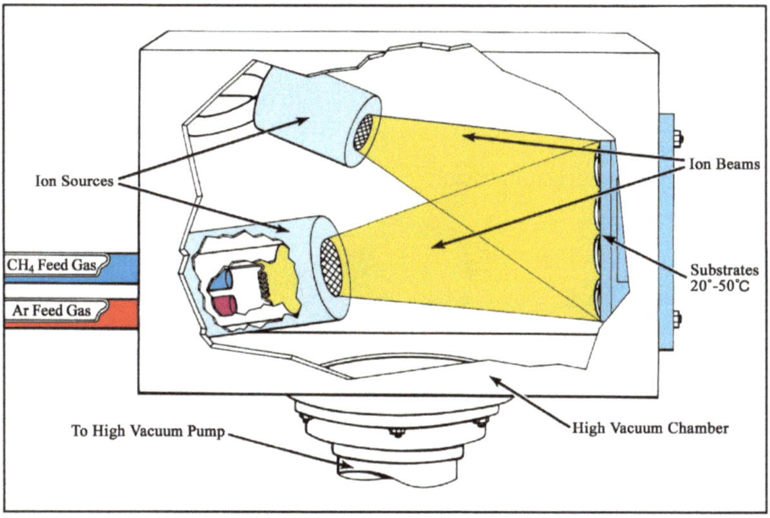

Figure 12.33 The arc deposition of DLC supplemented by the compaction of an argon ion beam.

During the arc deposition, the DLC coating may also be incorporated with non-carbon atoms (e.g., Si). The non-carbon atoms can be conveniently introduced by vaporizing the material with an electron gun directed by a magnet (Fig. 12.34). The electron gun evaporator is better than conventional resistor heater that will introduce the contaminant of the heater material (e.g., W).

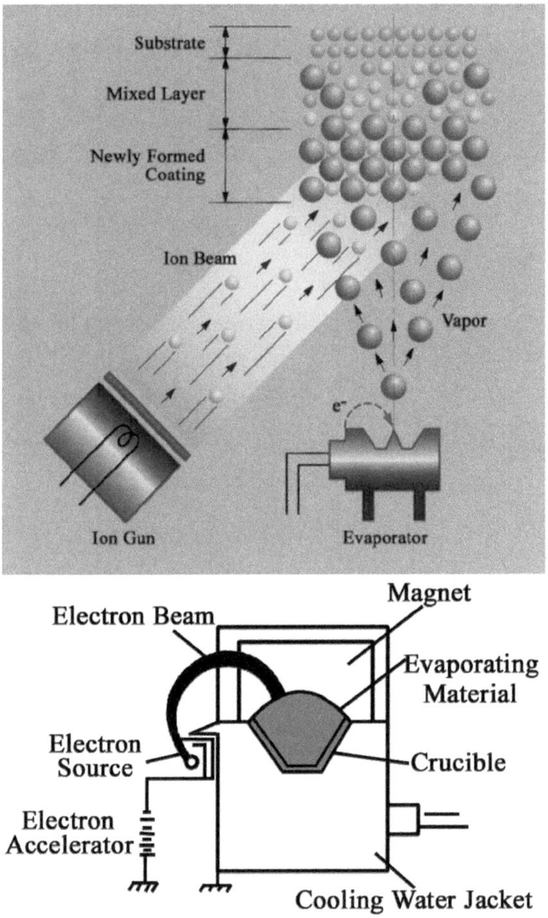

Figure 12.34 The incorporation of metal atoms in a DLC coating deposited by using an ion gun (top diagram). The metal atoms are evaporated by an electron gun that is bent by using a magnet (bottom diagram). A typical evaporator is operated at a voltage of 30 V with a current of 100 A. The evaporation rate is about 0.1 gram per minute.

12.9 Cathodic Arc Formation of Amorphous Diamond

Cathodic arc is commonly used to deposit a multitude of coating materials that may include metals (e.g., Ti, Zr), carbides (e.g., TiC, ZrC), nitrides (e.g., TiN, TiAlN), or oxides (e.g., TiO_2, Al_2O_3).

The deposited material is normally made into a cathode and it is mounted in a vacuum chamber. A pilot anode is then used to induce an electric arc from the cathode. The arc can vaporize the cathode material to form a plasma of cations and electrons (Fig. 12.35). Because the temperature may reach several thousand centigrade, droplets of cathode material may splash out and mingled in with the arc.

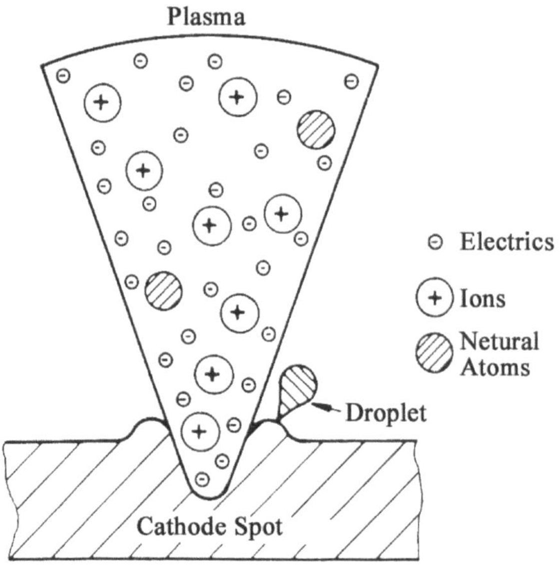

Figure 12.35 The formation of cathodic arc that may also contain clusters of solid atoms and droplets of molten material.

The cations are repelled and the electrons are attracted by an anode. The repelled cations are steered by a magnet to shoot toward a substrate that is kept at a negative bias. If the cathode is made of carbon material (e.g., graphite), the carbon cations will be attracted to the negative biased substrate where they will collide and form amorphous diamond (Fig. 12.36).

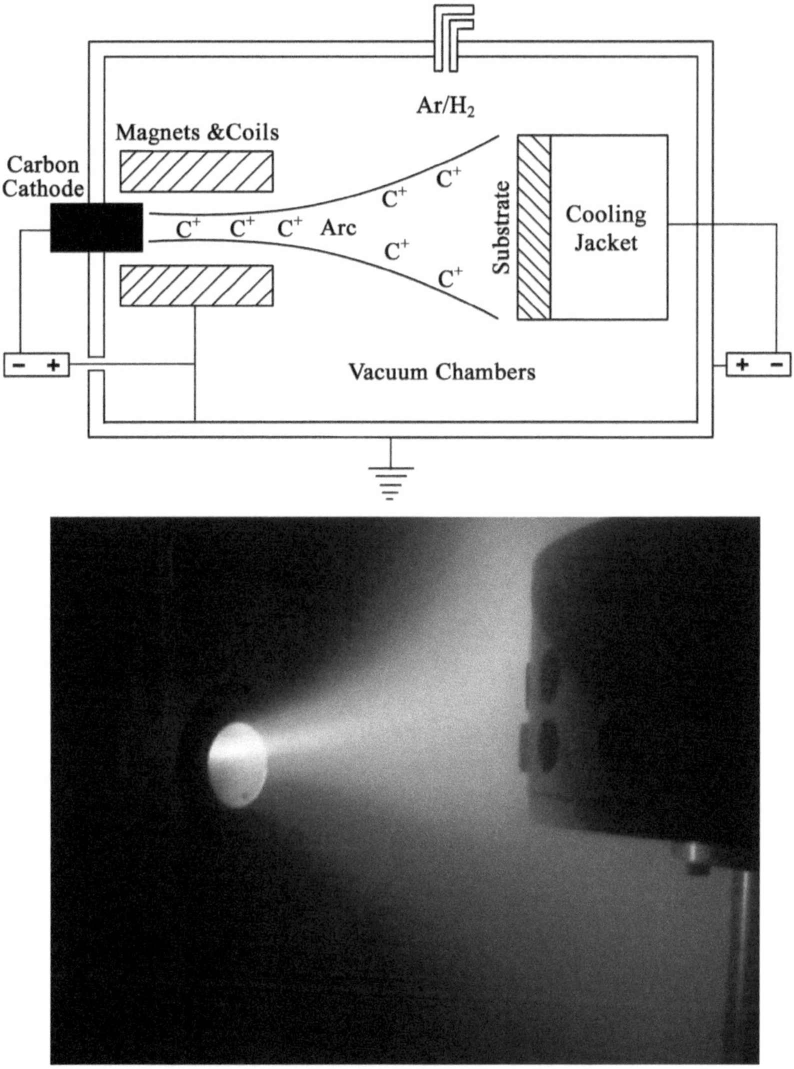

Figure 12.36 Cathodic arc deposition of amorphous diamond in principle (top) and in practice (bottom).

The cathodic arc may carry electricity as high as several million amperes per square centimeter with a power density reaching 10 million watt. The high-energy concentration has allowed a high degree of ionization (50–100%) with high kinetic energy (up to 100 eV) of carbon cations. The bombardment of such high dose of energetic cations on the substrate has allowed the deposition of amorphous diamond at high rate (up to 30 microns per hour) that is faster than most CVD of diamond films. Moreover, the amorphous diamond formed may contain high proportion of tetrahedral bonding (e.g. 85%) so the coating exhibits a hardness similar to diamond film as well.

However, the cathodic arc deposited amorphous diamond contains a high compression stress that is formed during the cooling of the substrate. As the substrate (e.g., high speed steel) tends to shrink several times more than amorphous diamond, the carbon atoms are squeezed considerably. The compressive stress will increase with the increasing coating thickness, so if the coating is more than 1 micron, it may sudden exfoliate after aging for some time (Fig. 12.37).

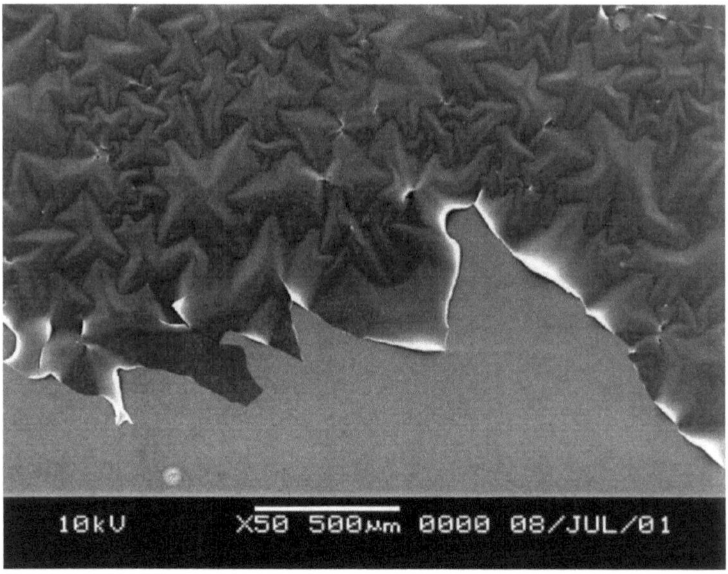

Figure 12.37 The wrinkling of an amorphous diamond squeezed by its large compression stress.

In order to cope with the attachment problem, the substrate for coating the amorphous diamond must be hard itself (e.g., cWC or tool steel). Moreover, the surface must be fully cleaned by both organic solvent and acidic solution. In addition, in process surface etching by glow discharge can effectively remove the residue contaminant (e.g., surface absorbed oxygen) that may still remain. One way to glow discharge is to apply a large negative bias (e.g., 1000 V) to the substrate in a vacuum that contains argon atoms. In this case, argon atoms will be ionized and bombard the surface of the substrate with high energy.

Once the surface is fully cleaned, the cathodic arc of carbon ions can be driven by the high bias (e.g., 500 V) of the substrate to implant beneath the substrate up to tens of nanometers. The high dose of implanted carbon atoms near the surface will harden the surface, so it is more receptive to the coating of amorphous diamond. Subsequently, the amorphous coating can be proceeded by delivering carbon cations at lower bias (e.g., 20 V). If properly coated, the amorphous diamond film can be very smooth and its surface finish may be comparable to that of the original surface of the substrate (Fig. 12.38).

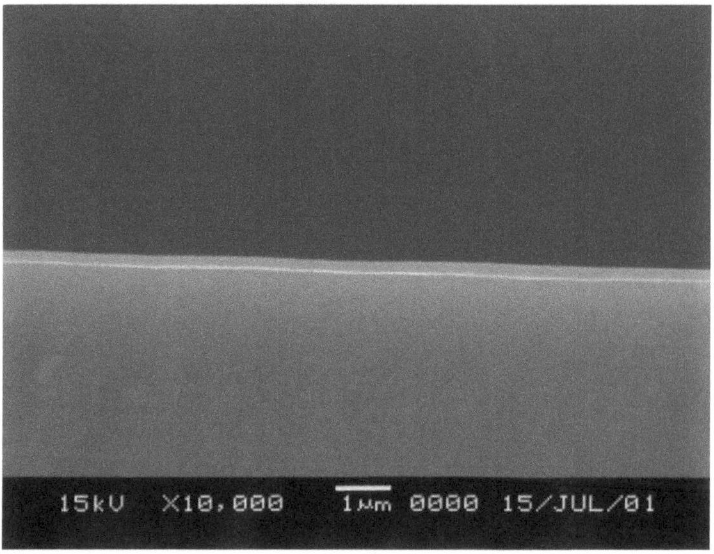

Figure 12.38 The coating of amorphous diamond on the substrate may be very uniform in thickness.

12.10 The Filtered Cathodic Arc Deposition

Although cathodic arc process has the advantage fast coating of superhard amorphous diamond with high uniformity, but this process has a problem associated with the high arc temperature. The evaporation of carbon atoms has inevitably fused carbon clusters to form droplets that may travel along with the arc, so the coating surface may invariably contain some adhered condensates of molten carbon (Fig. 12.39). These condensates are composed primarily of amorphous carbon with distorted sp^2 bonding.

Figure 12.39 The adherence of carbon clusters on the coating of otherwise smooth amorphous diamond.

The scattering of softer spots on the superhard amorphous carbon coating is prone to cause problems. As the softer spots wear out, the amorphous coating will begin to flake from the exposed region of the substrate. In addition to be the vulnerable points of the otherwise superhard coating, the protrusion of the carbon clusters can also increase the frictional coefficient so the amorphous diamond becomes undesirable for certain tribologic applications (e.g., as a protective coating of molds).

There are several ways to reduce the amount of carbon clusters or eliminate them completely from the surface of amorphous diamond. One method is to homogenize the heating on the surface of carbon cathode so that there are few or no local hot spots that are the source of forming molten carbon droplets.

The cathodic arc is triggered by a pilot anode that moves to the surface of the carbon electrode. Because of the intensification of the electrical field near the protrusion of the electrode surface, electrons will concentrate to the high point and emit toward the anode. This situation is similar to the principle of lightning rod that electrons are emitting from the sharp spike where the electrical field is the highest. As the result of the electricity concentration, the current density may approach hundred millions amperes per square centimeter. Due to the sudden heating effect, carbon will melt and explode into forming molten droplets. As soon as the molten carbon is expelled, the original protrusion point becomes a depressed area so electrons begin to diffuse toward other local high points where they will concentrate again. Some of the local high points may be created due to the condensation of molten droplets that fall back nearby. Hence, electrical spark may dance around the surface of the carbon electrode (Fig. 12.40). The erosion of the electrode will aggravate the surface relief so eventually the arc becomes so uneven that the electrode must be replaced.

Since the carbon droplets are formed by the jumping sparks on the surface of carbon electrode, one way to reduce the scale of carbon droplets is to keep the surface of the carbon electrode smooth so the electrical sparks will be more uniformly distributed. In order to minimize the surface unevenness, the carbon material must contain fine particles with very low porosity. Moreover, the electrode must be cooled efficiently to avoid overheating.

Another way is to use magnetic field to steer the random sparks on the electrode surface and sweep them rapidly across the entire surface of the carbon electrode. This use of steering arc can avoid the dwelling of sparks that tend to overheat a local region. The over concentration of the heat can create a large pool of molten carbon that may splash to form oversized droplets.

An effective method to eliminate the molten carbon droplets is to use magnets to bend the arc that contain charged particles with various mass (Fig. 12.41), similar to the principle used to separate isotopes by a mass spectrometer. The light carbon cations will turn smoothly following the magnetic field, but the oversized carbon clusters will have a much larger turning radius so they will hit the side filter. In this way, only monoatomic carbon cations can deposit the substrate so no carbon droplets will land on the amorphous diamond coating.

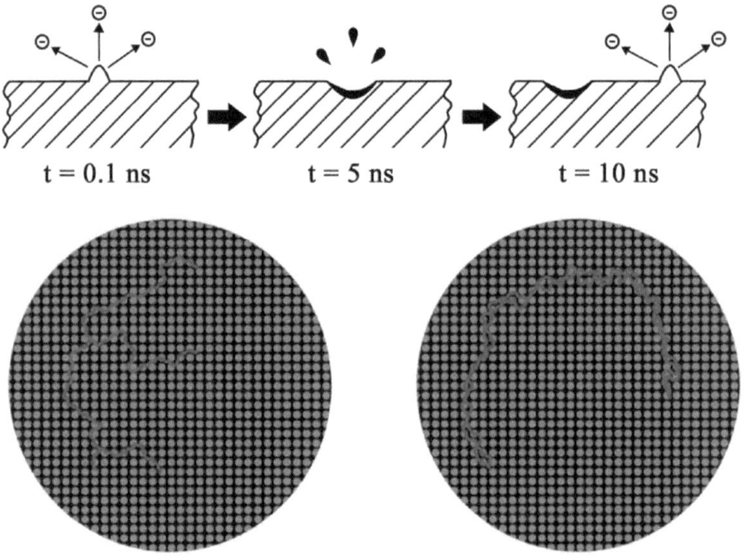

Figure 12.40 Electron emission may concentrate at local high point on the surface of the carbon electrode. The ensuing heating will splash molten carbon droplet to create a local depression. Subsequently to electron emission will move to the nearby local high point that may be formed by the falling back of carbon droplets (top). The jumping electron emission can form a dancing spark that fluctuated along the high regions of the surface of the carbon electrode (bottom).

The filtered cathodic arc may deposit a very smooth surface of amorphous diamond (Fig. 12.42). However, as the flux of the arc is diminished by the turning filter, the coating speed is greatly reduced (e.g., to only 1/5 of the original rate). Moreover, because of the travel around a longer path, the energy of the carbon cations is attenuated. As a consequence, the proportion of carbon atoms in tetrahedral bonding is lower (e.g., to <50%). But it is still much harder than most DLC films that are deposited by sputtering methods or that contain hydrogen.

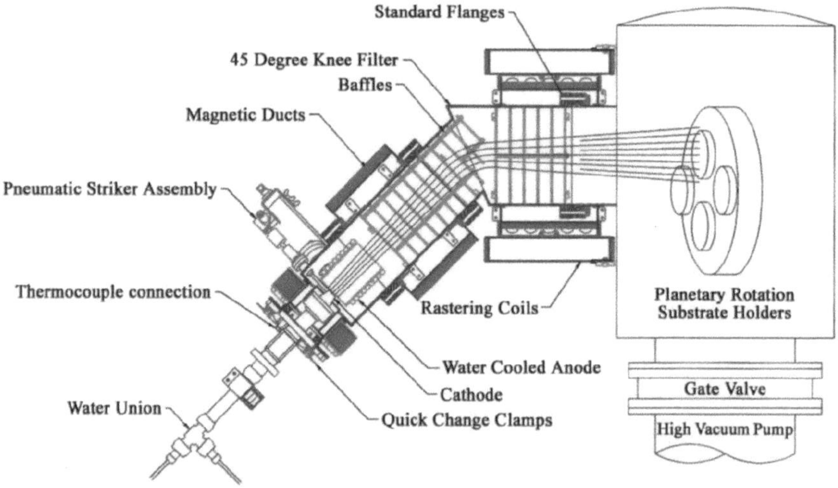

Figure 12.41 The bending of charged particles in an arc by magnets to filter out oversized carbon clusters (Commonwealth Scientific brochure).

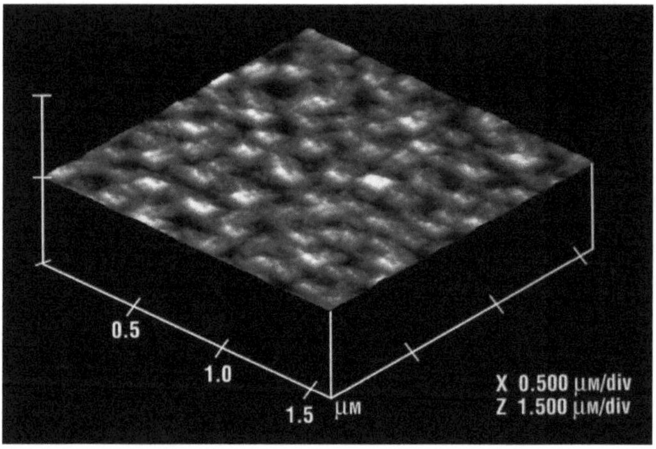

Figure 12.42 The highly smooth surface of amorphous diamond that contains no carbon clusters.

One important application of smooth amorphous diamond deposited by filtered cathodic arc is to coat the hard drive, which is loaded with ever-increasing density of magnetic signals (e.g., 10 Gb/in^2) (Fig. 12.43).

The trend of increasing the memory density has been even more spectacular than that dictated by Moore's Law for transistors. In fact, while transistors are entering the nanometer range (e.g., 90 nm for 2005), memory storage is now planning for atomic size using diamond as the substrate for the ultimate data storage (Fig. 12.44).

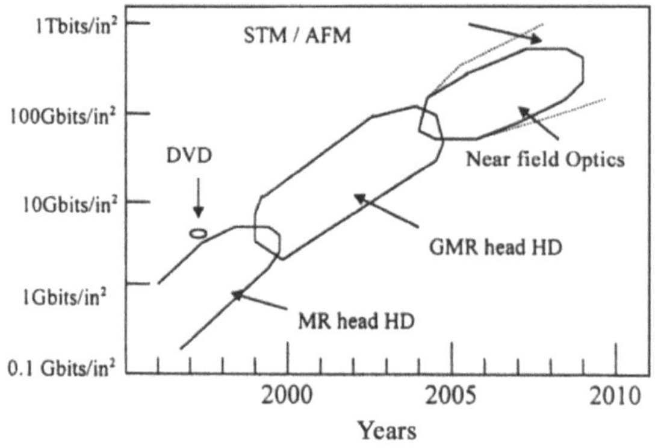

Figure 12.43 The future trend of information storage density.

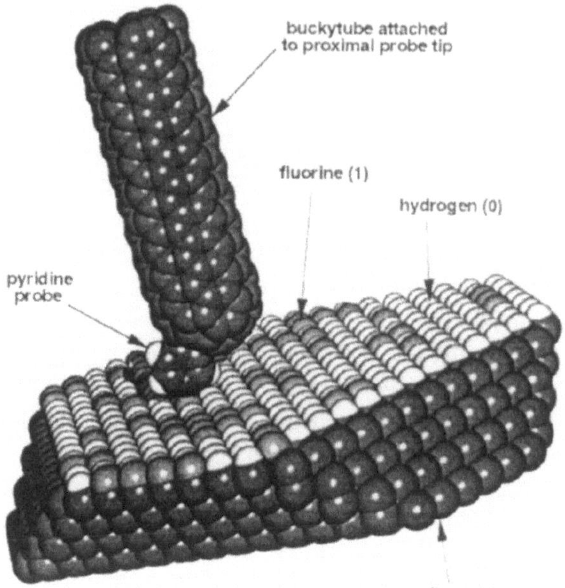

Figure 12.44 A novel data storage system capable of 10^{15} bytes/cm^2 is being explored by using diamond as the storing house. In this case, H atoms would be designated as 0 and F atoms as 1. A SWNT tip is used to distinguish between 0 and 1 rapidly and unambiguously.

For the present technology of using magnetic elements as the binary code, with a trend of miniaturizing memory space, the surface smoothness will also approach 3 Å, which is only about two layers of atoms. Moreover, the flying height of the read head based on giant magnetic resistivity (GMR) is narrowing down to about 1 nm (Fig. 12.45).

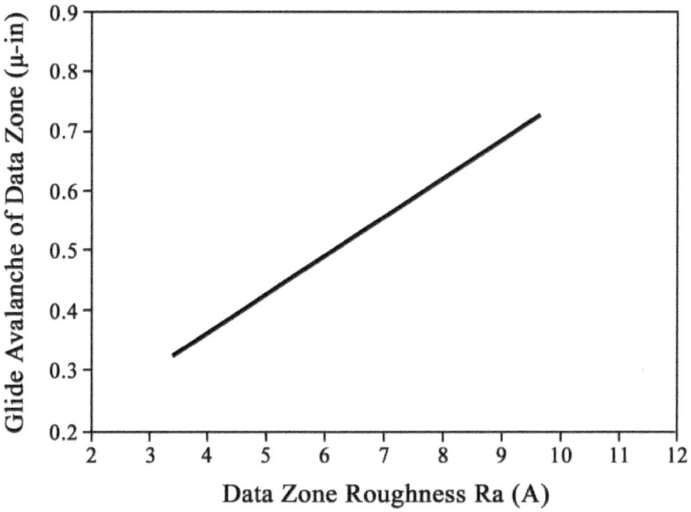

Figure 12.45 The dependence of flying height of the read head on the surface roughness of hard drive.

In order to protect the magnetic memory from the catastrophic scratches of the read head, the surface of the magnetic memory is coated with a thin layer of DLC (Fig. 12.46).

Normally, the DLC coating on hard drive is hydrogenated, which is deposited by sputtering. However, because such DLC is not superhard, it is required to have sufficient thickness to avoid being worn through. But as the flying height is no longer sufficient to allow such a coating thickness, harder DLC must be used. Consequently, CN films have been adapted as they are harder and also more slippery. However, the CN film contains a significant amount of graphitic bonds. Hence, the amorphous diamond deposited by filtered cathodic arc would be a better substitute, particularly if the amorphous diamond is also doped with nitrogen to increase the hardness, to reduce the friction, and to enhance the inertness.

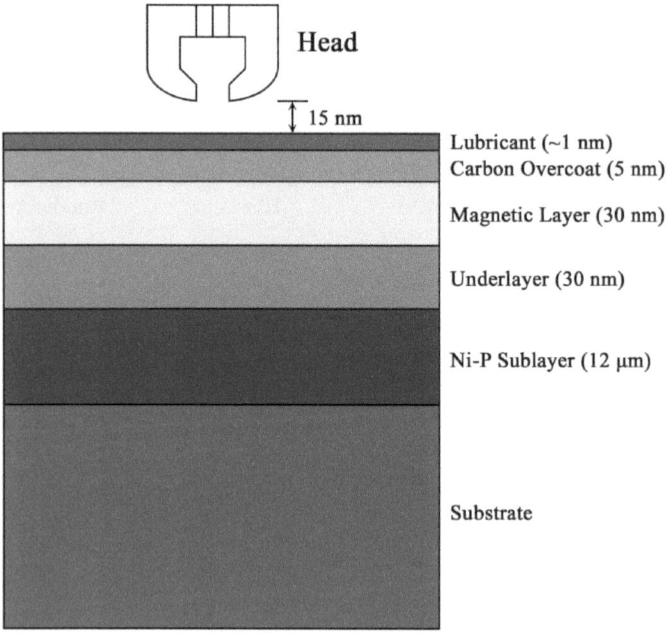

Figure 12.46 An example of the hard drive structure that shows the closeness of the read head to the top of the hard drive. The read head is looming above the top surface of the hard drive with conditions that may be compared to flying a Boeing 747 jumbo airplane just 10 m above a flat ground.

12.11 RF CVD Deposition of DLC

In addition to PVD method, very-low-energy (e.g., RF) plasma CVD may also be used to coat DLC. For example, hydrogenated DLC may be deposited by the following process (Table 12.8).

Table 12.8 The coating parameters of H-DLC deposited by RF CVD

Plasma	13.56 MHz
Substrate	Capacitively coupled
Pressure	10^{-1}~10^{-4} mbar
Gas	Ar, H_2, C_2H_2
Growth rate	0.5–6 μm/h
Coating thickness	0.5–5 μm

The properties of DLC deposited by RFCVD are compared with DLC deposited by PVD methods in Table 12.9.

Table 12.9 The comparison of DLC properties

	H-DLC	Si-DLC	C-DLC
Process	RFCVD	10n Gum	Cathodic Arc
Hardness ($HV_{0.025}$)	4000–6000 kg/mm^2	1200–1600 kg/mm^2	>6000 kg/mm2
Nanohardness	20–40 GPa	10–15 GPa	>70 GPa
Young's modulus	~250 GPa	~150 GPa	~800 GPa
Adhesion	Medium	Strong	Weak
Stress	1 GPa	0.5 GPa	2 GPa
Thermal Stability	300°C	250°C	500°C
Friction Coefficient	0.05	0.03	0.1

The H-DLC deposited by RFCVD is fairly uniform (Fig. 12.47). Moreover, the stress in the coating is low, so the adherence strength is high. Hence, the common problem of coating flaking is avoided.

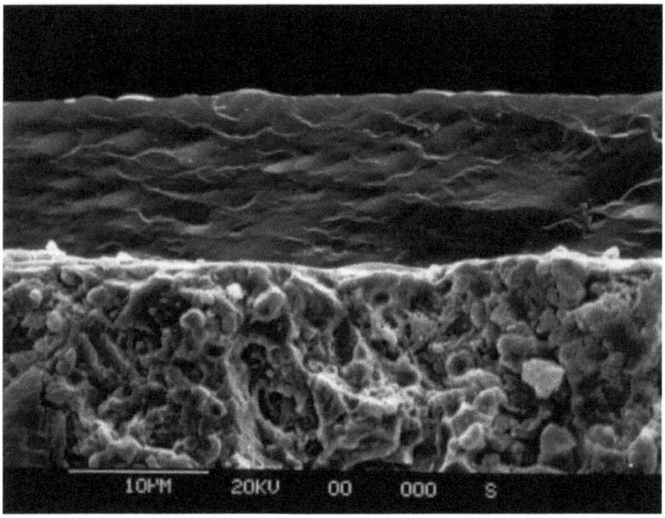

Figure 12.47 The amorphous appearance of hydrogenated DLC deposited by RFCVD.

RFCVD-deposited H-DLC is considerably harder than "soft" DLC deposited by ion plating, particularly, those that contain metal for reducing stress and friction (Fig. 12.48).

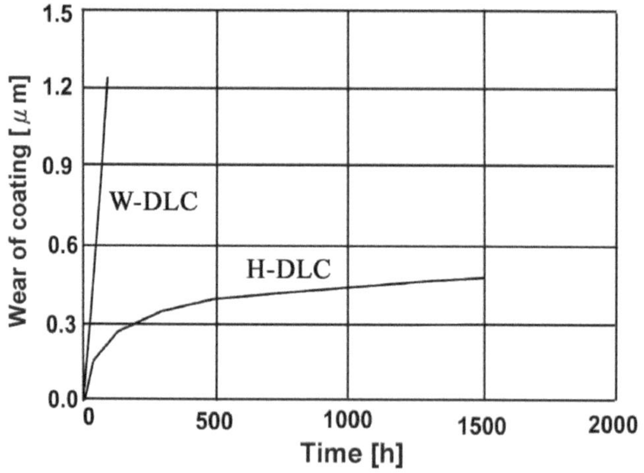

Figure 12.48 The wear of hydrogenated DLC deposited by RFCVD is much less than tungsten-impregnated DLC deposited by ion plating. Both DLCs are designed for coating on injector piston.

Chapter 13

Applications of Diamond-Like Carbon

Diamond-like carbon can be coated on almost everything from metal to plastic.

13.1 Diamond-Like Carbon

Although diamond has many extreme properties that may applied at coating material, but because of the incompatibility between these properties and the substrate materials, diamond cannot be coated on ordinary materials. In addition, diamond film is deposited by CVD processes that often require a high substrate temperature (e.g., >800°C), consequently, there are only a handful materials (e.g., Si, cemented WC, sintered SiC, Si_3N_4, or refractory carbide formers such as Ti, W) that may bond to diamond at such temperature. Thus, more typical industrial materials such as tool steels and aluminum silicon alloys cannot be coated with diamond. Hence, a diamond-like carbon (DLC) film that retains part of diamond properties may be used as a more general coating material.

Handbook of Industrial Diamonds: Diamond Films and Carbon Coatings
James Sung

ISBN TBA (Hardcover), TBA (eBook)
www.jennystanford.com

In order for the coating material to be less diamond (e.g., thermal expansion) in certain ways but still with some diamond attributes in other ways, the continuity of diamond lattice with tetrahedral bonds (sp^3) in the coating material must not be retained. Specifically, diamond properties may be partially retained by making the structure amorphous. Additionally, the coating material may contain non-diamond structures (e.g., planar or graphitic sp^2 bonds) or non-carbon elements (e.g., H). Thus, DLC falls into two categories: amorphous diamond (a-D), which contains pure carbon atoms held together by distorted tetrahedral bonds and distorted planar bonds, and hydrogenated DLC (a-C:H), which contains in addition hydrogen, and sometimes, other elements (e.g., Si, W). Hence, DLC is a composite material like alloy for metal or plastic for organic materials. Amorphous diamond is a physical composite of DLC; and hydrogenated DLC, a chemical composite (Fig. 13.1).

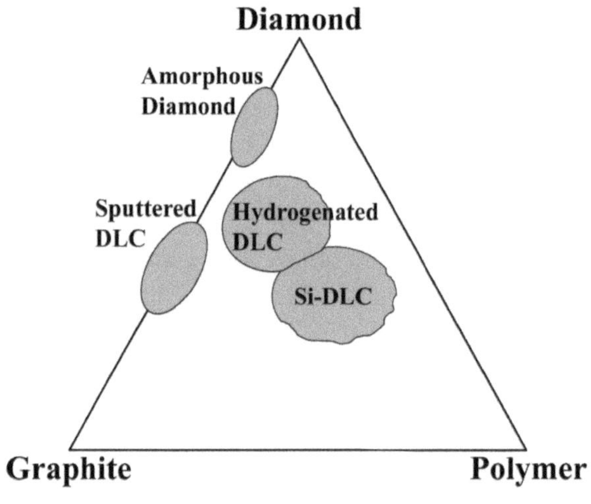

Figure 13.1 The composite nature of DLC and the compositional difference of amorphous diamond and hydrogenated DLC. The former contains pure carbon. The latter incorporates polymer-like hydrocarbons.

The composite nature of the DLC has compromised many diamond's extreme properties. For example, although amorphous diamond may retain much of the diamond hardness, but hydrogenated DLC is generally not superhard (Knoop micro hardness typically <4000 Kg/mm^2).

In contrast with hydrogenated DLC, amorphous diamond does not contain hydrogen, which makes DLC less hard. Such films may contain a higher proportion (>80%) of diamond bonds, so the hardness can be more than twice that of hydrogenated DLCs (Fig. 13.2).

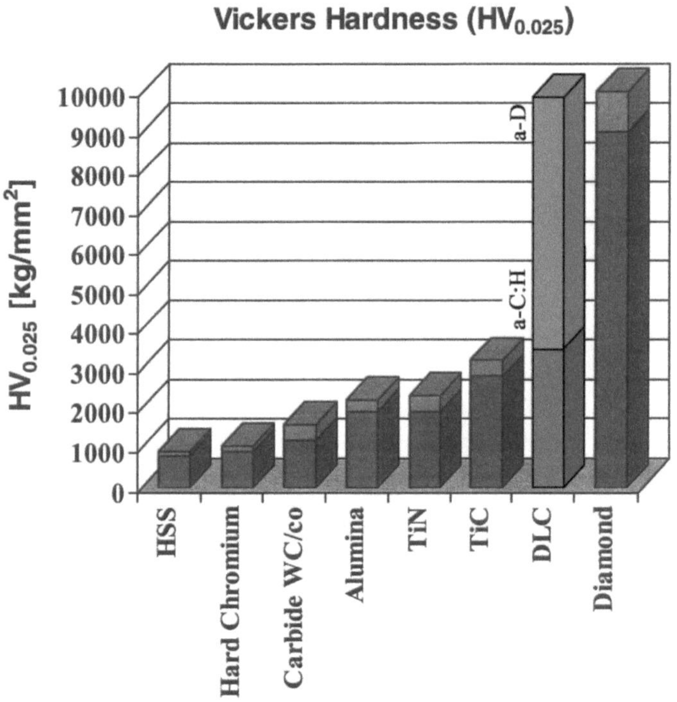

Figure 13.2 The comparison of hardness of various substrate and coating materials. The amorphous diamond with more than 80% in distorted tetrahedral bonding may be as hard as diamond itself because of its ability to prevent the sliding of dislocations.

Amorphous diamond and hydrogenated DLC differ not only in bonding types and thermal stability, but also in substrate adherence and coating properties, such as hardness (Fig. 13.3).

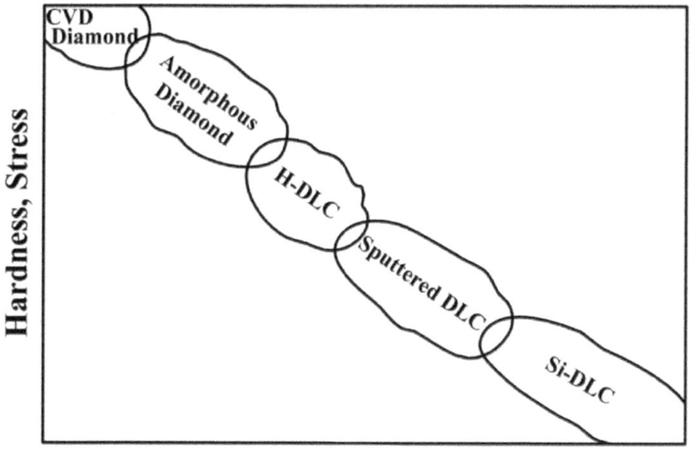

Figure 13.3 The trade-off of DLC properties between hardness and smoothness.

13.2 Coating Thickness

One of the diamond attributes that is always desirable is hardness. However, a harder material tends not to respond to temperature changes, so the thermal expansion coefficient of DLC is generally much lower than that of its substrate material. Consequently, the interface is likely stressful and it is prone to delamination. Most DLC films are deposited under ion bombardments. The impinging force tends to squeeze the structure. Moreover, the larger thermal contraction of the substrate material when it is cooled from the coating condition also tends to compress the joined DLC. As a result, a compressive stress is developed within the film. This compressive stress will increase with the increasing of coating thickness. Therefore, most DLC coatings are relatively thin (e.g., <3 microns) in order to avoid the accumulation of high stress.

DLC is the thinnest diamond-containing layer of various products (Fig. 13.4). As a result, it can be coated on a great varieties of substrate materials (metals, glass, polymers), even with complicated shapes.

DLC is formed at a lower temperature than CVD diamond film, so its thermal stability is much lower. Amorphous diamond is stable at temperatures up to 500°C. However, the presence of a significant amount of hydrogen (e.g., about 1/3) in hydrogenated DLC has made it unstable at temperatures above 250°C.

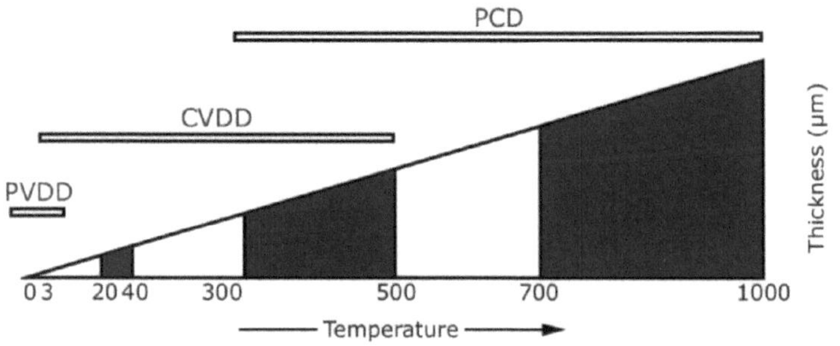

Figure 13.4 The temperature of formation and the thickness of diamond containing material formed. PVDD = Physical vapor deposited diamond (<200°C), CVDD = Chemical vapor deposited diamond (600–900°C), PCD = Polycrystalline diamond sintered under high pressure (>1200°C). The blackened bands represent industrial applications of diamond layers.

Figure 13.5 DLC coating may be much more smooth (e.g., Ra < 1 nm) than conventional ceramic coating (e.g., Ra > 10 nm).

As a thin coating, DLC can be much more smooth than conventional ceramic coatings (Fig. 13.5). This smoothness also aids to its low frictional coefficient that is critical for most tribologic applications.

13.3 Friction Force

One of the major applications of DLC is in tribology, which may require a very slippery surface. To make a surface slippery, the frictional coefficient must be low. The frictional coefficient can be reduced in two opposite situations: between soft materials as in the case of lubrication or between hard materials as in the case of sliding. The friction is reduced in lubrication because soft materials (e.g., teflon or graphite) will yield under stress. In this case, the smoothness of the interface is not essential. The friction is reduced in sliding because hard materials do not interact (e.g., diamond of DLC). In this case, the smoothness of the contacting surfaces is critical (Fig. 13.6).

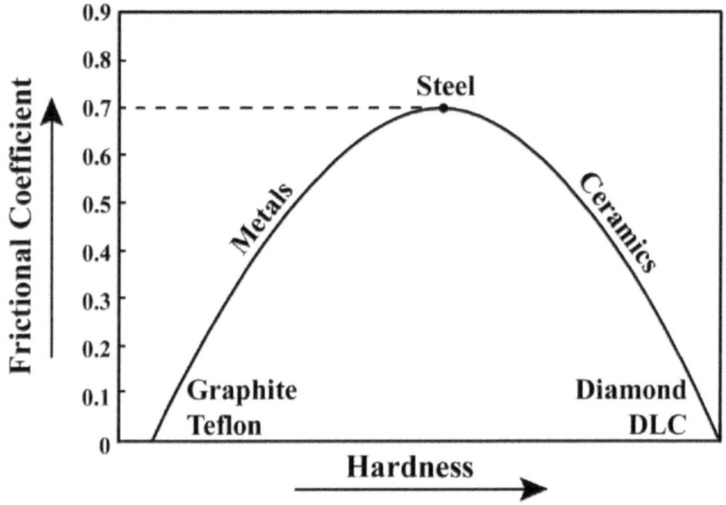

Figure 13.6 The frictional coefficient may be decreased by yielding of soft materials or sliding of hard materials. In the latter case, although DLC is less hard than diamond, it may have even lower frictional coefficient because of the surface treatment to eliminate dangling electrons.

The frictional coefficient of DLC is typically much lower than all other coatings including solid lubricants such as graphite, molybdenum disulfide. In fact, DLC can be made much more slippery than even teflon (Fig. 13.7).

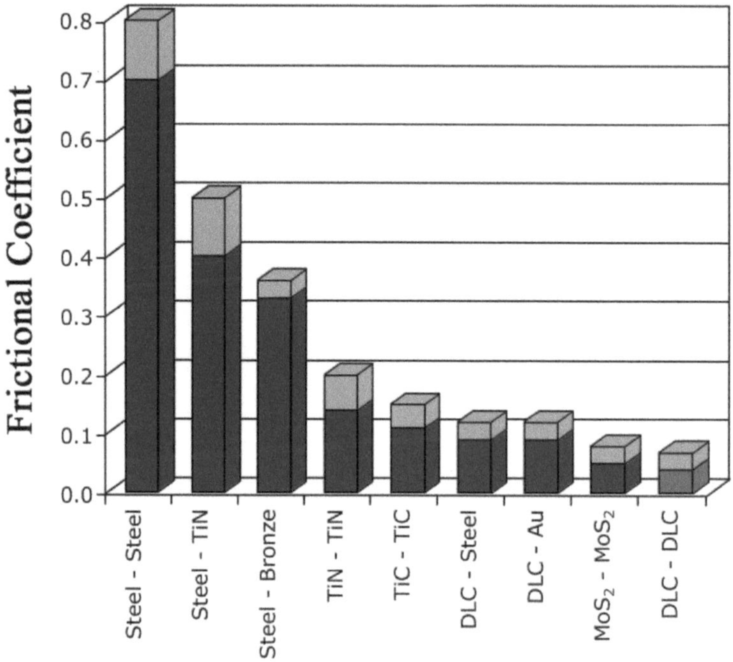

Figure 13.7 The frictional coefficient of steel and various coating materials. Note that DLC is the most slippery coating material available.

The frictional coefficient of diamond surface is dependent on several factors, the smoothness of the contact surfaces, the surface chemistry of absorbent species, and the atmospheric conditions.

The surface roughness may increase the friction between two contacting surfaces, particularly when they are under the pressure that would increase their contact area (Fig. 13.8).

The friction between two diamond surfaces is further complicated by the interaction of dangling electrons on the diamond surface. These dangling electrons may be shield in dry nitrogen or air so the frictional coefficient is low. But in humid air, the water molecules may be polarized between the two sliding diamond surfaces that contain dangling electrons. Such an "induction" resistance may increase the frictional coefficient. However, in vacuum, when no shield is present between two diamond surfaces, the frictional coefficient can suddenly surge because the drag of dangling electrons across the sliding surfaces (Fig. 13.9).

However, if one surface does not contain dangling electrons, then vacuum will actually decrease the dragging force at the interface, as in the case of sliding a polished diamond surface against a film of cubic boron nitride. On the surface of cubic boron nitride, there are no dangling electrons because boron has no unbonded electron and nitrogen has a stable lone pair of unbonded electrons (Fig. 13.10).

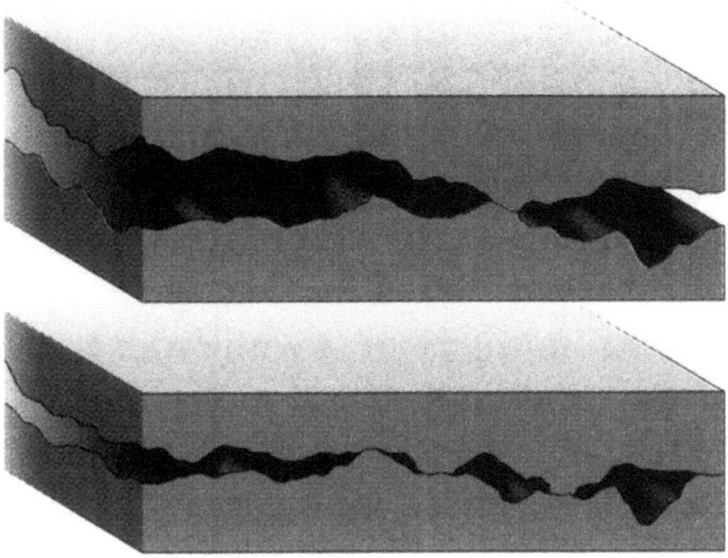

Figure 13.8 The contact between two rough diamond surfaces. The friction arises at contact points when they slide against each other (top diagram). If the pressure is applied so the contact area increases (bottom), the friction force will increase. It is this contact area, not the roughness, that determines the ultimate friction between the two surfaces.

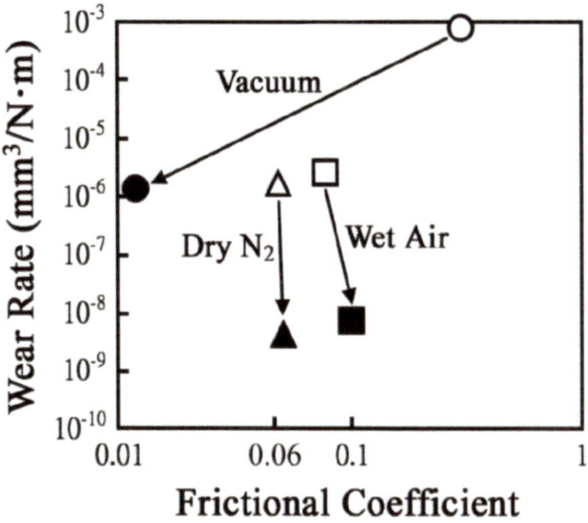

Figure 13.9 The friction between polished surfaces of CVD diamond and sintered Si_3N_4 (open circles), and between polished CVD diamond and cubic boron nitride film (black circles). Note that vacuum can increase the friction force in the former case, but it decreases the friction force in the latter case. The reason is that cubic boron nitride contains no dangling electrons on the surface so vacuum has no effect on the surface dragging force.

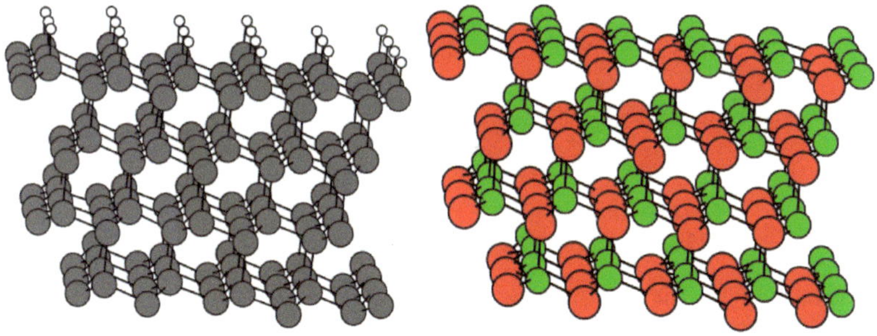

Figure 13.10 The dangling electrons (open circles) present on the carbon atoms (black balls) for the (111) face of diamond (left). The interaction of these dangling electrons across two sliding surfaces in vacuum can increase the friction force. The absence of dangling electrons on the same face of cubic boron nitride (right). In the latter case, boron atoms (red balls) do not possess unbonded valance electrons, whereas nitrogen atoms (green balls) contain unbonded electrons in a stable pair that does not interact.

For an ultimate slippery surface, the dangling electrons on diamond surface, e.g., (111) plane, may be sealed by hydrogen termination. This can be achieved by heating diamond under hydrogen gas to a high temperature (e.g., 800°C), or by bombarding its surface with hydrogen plasma in vacuum. The hydrogen-terminated diamond surface show partial positive charge due to the presence of polarized hydrogen protons so the contact surface is actually under repulsive electrical force. The frictional coefficient between such contactless surface is less than 0.01 that is much slippery than even grease (Fig. 13.11).

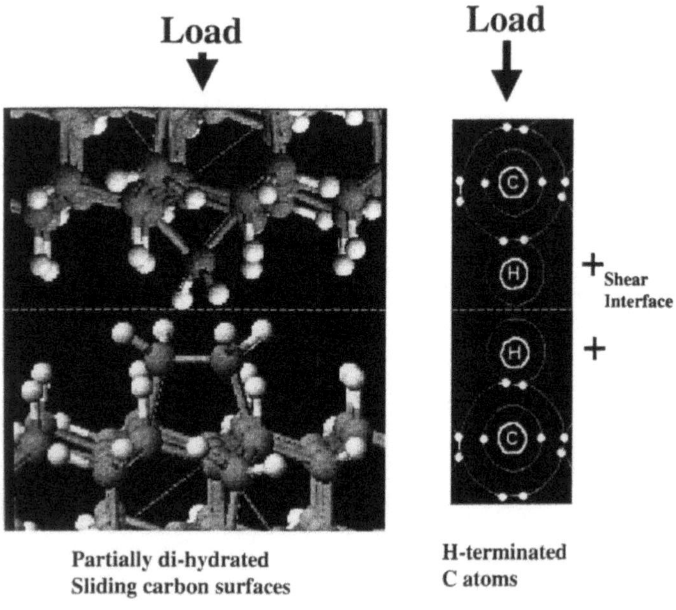

Figure 13.11 The hydrogenated surfaces of DLC (left) are separated by the mutual repulsion of opposing hydrogen atoms that show positive polarization.

If the two opposing diamond surfaces are fluorinated, then he repulsive force between them will be much larger than hydrogen-terminated diamond surfaces because each fluorine atom contains three lone pairs of electrons. In this case, the separation between the two fluorinated diamond surfaces are much farther than that for hydrogen-terminated ones. Such an action at distance between two fluorinated diamond surfaces, like two magnets suspended by mutual repulsion, would make the sliding of the two opposing surfaces virtually frictionless. This would be the lowest amount of interaction achievable between any pair of opposing surfaces.

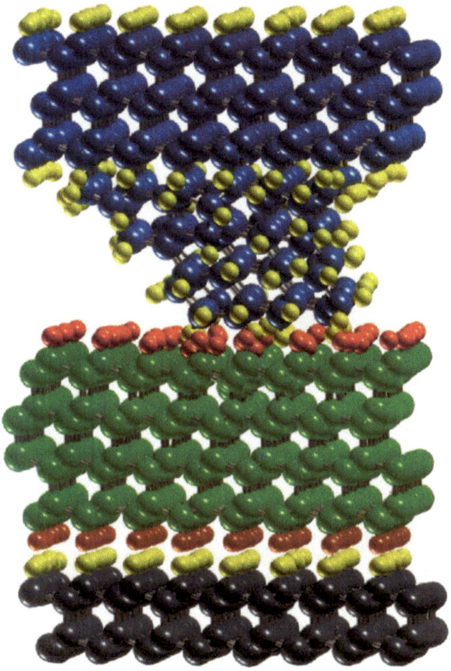

Figure 13.12 The sliding of a hydrogen-terminated diamond indenter over another hydrogen-terminated diamond surface may introduce physical deformation due to the mutual repulsion of electrostatic force. In this case, no chemical interaction is involved (upper interface). The sliding between two fluorine terminated diamond surfaces will be virtually frictionless because the opposing fluorine atoms will keep the two surfaces widely separated (lower interface). In the above illustration, large atoms are carbon, middle ones are fluorine, and small bunch are hydrogen.

In addition to the possible mechanical interference and electrical interaction between two diamond surfaces, chemical reactions may also occur at the interface (Fig. 13.13).

Since hydrogen termination may plug the dangling electrons of carbon atoms; hence, higher amount of hydrogenation of DLC may reduce the frictional coefficient (Fig. 13.14).

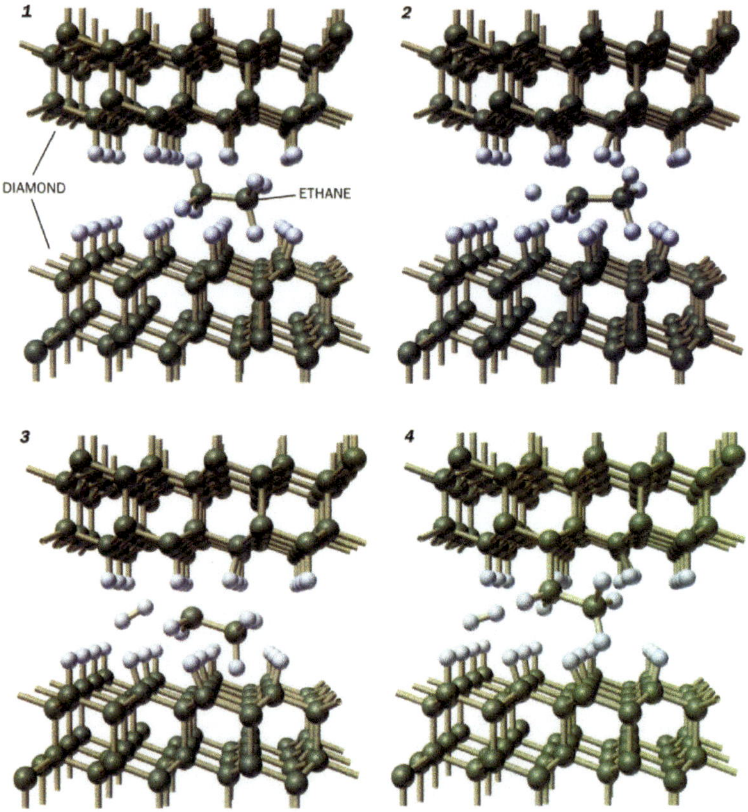

Figure 13.13 Chemical reaction may occur between two sliding surfaces of diamond. In this illustration, an ethane molecule, composed of two carbon atoms (green) and six hydrogen atoms (blue), is sandwiched between two diamond surfaces, which are terminated with hydrogen atoms (1). As the surfaces slide, the ethane loses a hydrogen atom (2), becoming an ethyl radical. The free hydrogen subsequently removes a hydrogen atom from the diamond and bonds with it to form a molecule of hydrogen gas (3). The ethyl radical eventually becomes chemically bound to one of the diamond surfaces (4). The diagram is based on computer simulations conducted by Judith A. Harrison and her co-workers at the U.S. Naval academy.

Although hydrogenation may reduce the frictional coefficient, there are undesirable properties coming along with hydrogen. In addition to have a lower hardness of the DLC, its property may also be unstable as hydrogen is interactive with moisture and oxygen in air. Hence, the low frictional coefficient of hydrogenated DLC may change in air with high relative humidity (Fig. 13.15).

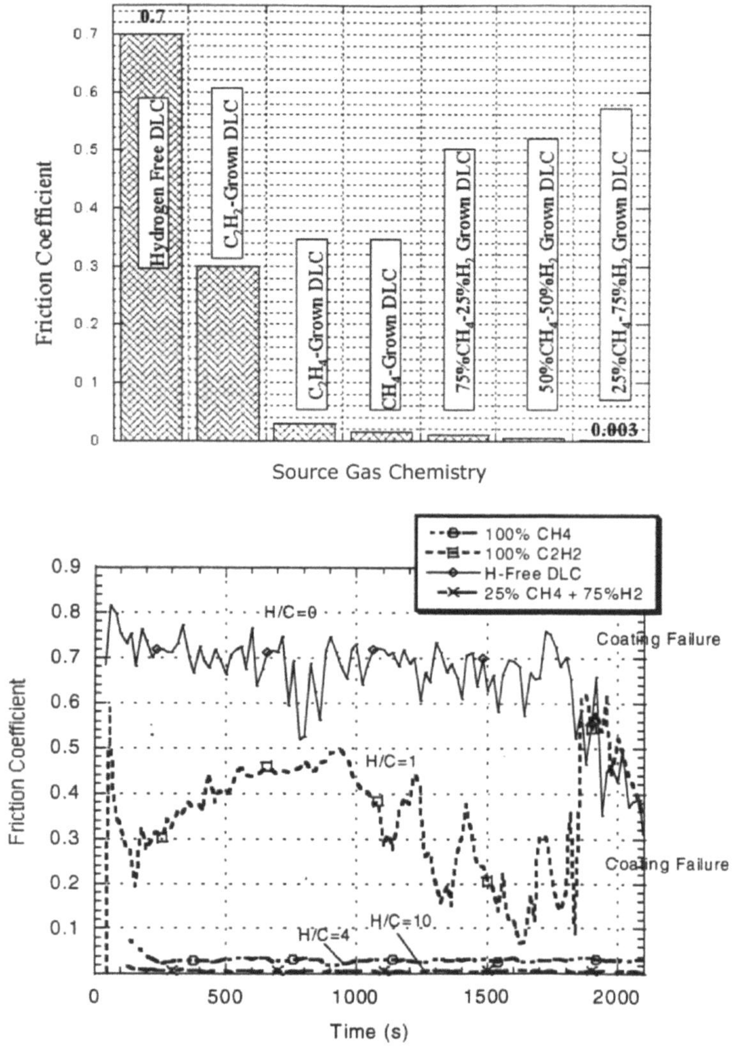

Figure 13.14 The increased hydrogenation in DLC can reduce the frictional coefficient of the coating (top). The frictional coefficient of DLC deposited from carbonaceous gas plasma (bottom). The gas plasma used is carbon cations only for H/C = 0, C_2H_2 plasma for H/C = 1, CH_4 plasma for H/C = 4, the plasma of 35% CH_4 + 75% H_2 for H/C = 10.

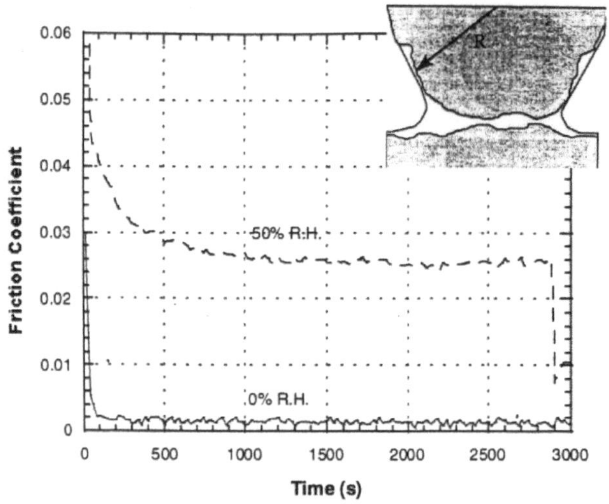

Figure 13.15 The moisture gathered by hydrogen in DLC can increase the drag force and the frictional coefficient when exposed to high humidity.

The moisture interactive hydrogenated DLC may increase the friction force with time in humid air. In contrast, the moisture inert amorphous diamond, although having a higher initial frictional coefficient, may actually become more slippery than hydrogenated DLC when exposed in humid air (Fig. 13.16).

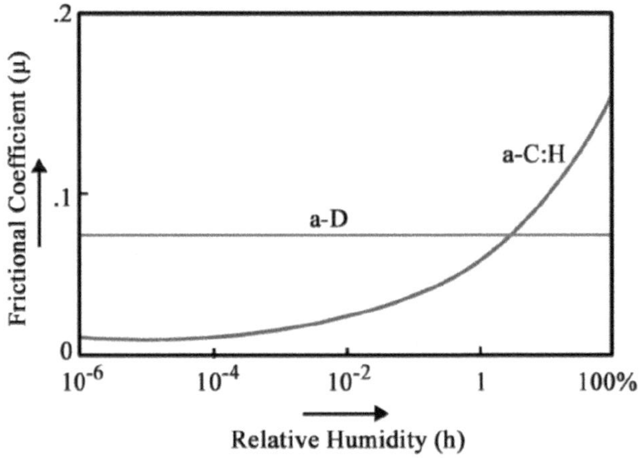

Figure 13.16 The increase of frictional coefficient with relative humidity for hydrogenated DLC and the more stable tribologic property of amorphous diamond in humid air.

However, if the moisture is not present in the environment, hydrogenated DLC can have extremely low frictional coefficient due to its superior surface smoothness and the reduced dangling electrons (Fig. 13.17).

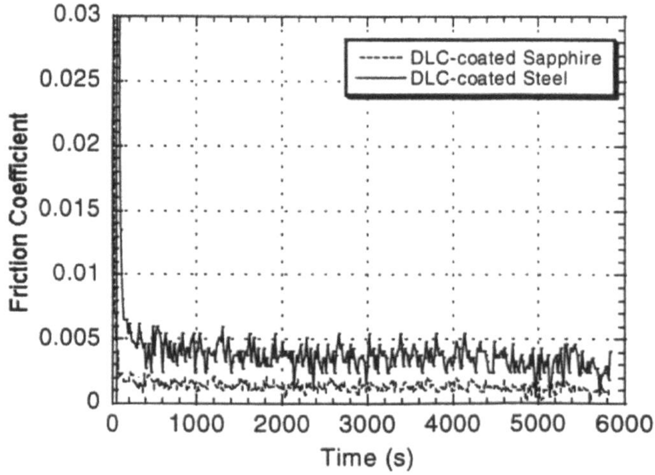

Figure 13.17 The extremely low frictional coefficient of hydrogenated DLC in dry nitrogen. The higher friction force associated with steel is due to the interaction of carbon's dangling electrons with iron's d-orbital electrons. Test conditions: load 10N, speed 0.3 m/sec, room temperature.

13.4 Applications of Amorphous Diamond Coating

Amorphous diamond's properties are not environment sensitive as hydrogenated DLC. Some of amorphous diamond's unique properties include the low frictional coefficient (<0.1) and the high corrosion resistance (un-reactive even to aqua regia). These properties are highly desirable for many industrial applications.

Amorphous diamond can be used to coat a variety of cermet (e.g., cemented WC) or metal (e.g., high-speed steel) substrates. During the coating process, the work piece is maintained at a temperature of less than 150°C. Hence, the properties of the underlying substrate may not be altered. Moreover, the coating can be applied to large areas (up to 1 m across). The thickness of the coating can be from 300 Å to 3 μm. The thickness variation is typically within 10%. The surface smoothness is basically maintained with an increase of Ra of about 0.03 μm.

With such a versatility in coating applications, amorphous diamond has been applied to many fixtures and devices as wear resistance coatings, tribologic slippery surfaces, chemically inert barriers, etc.

The most common application of amorphous diamond is to provide a superhard protective coating for cutting tools such as twist drills (Fig. 13.18).

Figure 13.18 Amorphous diamond coated twisted drills and bearing surfaces.

Some examples of the tools performance after coating with amorphous diamond are listed in Table 13.1:

Table 13.1 Amorphous diamond coating performance

Tool	Material	Life	Comparison
Carbide drills	RF 38 brake lining	4×	Uncoated
Carbide drills	Al-18%Si	16×	Uncoated
Carbide drills	G-10 fiberglass	40×	Uncoated
HS center drills	RF 38 brake lining	36×	Uncoated

Tool	Material	Life	Comparison
Carbide micro drills	Printed circuit board	5×	Uncoated
HS end mills	RF 38 brake lining	9×	4.5 × TiN
HS end mills	Graphite	3×	TiN
Carbide inserts	Al-10%Si	2×	Uncoated
Carbide inserts	Al-6061T	3×	Uncoated
Carbide reamers	Al-16%Si30%+	5×	TiCN
Carbide ball end mill	P-20 mold steel	3×	Uncoated

Cemented tungsten carbide of micro drills (0.5 mm diameter or smaller) are widely used to drill miniature holes through printed circuit boards. As the holes become smaller, the cutting debris is more difficult to extrude. Hence, micro drills tend to jam in the hole and break frequently. In order to increase the shear strength of the drill, the cobalt content has to increase. As the result, the hardness of the drill decreases, which shortens the life of the drill. Moreover, the high rotation speed (e.g., hundreds thousands turns per minute) of the micro drill in a tiny hole (e.g., 0.3 mm) may heat up the cutting tip to over 400°C that may soften the drill with high cobalt content. Hence, the vicious cycle has impeded the productivity of drilling holes through printed circuit boards.

With the application of a thin (e.g., <1 micron) layer of amorphous diamond on the flute of micro drills (Fig. 13.19), not only the wear of cutting edge is reduced, but also the heat generated is less. In addition, the debris can be discharged smoothly. As a result, the number of holes drilled can be increased from less than a thousand to over five thousands. The increase of service life not only reduces the cost of drilling, but also the throughput of production. Furthermore, the wall of drilled holes is smoother. In addition, there is less smearing of the copper layer embedded in the fiber reinforced board due to the less drag force exerted.

The mold made of high-speed steel or cemented carbide for die forging of electrical contacts can be worn out of shape quickly. Amorphous diamond has been applied to improve the life of such molds (Fig. 13.20). As a result, the service life of the mold can be increased from less than 10,000 electrical contacts produced per mold to over one million. Moreover, the electrical contacts so produced are much smoother; hence, the electricity loss due to the contact resistance is also reduced.

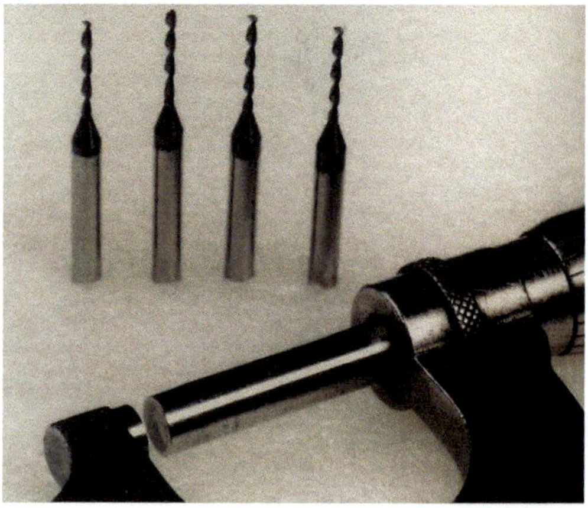

Figure 13.19 Amorphous diamond-coated micro drills and a micrometer.

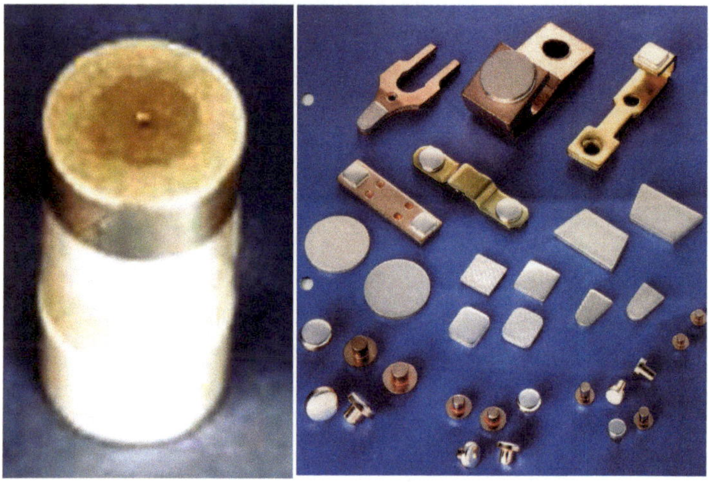

Figure 13.20 The amorphous diamond-coated mold (left) and a variety of electrical contacts formed from such a mold (right).

IC forming plungers are widely used to bend the lead frame of packaging devices. During the bending, the soldering tin often sticks to the plunger, so the bent lead cannot be released during the retraction of the plunger. This problem can cause frequent down time of the high throughput machine. Hydrogenated DLC coating has been applied to the leading face of the plunger (Fig. 13.21). As a consequence, the yield of production has been increased, and the machine down time has been reduced.

Figure 13.21 Nanocrystalline coated lead bending plungers for packaging IC.

DLC has been coated on molds for making ophthalmic lenses (Fig. 13.22). The coated mold not only eliminates the requirement of applying a lens-separating medium (e.g., sprayed boron nitride) but also makes the lens surface shiny. Moreover, the mold life and the product yield are both increased.

Figure 13.22 DLC-coated mold for forming ophthalmic lenses (Kinik brochure).

DLC has also been applied to molds for marking aluminum cans with dramatically increases of service life (Fig. 13.23).

Figure 13.23 DLC coated mold for marking aluminum cans.

DLC is also used to coat wire-supporting pulleys with a much lower drag force of the wire (Fig. 13.24).

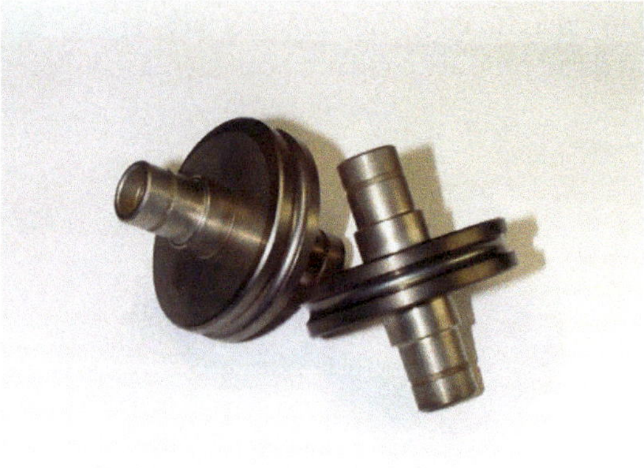

Figure 13.24 DLC coated wire-drawing pulleys to reduce the stress and wear of the passing wire.

Chemical mechanical planarization (CMP) of silicon wafers is a common practice to manufacture integrated circuitry (IC) and other sophisticated components. During CMP, a rotating pad that is covered with a slurry is used to polish the wafer. In order to continual the process, a diamond dresser must be used to scrape the debris away from the pad and also to renew its surface for subsequent polishing. However, many slurries used in CMP are highly acidic (e.g., pH = 3). If the diamond dresser is used to dress the pad while the acidic slurry is used, the metal bond that used to hold diamond grits on the pad dresser may be dissolved away. As a consequence, diamond grits may fall out of the dresser. The lost diamond can easily cause major scratches on expensive wafers being polished.

Amorphous diamond coating has been applied to diamond pad dressers that are employed to condition a polishing pad (Fig. 13.25). Conventional pad dressers can only dress the pad ex situ, after the acidic slurry is washed away. Amorphous diamond coated diamond dressers can dress in situ while the silicon wafer is being polished by the acidic slurry. As a result of the non-stop polishing, the throughput of the CMP production can be significantly increased.

Figure 13.25 DLC-coated diamond pad dresser (Kinik brochure).

One of diamond's unique properties is that it can form the sharpest edge of all materials. This is due to its small atoms and the strong covalent bonding so they can support the sharp radius with few atoms (Fig. 13.26). In contrast, the edge of a steel razor blade cannot be very sharp because the iron atoms are too big to define a small radius on the edge.

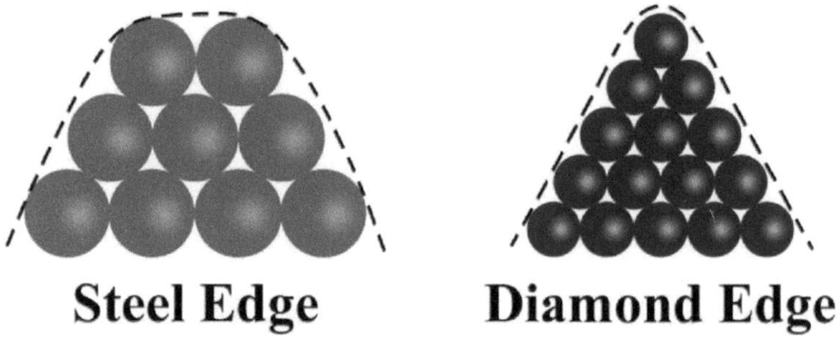

Figure 13.26 The dull edge that is supported by a pile of large iron atoms held together loosely (left) and the sharp edge that is formed by a lattice of smaller carbon atoms bonded together strongly (right).

In addition to be sharp, diamond's blade edge can also be the straightest. In comparison, the weak metallic bonding of a few iron atoms cannot maintain a straight line. The apparent straight line of a sharp razor blade is actually zigzag on the ridge when examined under an electron microscope (Fig. 13.27).

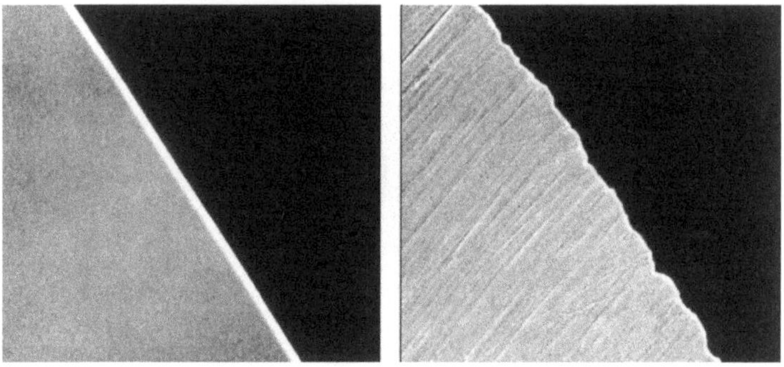

Figure 13.27 The contrast of sharpness and straightness of a diamond knife (left) and a razor blade (right) (Drukker brochure).

Because of its sharpness, diamond knives have been used as surgical knives for eye operation or for other industrial applications (Fig. 13.28).

The sharpness of a cutting edge is formed by polishing. If amorphous diamond is applied as a coating, it would round off the radius so the edge becomes dull. However, if the coating is applied with a line-of-sight projectile of carbon cations, as in the case of spraying with a cathodic arc, carbon cations can be blasted from the leeway so they converge in the front of the edge. In this way, very sharp edge can be formed (Fig. 13.29).

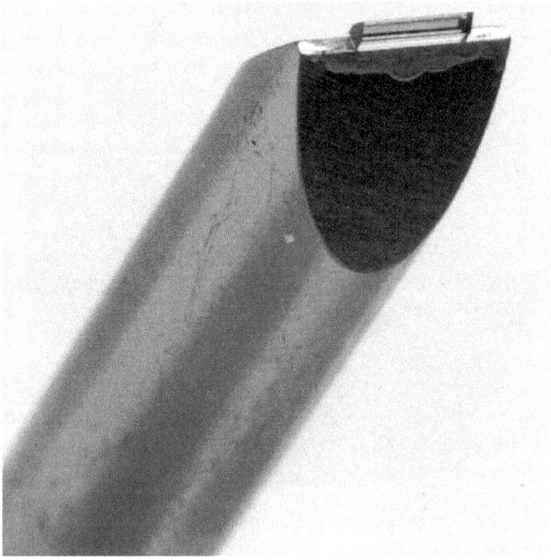

Figure 13.28 The diamond knife used for cleaving hard optical fibers made of silica. After cleaving, the optical fiber may be broken with a very straight fracture plane that is suitable for joining with other fibers.

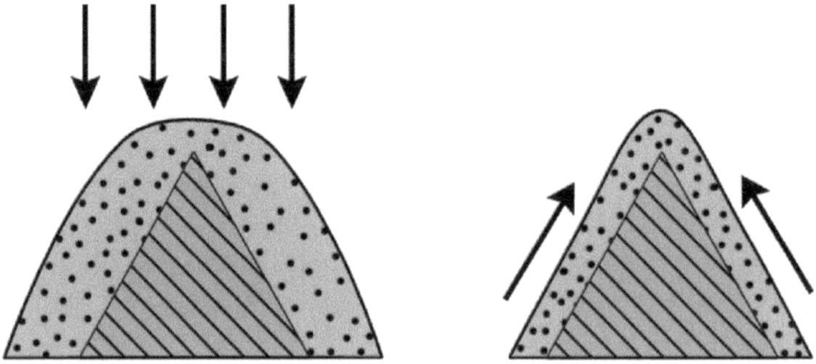

Figure 13.29 The front coating of a sharp edge will make it dull (left). The backward coating of a sharp edge can make it sharper (right).

The common steel razor blade may be over coated with a thin layer of chromium that is harder than iron with a straighter edge. Amorphous diamond or hydrogenated DLC of nanometer thickness can be applied over the chromium base. As the result of sharp edge and straight ridge, cleaner shaving can be achieved with a lower risk of scratching the skin. The slow wear of amorphous diamond will also make the razor blade last for a long time (Fig. 13.30).

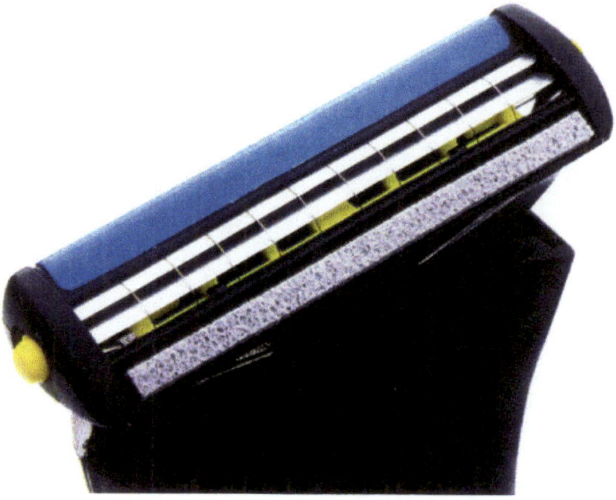

Figure 13.30 The amorphous diamond coated razor blade (Schick and Wilkinson brochure).

Normally amorphous diamond coating is applied with high smoothness to suit mechanical and tribologic applications. But for other applications, rougher surface may be more desirable. One of such application is to use amorphous diamond as electron emitting surface. For example, amorphous diamond is applied to coat electrodes (e.g., over platinum or glassy carbon) for electrolysis. Not only the coating can avoid the corrosion of the electrode, but also it will stabilize the electrical current so the current signal is better resolved (Fig. 13.31).

Another application of amorphous diamond for electron emission is to coat electrodes of lighting devices. For example, miniature fluorescence light has been widely used in LCD displays, scanners, PDA and copying machines, etc. The device contains a pair of cold cathode (e.g., W/Ni) that emit electrons to a gas mixture of argon and mercury. These gases are sealed in a glass tube coated with photofluorescent materials. The emitting electrons will excite the mercury atoms. The return of the electron to the ground state will emit a photon that will excite the fluorescence light. By coating with the cold cathode with amorphous diamond, more electrons can be emitted with lower power required (Fig. 13.32). The temperature of the electrode also reduced so the service life of the light can be extended considerably.

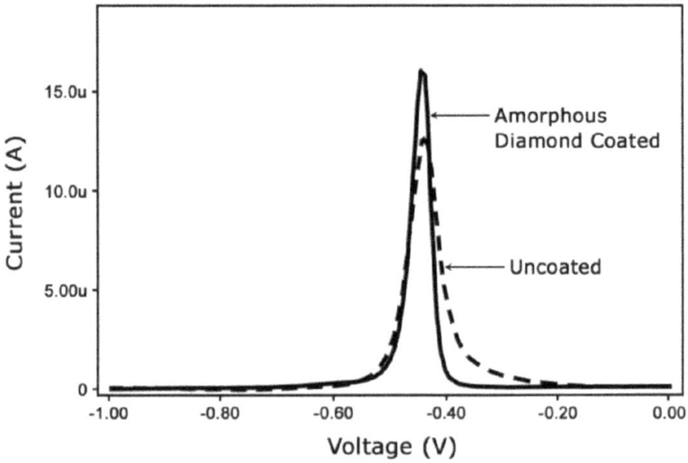

Figure 13.31 The amorphous diamond coated electrode can produce a more clear signal of electrical current during the electrolysis.

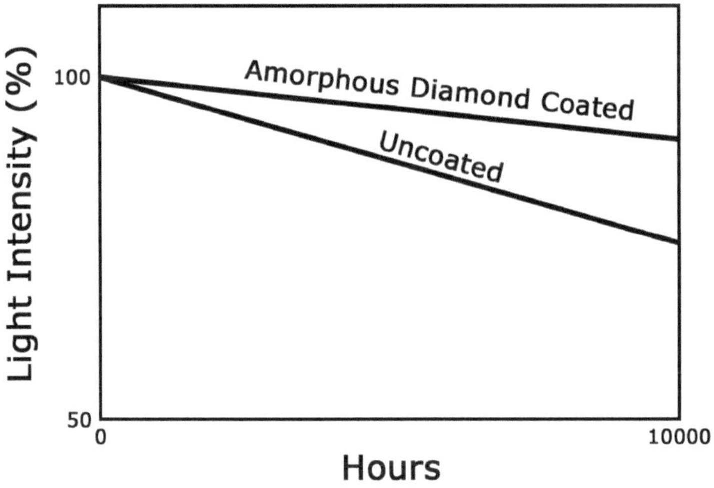

Figure 13.32 The lower decay rate of light luminosity by coating cold cathode with amorphous diamond.

13.5 Application of Hydrogenated DLC

In many applications, the strong internal stress associated with hardness of amorphous diamond may cause severe problems, such as flaking of the coating due to poor substrate adherence and the higher frictional coefficient due to the rougher surface conditions. In this case, less harder hydrogenated DLC coating is needed to improve the adherence and also the smoothness of the surface.

One of the applications that require high adherence and super smoothness of hydrogenated DLC is in coating the magnetic tapes that contains evaporated metal (Fig. 13.33).

Cassette discs are mass-produced by injection molding. Hydrogenated DLC has been applied to protect the outside mold that is used to support the injection mold (Fig. 13.35). The nickel stamper on the outside mold is easily scratched by the reciprocate movements of the injection mold due the expansion and contraction of each thermal cycle. As a result, the service life of coated outside mold has increased from less than two months to over eight months.

The coating of hydrogenated DLC can drastically reduce the frictional coefficient and greatly extend the service life of the magnetic tape.

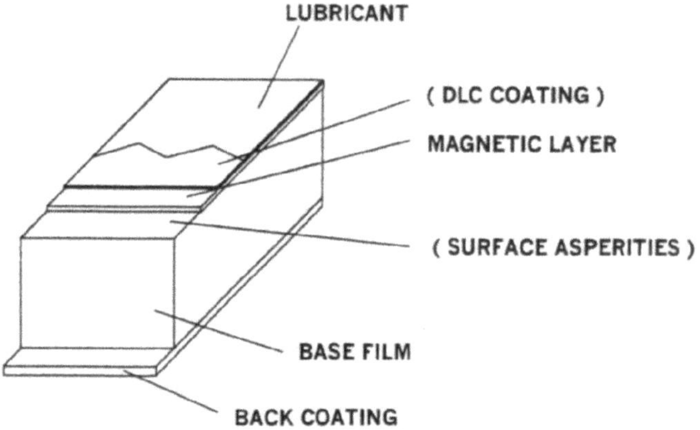

Figure 13.33 Typical structure of a metal evaporated magnetic tape.

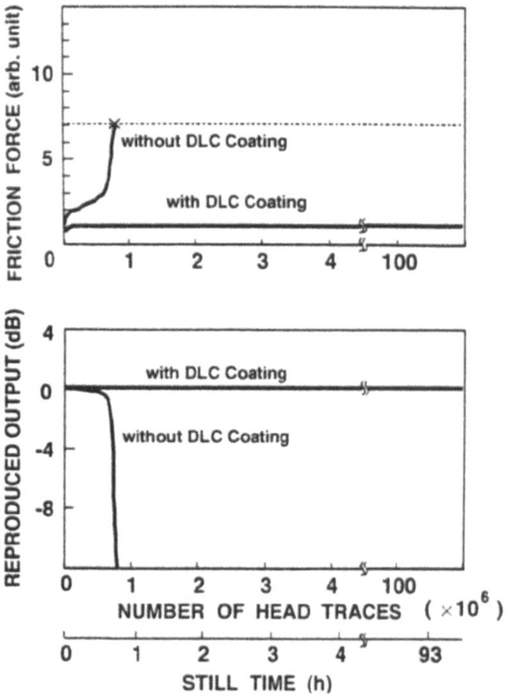

Figure 13.34 Hydrogenated DLC coating can improve dramatically the durability of magnetic tape in still mode. The testing conditions are: relative speed 10 m/sec, tape tension 7 gram, head number 2, and with lubricant.

Figure 13.35 DLC coated CD molds (Kinik brochure).

Hydrogenated DLC is also coated on inside of air bearing for open-end spinning at 100,000 rpm. The result is a dramatic increase of start and stop times before failure (Fig. 13.36).

Hydrogenated DLC may also protect the gap between piston and cylinder in a compressor for gas liquefaction (Fig. 13.37).

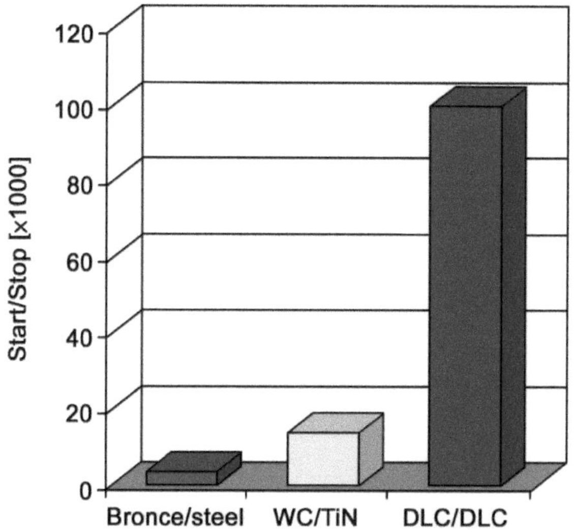

Figure 13.36 The air bearing's service life can be greatly lengthened by applying coating of hydrogenated DLC.

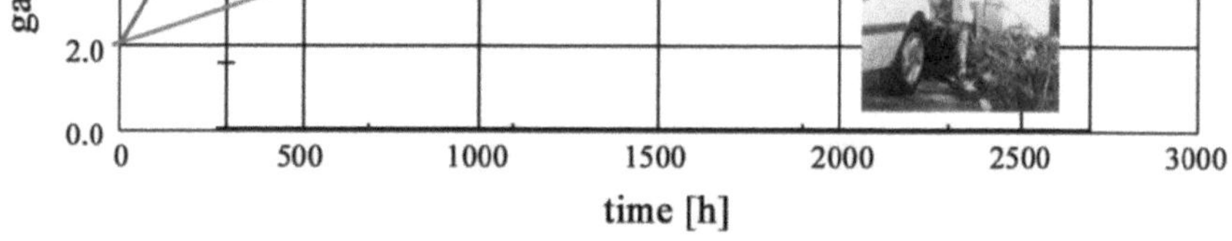

Figure 13.37 The enlargement of the gap between piston and cylinder can be slowed down by applying coating of hydrogenated DLC. The testing was conducted at 200 bar in dry conditions.

Hydrogenated DLC coating is also applied to golf clubs heads of the driver (Fig. 13.38). The slipper surface can reduce the spin of golf balls during hitting. As a result, the energy of swinging momentum can be fully transferred to the departing ball so it may fly straighter and farther.

Figure 13.38 DLC coated golf club heads (Kinik brochure).

Figure 13.39 The DLC protected church dome may preserve the gold color of TiN coating.

Hydrogenated DLC may also be applied to prevent the weathering of outdoor metals such as gold appearing TiN coating on the dome of a cathedral (Fig. 13.39). The TiN coating with its gold color may tarnish when exposed to wind and rain.

There are new applications of DLC coating that are tried from time to time; some are in unexpected areas. For example, DLC has been coated at the bottom of fiberglass yacht to prevent barnacles from building up. The result is the reduction in the drag force and the avoidance of corrosion of the boat.

13.6 Medical Applications of DLC Coating

In addition to mechanical and tribologic applications, hydrogenated DLC may also be used as a chemical barrier to prevent corrosion or other reactions, such as hydrogen embitterment (Fig. 13.40).

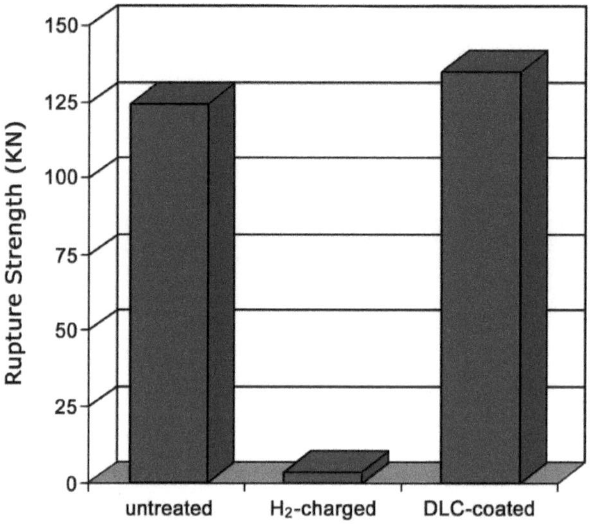

Figure 13.40 Hydrogen permeation in steel can reduce its rupture strength by hydrogen embitterment. However, the coating of hydrogenated DLC can prevent hydrogen permeation and avoid the embitterment.

Microtome scalpels are used to strip biological tissues for instrumental analysis (e.g., electron microscopy). Each slice may be as thin as 20 nm that may be separated by submerging in water. DLC may be used to coat these scalpels to improve the sharpness and durability (Fig. 13.41). However, in order for the scalpel to be wetted by water, its surface needs to be treated with hydroxyl radicals to make it hydrophilic. Normal DLC deposited with hydrogen-terminated surface is hydrophobic so it can repel water. The hydrophobic surface of a DLC surface may be converted to hydrophilic one by heating treatment under oxygen or bombardment with oxygen plasma.

DLC may also be used for coating other biological materials such as stents for opening blood vessels, valves for keeping heart pumping, or artificial joints for supporting hipbone (Fig. 13.42).

As a biological implant, DLC is ideal in many ways. First, about 1/4 of human body weight is made of carbon, the same element that defines DLC. Hence, DLC is not toxic in human body even with prolonged use. Second, DLC is hard and slippery so it does not wear out as fast as other implant materials. Third, DLC does not interact with blood or body fluids so the DLC coated implant will be like new all the time. This is particularly important for heart valves or stent coating as platelets may build up and cause the failure of the implant part (Fig. 13.43).

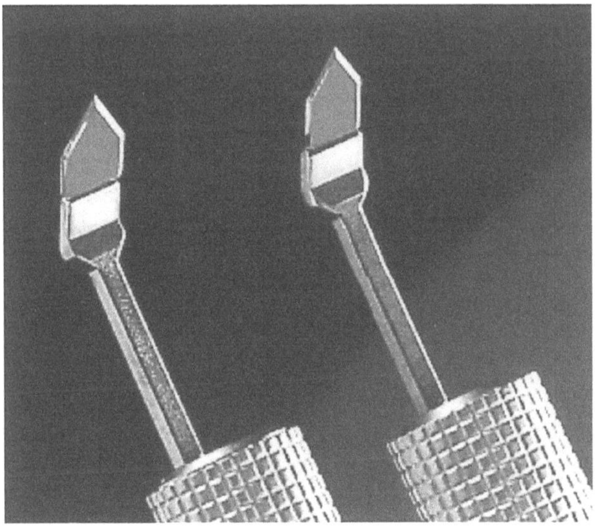

Figure 13.41 DLC-coated microtome scalpels for sampling biological tissues.

Figure 13.42 DLC coated hip joint can lubricate the interface between the metal (e.g., Co-Cr) ball and plastic (e.g., polyethylene) socket and prevent the accumulation of plastic debris that may damage the surrounding muscle.

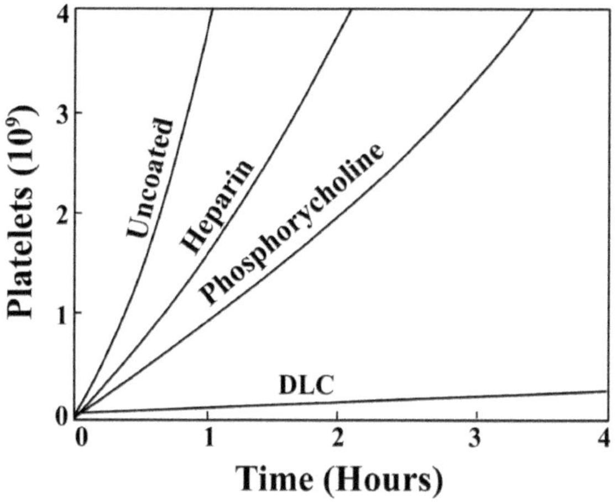

Figure 13.43 The accumulation of platelets on the implant materials.

Index